Stallcup's Electrical Grounding and Bonding Simplified

based on the nec®, codes and standards

National Fire Protection Association
Quincy, Massachusetts

Written by James Stallcup, Sr.
Edited by James Stallcup, Jr.
Design, graphics and layout by Billy G. Stallcup

Copyright ©2002 by Grayboy, Inc.

Published by the National Fire Protection Association, Inc.
One Batterymarch Park
Quincy, Massachusetts 02269

All rights reserved. No part of the material protected by this copyright notice may be reproduced or utilized in any form without acknowledgment of the copyright owner nor may it be used in any form for resale without written permission from the copyright owner and publisher.

Notice Concerning Liability: Publication of this work is for the purpose of circulating information and opinion among those concerned for fire and electrical safety and related subjects. While every effort has been made to achieve a work of high quality, neither the NFPA nor the authors and contributors to this work guarantee the accuracy or completeness of or assume any liability in connection with the information and opinions contained in this work. The NFPA and the authors and contributors shall in no event be liable for any personal injury, property, or other damages of any nature whatsoever, whether special, indirect, consequential, or compensatory, directly or indirectly resulting from the publication, use of or reliance upon this work.
 This work is published with the understanding that the NFPA and the authors and contributors to this work are supplying information and opinion but are not attempting to render engineering or other professional services. If such services are required, the assistance of an appropriate professional should be sought.

National Electrical Code® and *NEC*® are registered trademarks of the National Fire Protection Association, Inc.

NFPA No.: SEG02
ISBN: 0-87765-507-3
Library of Congress Card Catalog No.: 2002101442

Printed in the United States of America
03   04   05   06   07   5   4   3   2   1

Introduction

Without a good basic knowledge of the proper grounding and bonding of electrical systems, there is a strong tendency for designers, installers and inspectors to either be lax in certain code requirements or to overcompensate with excessive cost resulting. An effective grounding system must be designed like any other portion of the circuit. A reliable grounding scheme does not just happen, it must be well conceived and accurately calculated to perform just like any functional circuit.

Through this publication, the authors have attempted to put grounding and bonding as a subject into proper perspective, while emphasizing the National Electrical Code® and safety with practical applications. This book covers all of the general grounding and bonding of electrical installations, including specific requirements for the following topics:

- Circuit and system grounding
- Grounding electrode system and electrode conductor
- Enclosure, raceway and service cable grounding
- Bonding
- Equipment grounding and equipment grounding conductor
- Methods of equipment grounding
- Direct-current systems
- Instruments, meters and relays
- Grounding of systems and circuits of 1 kV and over
- Grounding information technology systems
- Lightning protection on outside overhead lines
- Lightning protection for buildings and structures
- Surge arresters

Because of the introduction of sophisticated technology and the many variables when installing electrical systems, grounding and bonding has become a major stumbling block for some designers, installers and inspectors. Through practical experience and skillful knowledge of the subject, combined with the ability to illustrate visually for the reader, the authors have endeavored to make this publication on grounding and bonding the standard for the electrical industry.

Table of Contents

Chapter 1:
General Requirements 1-1

Chapter 2:
Circuit and System Grounding 2-1

Chapter 3:
Grounding Electrode System and Grounding Electrode Conductor 3-1

Chapter 4:
Enclosure, Raceway and Service Cable Grounding 4-1

Chapter 5:
Bonding 5-1

Chapter 6:
Equipment Grounding and Equipment Grounding Conductors 6-1

Chapter 7:
Methods of Equipment Grounding 7-1

Chapter 8:
Direct-current Systems 8-1

Chapter 9:
Instruments, Meters and Relays 9-1

Chapter 10:
Grounding of Systems and Circuits of 1 kV and Over (High-Voltage) 10-1

Chapter 11:
Grounding Information Technology Equipment 11-1

Chapter 12:
Lightning Protection on Outside Overhead Lines 12-1

Chapter 13:
Lightning Protection for Buildings and Structures 13-1

Chapter 14:
Surge Arresters 14-1

Chapter 15:
Transient Voltage Surge Suppressors 15-1

Appendix: A-1

Abbreviations: A-19

Glossary: G-1

Topic Index: T-1

1

GENERAL REQUIREMENTS

This chapter deals with general requirements pertaining to grounding and bonding of electrical systems. Equipment and circuits that are grounded and bonded properly are protected from lightning strikes and excessive surges of voltages. Property is also protected as well as safety for people is provided. There are three grounding schemes in a well designed and installed grounding system and they are as follows:

- Circuit and system grounding
- Equipment grounding
- Bonding supply and load side

GENERAL REQUIREMENTS FOR GROUNDING AND BONDING
250.4

The following types of systems have general requirements that are to be applied for grounding and bonding:
- Grounded systems
- Ungrounded systems

GROUNDED SYSTEMS
250.4(A)

The following general requirements for grounding and bonding shall be accomplished for grounded systems to protect property as well as provide safety for people:
- Electrical system grounding
- Grounding of electrical equipment
- Bonding of electrical equipment
- Bonding of electrically conductive materials and other equipment
- Effective ground-fault current path

ELECTRICAL SYSTEM GROUNDING
250.4(A)(1)

Systems and circuits shall be grounded to limit voltage due to lightning, line surges or unintentional contact with higher voltage lines. Systems and circuits shall be grounded to stabilize the voltage-to-ground during normal operation. Systems and circuit shall be solidly grounded to facilitate overcurrent device operation in case of ground-faults. **(See Figure 1-1)**

Figure 1-1. This illustration shows the benefits of grounding an electrical system for safety on the supply side.

Grounding Tip 1: On a 480/277 volt system, if a 480 volt phase voltage is measured, the transformer is not defective. However, if the voltage-to-ground is measured and it is floating, the electrical system is losing its connection to earth ground.

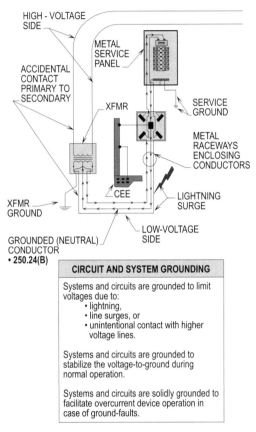

Grounding Tip 2: Circuit and system grounding consist of grounding the electrical system on the supplying side of the service equipment disconnecting means and OCPD.

GROUNDING OF ELECTRICAL EQUIPMENT
250.4(A)(2)

Noncurrent-carrying conductive materials that enclose electrical conductors or equipment, or forming part of such equipment, shall be connected to earth so as to limit the voltage-to-ground on these materials.

Circuits and equipment shall be grounded to facilitate overcurrent device operation in case of insulation failure or ground-faults. To insure this, the grounding path from circuit equipment and metal enclosures shall be continuous and not subject to damage. This path shall be capable of safely handling fault currents that may be imposed on it. Impedance shall be sufficiently low enough to keep the voltage-to-ground at a minimum and facilitate the opening of overcurrent protection devices ahead of the circuit. **(See Figure 1-2)**

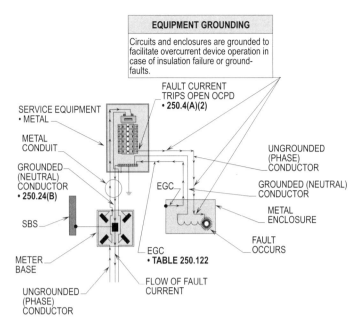

Figure 1-2. This illustration shows the benefits of equipment grounding on the load side

Grounding Tip 3: Equipment grounding conductors connecting the metal of enclosures housing electrical components and wiring keeps such enclosures as equal to ground potential as possible. This is accomplished by connecting them to the single-point ground of the grounding electrode system at the service equipment.

BONDING OF ELECTRICAL EQUIPMENT
250.4(A)(3)

Noncurrent-carrying conductive materials that enclose electrical conductors or equipment, or forming part of such equiopment, shall be connected together and to the electrical supply source in such a manner so as to establish an effective ground-fault current path. **(See Figure 1-3)**

Grounding Tip 4: All bonding jumpers, grounded (neutral) conductors and grounding conductors on the supplying side shall be sized based on the size of the ungrounded (phase) conductors. In other words, there is not an OCPD ahead of the conductors supplying the service equipment.

BONDING OF ELECTRICALLY CONDUCTIVE MATERIALS AND OTHER EQUIPMENT
250.4(A)(4)

Electrically conductive materials, such as metal water piping, gas piping and structural steel members that are likely to become energized shall be bonded together and to the electrical supply source in such a manner that an effective ground-fault current path is established. **(See Figure 1-3)**

EFFECTIVE GROUND-FAULT CURRENT PATH
250.4(A)(5)

An effective ground-fault current path from circuits, equipment and conductor enclosures shall be installed in a manner that creates a permanent, low-impedance circuit capable of safely carrying the maximum ground-fault current likely to be imposed on it from any point on the wiring system. **(See Figure 1-4)**

Grounding Tip 5: To provide the means for keeeping the path permanent and continuous, see **300.10, 300.12** and **300.15(A)**.

Grounding Note: The earth shall not be used as the sole equipment grounding conductor or effective ground-fault current path.

Figure 1-3. This illustration shows methods of how to bond electrically conductive materials (other than metal water piping) and other equipment.

Grounding Tip 6: All bonding jumpers and grounding conductors used to ground the metal of enclosures in an electrical circuit on the load side of the service equipment shall be sized based on the size OCPD ahead of such circuit.

Grounding Tip 7: Equipment grounding conductors in circuit raceways or cables shall be permitted be used to ground the metal piping that their ungrounded (phase) circuit conductors may energize.

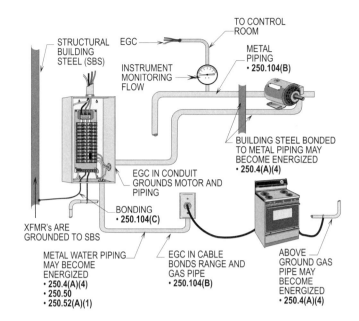

BONDING OF ELECTRICAL EQUIPMENT
BONDING OF ELECTRICALLY CONDUCTIVE MATERIALS AND OTHER EQUIPMENT
NEC 250.4(A)(3)
NEC 250.4(A)(4)

Figure 1-4. This illustration shows that an effective ground-fault current path from circuits, equipment and conductor enclosures shall be installed in a manner that creates a permanent, low-impedance circuit capable of safely carrying the maximum ground-fault current likely to be imposed on it from any point on the wiring system.

Grounding Tip 8: To size the grounded (neutral) conductor and equipment grounding conductor large enough to clear a ground-fault, size grounded (neutral) conductors based on the size of the service conductors (ungrounded conductors) per **Table 250.66** and equipment grounding conductors based on the size OCPD's ahead of the circuit conductors per **Table 250.122**.

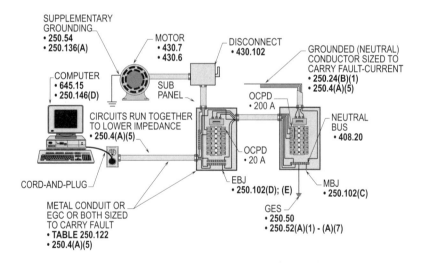

EFFECTIVE GROUND FAULT CURRENT PATH
NEC 250.4(A)(5)

UNGROUNDED SYSTEMS
250.4(B)

The following general requirements for grounding and bonding shall be accomplished for ungrounded systems to protect property as well as provide safety for people:

- Grounding electrical equipment
- Bonding of electrical equipment
- Bonding of electrically conductive materials and other equipment
- Path for fault-current

General Requirements

GROUNDING ELECTRICAL EQUIPMENT
250.4(B)(1)

Noncurrent-carrying conductive materials that enclose electrical conductors or equipment, or forming part of such equipment, shall be connected to earth in such a manner so as to limit the voltage imposed by lightning or unintentional contact with higher voltage lines and limit the voltage-to-ground on these materials. **(See Figure 1-5)**

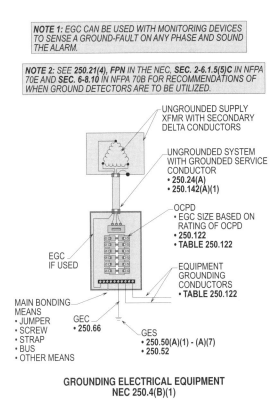

Figure 1-5. This illustration shows the grounding and bonding requirements for an ungrounded system.

BONDING OF ELECTRICAL EQUIPMENT
250.4(B)(2)

Noncurrent-carrying conductive materials that enclose electrical conductors or equipment, or forming part of such equipment, shall be connected together and to the supply system grounded equipment in such a manner so as to create a permanent, low-impedance path for ground-fault current that is capable of carrying the maximum fault current likely to be imposed on it.

BONDING OF ELECTRICALLY CONDUCTIVE MATERIALS AND OTHER EQUIPMENT
250.4(B)(3)

Electrically conductive materials, such as metal water piping, gas piping and structural steel members that are likely to become energized shall be bonded together and to the supply system grounded equipment in such a manner to create a permanent, low-impedance path for grounding current that is capable of carrying the maximum fault current likely to be imposed on it.

PATH FOR FAULT CURRENT
250.4(B)(4)

Electrical equipment, wiring and other electrically conductive material likely to become energized shall be installed in a manner that creates a permanent, low-impedance circuit from any point on the wiring system to the electrical supply source so as to facilitate the operation of overcurrent devices should a second fault develop on the wiring system.

Grounding Note: The earth shall not be used as the sole equipment grounding conductor or effective ground-fault current path.

OBJECTIONABLE CURRENT OVER GROUNDING CONDUCTORS
250.6

Grounding Tip 9: To provide a low-impedance, run grounded (neutral) conductors and equipment grounding conductors in the same raceway, cable, cable tray, etc., per **300.3(B), 300.5(I), 300.20** and **250.134(B)**.

Unbalanced loads along with multiple grounding points may have objectionable current flowing in grounding conductors within the circuits of the wiring system. When the grounded (neutral) conductor is intentionally connected to earth ground at the supply transformer and the service equipment of the building, there will be two parallel paths present for unbalanced current to travel over. The two parallel paths of unbalanced current will return back to the utility supply transformer through the grounded (neutral) conductor and through the ground path of the earth soil. More current returns from the transformer to the service equipment over the faulted ungrounded (phase) conductor and trips open the OCPD of the faulted circuit conductor. The following conditions of use shall be considered for objectionable current over grounding conductors: **(See Figure 1-6)**

- Arrangement to prevent objectionable current
- Alterations to stop objectionable current
- Temporary currents not classified as objectionable currents
- Limitations to permissible alterations
- Isolation of objectionable direct-current ground currents

Grounding Tip 10: The grounded (neutral) conductor only carries the unbalanced current between the ungrounded (phase) conductors per **220.22**. The grounded (neutral) conductor shall be permitted to be increased in size due to nonlinear loading on the ungrounded (phase) conductors per **220.22, FPN 2**. For transformer requirements pertaining to nonlinear loading, see **450.3, FPN 2**.

ARRANGEMENT TO PREVENT OBJECTIONABLE CURRENT
250.6(A)

Depending on the resistance of the ground path of the grounded (neutral) conductor and earth soil, about 10 percent of the current will flow through the ground and approximately 90 percent will flow over the grounded (neutral) conductor. Balancing the load of the system as well as possible reduces the unbalanced current (objectionable) so that excessive amounts will not flow over the grounded (neutral) conductor. **(See Figure 1-7)**

General Requirements

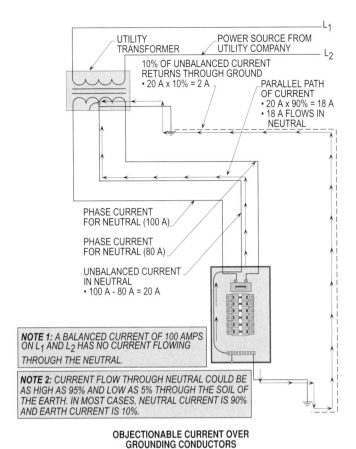

Figure 1-6. This illustration shows a condition where objectionable current flow in the earth will not be a problem.

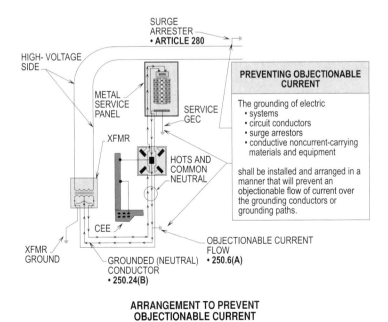

Figure 1-7. This illustration shows a connection to ground of the grounded (neutral) conductor that is not considered an objectionable current flow.

Grounding Tip 11: On a three-phase, four-wire wye system, the grounded (neutral) conductor may be current-carrying per **310.15(B)(4)(c)**. For the possibility of control circuits becoming current-carrying, see **725.28(B)(1)** and **(B)(2)**.

Grounding Tip 12: Electrical systems shall be installed to prevent objectionable current flow over certain elements. For example, the grounded (neutral) conductor shall not be bonded to the metal enclosure of a subpanel, this causes objectionable current flow over enclosures, conduits, etc.

1-7

ALTERATIONS TO STOP OBJECTIONABLE CURRENT
250.6(B)

Grounding Tip 13: Warning, it is never permissible to disconnect a grounding conductor to a grounding electrode and allow such electrode to float. Reconnect it into the electrode system somewhere

Objectionable current could occur from multiple grounds being installed on the same system. Objectionable current could flow from one ground connection to another, after such current has entered from one or more of the grounding points in the grounding system. If objectionable current is present, one or more of the following steps will help solve this problem:

- One or more grounding connections should be disconnected, but do not disconnect all of them.
- Placement of grounds should be altered.
- Break the conductive path of interconnecting grounding connections, so as to eliminate the continuity between grounding connections.
- Consult the authority having jurisdiction for suitable methods which are acceptable to help solve the problem. **(See Figure 1-8)**

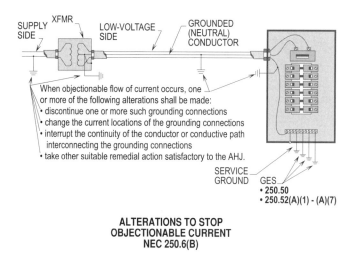

Figure 1-8. This illustration shows methods used to help eliminate objectionable current flow.

Grounding Tip 14: There are conditions where a grounding conductor can be removed without disrupting the safety of the grounding system. For example, a grounding conductor connecting the metal of an electrical enclosure to the structure steel for the diversion of static electricity build-up on such equipment.

TEMPORARY CURRENTS NOT CLASSIFIED AS OBJECTIONABLE CURRENTS
250.6(C)

Temporary currents develop from accidental shorts such as ground-fault currents that will flow over grounding conductors shall not be considered objectionable currents. **(See Figure 1-9)**

LIMITATIONS TO PERMISSIBLE ALTERATIONS
250.6(D)

Grounding Tip 15: Section 250.30(A)(1), Ex. 1 clearly states that the earth shall not be considered as providing a path for parallel current flow.

Currents generated from electronic equipment shall not be considered objectionable currents where such currents introduce noise or data errors. Currents that have such electrical noise or voltage fluctuations that produce dirty power conditions over the circuits shall also not be considered and objectionable current flow. **(See Figure 1-10)**

General Requirements

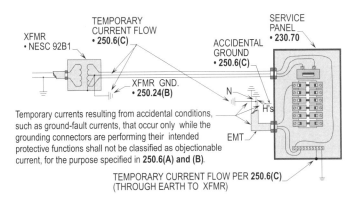

**TEMPORARY CURRENT NOT CLASSIFIED AS OBJECTIONABLE CURRENTS
NEC 250.6(C)**

Figure 1-9. This illustration shows examples of temporary currents that are not considered objectionable currents.

Grounding Tip 16: Currents that flow over equipment grounding conductors, bonding jumpers and grounded (neutral) conductors are considered temporary currents and shall not be defined as objectionable current flow per **250.6(C)** for they will only flow until the OCPD of the circuit trips open.

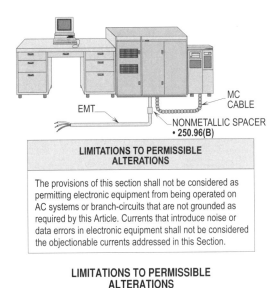

LIMITATIONS TO PERMISSIBLE ALTERATIONS

The provisions of this section shall not be considered as permitting electronic equipment from being operated on AC systems or branch-circuits that are not grounded as required by this Article. Currents that introduce noise or data errors in electronic equipment shall not be considered the objectionable currents addressed in this Section.

**LIMITATIONS TO PERMISSIBLE ALTERATIONS
NEC 250.6(D)**

Figure 1-10. This illustration shows a conduit system that is altered to help reduce electrial noise generated from the operation of the electronic equipment.

Grounding Tip 17: To correct noise problems for sensitive electronic equipment, it shall be permitted to isolate the supply conduit with a nonmetallic spacer. This spacer shall be located at the sensitive electronic equipment and cannot take the place of the safety ground (equipment grounding conductor).

ISOLATION OF OBJECTIONABLE DIRECT-CURRENT GROUND CURRENTS
250.6(E)

Where isolation from undesirable DC ground current, such as in the area of cathodic protected systems, a listed solid state AC coupling/DC isolating device shall be permitted to be placed in the grounding path to provide an effective return path for AC ground-fault current and also blocking DC current. **(See Figure 1-11)**

Note: It is necessary to restrict the flow of DC cathodic protection current to only the metallic parts to be protected. Such listed AC coupling and DC blocking devices permits flow of the AC current while blocking DC. These devices are evaluated for their ability to withstand the rated fault-currents. They are also evaluated for connections in the electrical grounding system in accordance with the instructions for the equipment.

Figure 1-11. This illustration shows an AC coupling and DC blocking device used for cathodic protection.

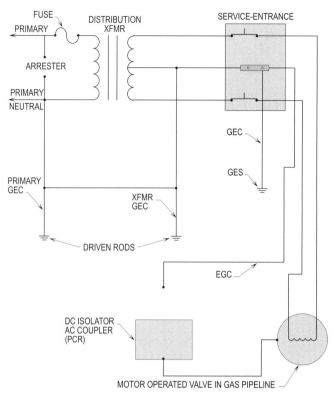

Grounding Tip 18: Equipment installed in series with the equipment grounding conductor shall be approved by the AHJ or be third party approved by a **UL (type) 508** shop. Note that the PCR has been tested and found to comply as an effective grounding path per **250.4(A)(5)**.

CONNECTION OF GROUNDING AND BONDING EQUIPMENT
250.8

Grounding Tip 19: It is most important that the connection device used to tie electric equipment to the grounding electrode system be approved for such use. For example, a device used to tie the grounding electrode conductor to a driven rod in the earth shall be listed for such use.

Pressure connectors, clips, clamps, lugs and devices shall be approved for the installation when used to connect grounding conductors to metal enclosures, electrodes and other types of equipment requiring grounding.

The grounding conductors and bonding jumpers shall be attached to circuits, conduits, cabinets, equipment, etc. which shall be grounded, by means of suitable lugs, listed pressure connectors, clamps, exothermic welding or other approved means. **(See Figure 1-12)**

Grounding Tip 20: Section **4 in UL 467** lists requirements for devices connecting the grounding electrode conductor to electrodes.

Grounding Note: The device used shall be identified as being suitable for such purpose. Connection devices or fittings that depend solely on solder shall not be permitted to be used. Sheet-metal screws shall not be permitted to be used to connect grounding conductors to enclosures.

PROTECTION OF GROUND CLAMPS AND FITTINGS
250.10

Grounding Tip 21: If the connection of the grounding electrode conductor to a driven rod, concrete-encased electrode or ground ring is below earth grade, the connecting device which connects the grounding electrode conductor to such electrode shall be listed for direct burial use and be protected from physical damage.

Unless approved for general use without protection, ground clamps and fittings shall be installed as follows:

• Be located so that they will not be subject to damage, or

• Be enclosed in metal, wood or equivalent protective covering.

With exothermic welding (cad-welding), checked for the resistance of the connection with a Ductor. If a low-resistance is measured, the problems of a high-resistance connection should be eliminated. **(See Figure 1-13)**

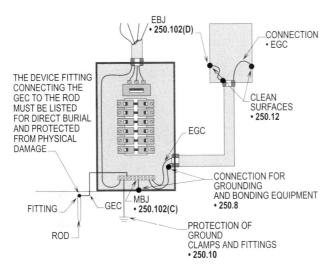

Figure 1-12. This illustration shows that the devices used to connect the grounding electrode conductor to the electrodes shall be listed for such use.

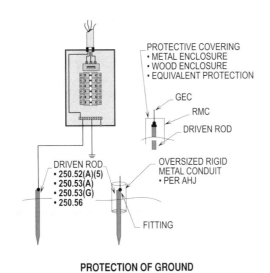

Figure 1-13. This illustration shows the methods that shall be permitted to be used to protect the grounding clamp from physical damage.

Grounding Tip 22: For further information pertaining to rules for protection to the connecting clamp, See **4.1.2 in UL 467**.

CLEAN SURFACES
250.12

Nonconductive coatings such as paint, lacquer and enamel shall be removed from threads and other contact surfaces to ensure good electrical continuity or by means of fittings designed to make such removal unnecessary. **(See Figure 1 - 12)**

Grounding Tip 23: If the device fitting used to connect the grounding electrode conductor to the grounding electrode does not have the capability to self clean the surface being connected to, the installer shall clean the surface in such a manner so a reliable and dependable connection is made.

Name _____   Date _____

Chapter 1
General Requirements

	Section	Answer

1. Systems and circuits shall be grounded to limit voltage due to lightning, line surges or unintentional contact with higher voltage lines. _____ T   F

2. Circuits and enclosures shall be grounded to facilitate overcurrent device operation in case of insulation failure or ground-faults. _____ T   F

3. The earth shall be permitted to be used as the sole equipment grounding conductor. _____ T   F

4. Connection devices or fittings that depend solely on solder shall be permitted to be used for connection of grounding and bonding of equipment. _____ T   F

5. Objectionable current can occur from multiple grounds being installed on the same system. _____ T   F

6. Temporary currents which develop from accidental shorts such as ground-fault currents that will flow over grounding conductors shall be considered objectionable currents. _____ T   F

7. Currents generated from electronic equipment shall be considered objectionable currents. _____ T   F

8. Pressure connectors, clips, clamps, lugs and devices shall be approved for the installation when used to connect grounding conductors to metal enclosures, electrodes and other types of equipment requiring grounding. _____ T   F

9. Ground clamps and fittings (not approved for general use without protection) shall to be enclosed in metal, wood or equivalent protective covering. _____ T   F

10. Nonconductive coatings such as paint, lacquer and enamel shall not be required to be removed from threads and other surfaces to ensure good electrical continuity. _____ T   F

2

CIRCUIT AND SYSTEM GROUNDING

This chapter presents basic system and equipment grounding requirements and grounding arrangements for service equipment, separately derived systems, emergency and standby power systems.

The purpose of the material is to define the fundamental grounding requirements and to illustrate grounding arrangements for several types of system grounding techniques that are most often selected for residential, commercial and industrial installations.

System and circuit conductors are grounded to limit voltage due to lightning, line surges or unintentional contact with higher voltage lines and to stabilize the voltage-to-ground during normal operation. Systems and circuit conductors are solidly grounded to facilitate overcurrent device operation in case of ground-faults.

AC CIRCUITS AND SYSTEMS TO BE GROUNDED
250.20

The following AC circuits and systems shall be grounded:

- AC circuits of less than 50 volts.
- AC circuits of 50 to 1000 volts.
- AC systems of 1 kV and over.
- Separately derived systems.

See Figure 2-1 for a detailed discription when applying these requirements.

Figure 2-1. This illustration shows electrical systems which are required to be grounded under certain conditions of use.

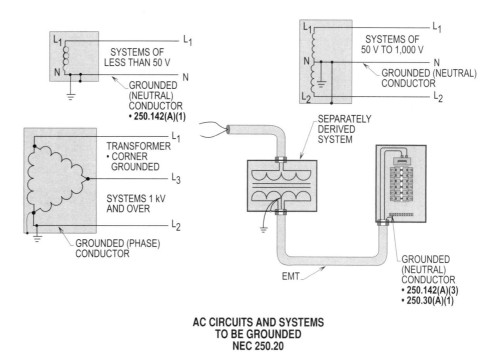

AC CIRCUITS AND SYSTEMS
TO BE GROUNDED
NEC 250.20

AC CIRCUITS OF LESS THAN 50 VOLTS
250.20(A)

Circuits and equipment operating at less than 50 volts are found in **Article 720**. Class 1, 2 and 3 circuits including rules pertaining to their installation procedures are listed in **725.21(A)** and **725.41**.

AC circuits of less than 50 volts shall be grounded under the following conditions:

- A transformer installed to supply low-voltage receives it's supply from a transformer exceeding 150 volts-to-ground.

For example, a circuit of 277 volts-to-ground supplying the primary side of such a transformer is considered a circuit of over 150 volts-to-ground and its secondary shall be grounded.

- A transformer installed to receive its supply from a transformer with an ungrounded system. This condition includes a supply voltage obtained from a transformer which has an ungrounded secondary.

- Where low-voltage overhead conductors are installed outside and not inside.

For example, AC conductors of 50 volts which are run outside overhead shall have one conductor grounded.

See Figure 2-2 for a detailed illustration for AC circuits of less than 50 volts.

Grounding Tip 24: Systems of less than 50 volts are systems such as Class 1, Class 2 and Class 3 circuits found in **725.21(A), 725.41(A)** and **Tables 11(A)** and **(B) in Ch. 9** in the NEC. **Article 720** in the NEC covers the rules for installing circuits and equipment operating at less than 50 volts.

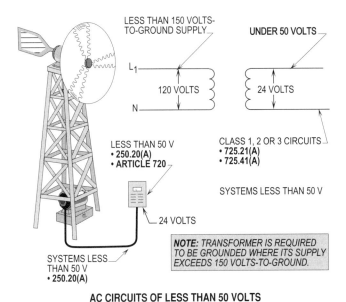

Figure 2-2. This illustration shows examples of systems less than 50 volts.

AC CIRCUITS OF 50 TO 1000 VOLTS
250.20(B)

AC circuits of 50 to 1000 volts shall be grounded under any one of the following conditions:

- The maximum voltage-to-ground on the ungrounded (phase) conductors does not exceed 150 volts. This voltage-to-ground circuit is usually a 120 volt circuit derived from a 120/240 or 120/208 volt system.

- The system is nominally rated as 120/208 and 277/480 volt, three-phase, four-wire wye and is connected so that the neutral can be used as a circuit conductor. The voltage-to-ground and between phases for 50 to 1000 volts can be any level. Wye systems of 2400/4160 volts are also used to supply circuits requiring a higher voltage served from a wye hookup.

- The system is nominally rated as 240/120 volt, three-phase, four-wire delta connected in which the midpoint of one phase is used as a circuit conductor with one phase conductor having a higher voltage-to-ground than the other two. **(See Figure 2-3)**

The following types of AC systems which are rated 50 volts to 1000 volts shall be grounded:

- 120 volt, two-wire, single-phase
- 120/240 volt, three-wire, single-phase
- 120/208 volt, four-wire, three-phase wye
- 277/480 volt, four-wire, three-phase wye
- 480 volt corner grounded, three-phase delta
- 240 volt, three-wire, three-phase delta
- 480 volt, three-wire, three-phase delta
- 600 volt, three-wire, three-phase delta

Grounding Tip 25: The requirements in **250.20(B)** cover the rules for grounding utility related transformers, usually located outside on a pad or pole.

Grounding Tip 26: The electrical power systems in **250.20(B)** are normally rated (if grounded) as follows:

- 120 V, 2-wire, 1Ø
- 120/240 V, 3-wire, 1Ø
- 120/208 V, 4-wire, 3Ø wye
- 277/480 V, 4-wire, 3Ø wye
- 480 V corner grounded, 3Ø delta
- 240 V, 3-wire, 3Ø delta
- 480 V, 3-wire, 3Ø delta
- 600 V, 3-wire, 3Ø delta

Figure 2-3. This illustration shows AC circuits of 50 volts to 1000 volts that shall be grounded when the grounded (neutral) conductor or ungrounded (phase) conductor is used as an equipment grounding conductor during a ground-fault condition.

Grounding Tip 27: For rules and regulations pertaining to grounding electrodes, see **Sec. 9 in UL 467**.

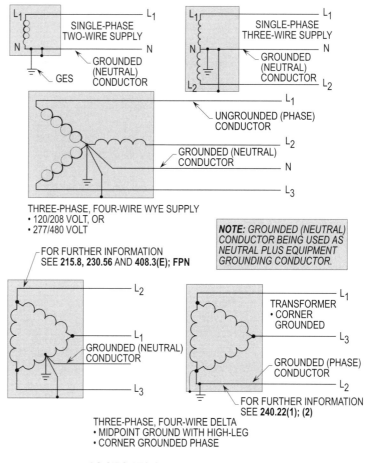

AC SYSTEMS OF 1 kV AND OVER
250.20(C)

AC systems of 1 kV and over shall be grounded if supplying mobile or portable equipment per **250.188**. Other type of AC systems of 1 kV and over that are installed do not have to be grounded. Where such systems are grounded, they shall comply with the applicable provisions of **Article 250**. **(See Figure 2-4)**

SEPARATELY DERIVED SYSTEM
250.20(D)

A separately derived system is derived from an generator, converter windings or transformer to reduce the voltage from high-voltage to low-voltage levels or vice versa in a building. Transformers of 4160 or 13,800 volts are voltages that may be reduced to utilization voltage of 480 volt or 120/208 volt to supply loads that are located at different floor levels within a building. **(See Figure 2-5)**

Circuit and System Grounding

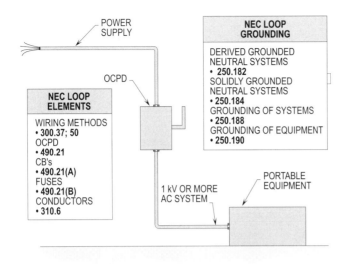

Figure 2-4. This illustration shows AC systems of 1 kV and over shall be grounded if supplying mobile or portable equipment.

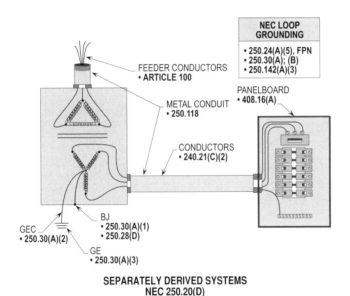

Figure 2-5. This illustration shows a separately derived system is derived from a generator, converter windings or transformers to reduce the voltage from high-voltage to low-voltage or vice versa in a building.

AC SYSTEMS OF 50 VOLTS TO 1000 VOLTS NOT REQUIRED TO BE GROUNDED
250.21

The following AC systems of 50 volts to 1000 volts shall not be required to be grounded:

- Circuits installed for industrial electric furnaces or any other means of heating metals for refining, melting or tempering.

- Separately derived systems used exclusively for rectifiers supplying only adjustable speed industrial drives. Such systems are used for speed control in industrial facilities and shall not be permitted to be utilized for anything else.

- Separately derived systems installed with primaries not exceeding 1000 volts used exclusively for secondary control circuits where conditions of maintenance and supervision ensure that only qualified persons will service the installation. However, the continuity of control power is required with ground detectors installed on the control system to sound an alarm if one phase should become grounded and a ground-fault condition occurs. **(See Figure 2-6)**

Grounding Tip 28: This section along with **240.21(C)(4)(1) through (4)** allows a separately derived system (transformer) to be installed outside and the secondary conductors are not limited in length until they enter the building and terminate in the control cabinet supplying an adjustable speed industrial drive system.

Figure 2-6. This illustration shows a system shall not be required to be grounded due to its supplying rectifiers powering and adjustable speed industrial drive.

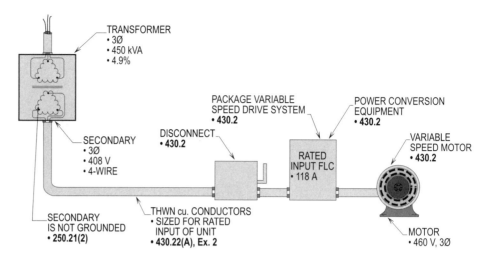

AC SYSTEMS OF 50 VOLTS TO 1000 VOLTS NOT REQUIRED TO BE GROUNDED
NEC 250.21

CIRCUITS NOT TO BE GROUNDED
250.22

The following circuits shall not be grounded due to their conditions of use:

- Circuits for electric cranes operating over combustible fibers in Class III locations per **503.13**. Combustible fibers are easily ignited with sparking or arcing devices. This rule was designed to eliminate this problem and hazardarous condition.

- Circuits operating in health care facilities such as anesthetizing locations per **517.160(A)(2)**. In operating rooms, circuits shall not be permitted to be grounded to ensure against ground-faults.

- Circuits for electrolytic cells shall not be permitted to be grounded per **Article 668**.

- Secondary circuits for lighting systems shall not be permitted to be grounded per **411.5(A)**.

See Figure 2-7(a), (b) and **(c)** for a detailed illustration when applying these requirements.

Grounding Tip 29: Section **250.22** usually requires a separately derived system to be installed, without grounding it, so as to provide a power supply that is not grounded for cranes operating in specific areas and operating rooms in hospitals.

Grounding Tip 30: The requirements in 250.24(A) parallels with the rules in **300.3(B), 300.5(I)** and **250.134(B)** which requires the grounded (neutral) conductor and equipment grounding conductor (if used) to be run with the ungrounded (phase) conductors. In other words, the equipment grounding conductor shall not be permitted to be laid in a ditch beside the conduit containing the ungrounded (phases) conductors and grounded (neutral) conductor. An equipment grounding conductor shall not be permitted to be wrapped to the outside of a cable.

SYSTEM GROUNDING CONNECTIONS
250.24(A)

The grounding electrode conductor shall be installed to bond and ground the service equipment terminal neutral bus to the grounding electrode system per **250.52(A)(1) thru (A)(6)**. The grounding electrode conductor is installed to connect equipment grounding conductors, neutral conductors and the service equipment enclosure to a reliable earth ground. Bonding jumpers shall be installed by using a wire, bus, screw or other suitable conductor per **250.28(A)**. Grounding electrodes bonded together for two or more services shall be considered a single grounding electrode system which will bond the metal enclosures of equipment of all sources together to create equipotential planes. Note that the head of a screw used to bond in the above items shall be green in color and be visible after the installation per **250.28(B)**. **(See Figure 2-8)**

Circuit and System Grounding

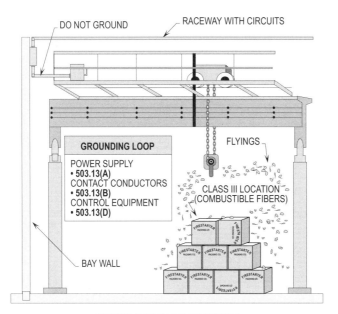

Figure 2-7(a). This illustration shows a system that is not permitted to be grounded.

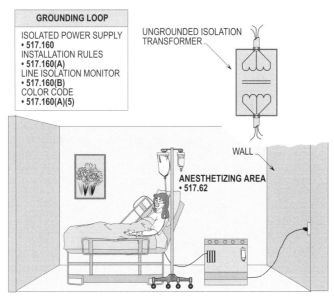

Figure 2-7(b). This illustration shows a system that is not permitted to be grounded.

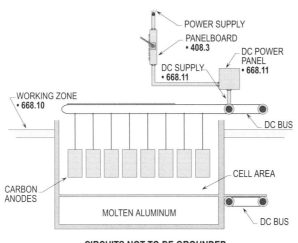

Figure 2-7(c). This illustration shows a system that is not permitted to be grounded.

Figure 2-8. The grounding electrode conductor shall be installed in such a manner so as to ground the grounded busbar, the equipment grounding conductors and grounded (neutral) conductor to the grounding electrode system to form a single-point ground connected to earth ground.

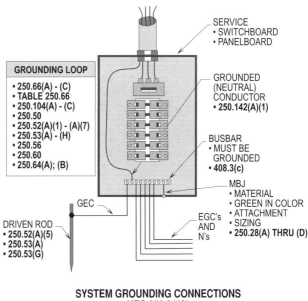

GROUINDING SERVICE-SUPPLIED AC SYSTEMS
250.24(A)(1) THRU (A)(5)

Grounding Tip 31: Unless otherwise permitted, the grounding electrode conductor shall be connected to the same grounded busbar that the grounded (neutral) conductor is terminated to.

AC grounded systems shall be grounded at each service by a grounding electrode conductor which is usually from the neutral bus terminal to a grounding electrode system which could be a cold water pipe, driven rod or other electrode, or any combination per **250.52(A)(1) thru (A)(6)**. The grounding electrode conductor shall be connected to the grounded service conductor when applying one or more of the following grounding requirements:

- General
- Outdoor transformer
- Dual fed services
- Main bonding jumper as wire or busbar
- Load-side grounding connections

GENERAL
250.24(A)(1)

The grounding electrode conductor shall be installed to bond and ground the service equipment terminal neutral bus to the grounding electrode system per **250.52(A)(1) thru (A)(6)**. The grounding electrode conductor shall be installed to connect equipment grounding conductors, neutral conductors and the service equipment enclosure to a reliable earth ground. **(See Figure 2-9)**

OUTSIDE TRANSFORMER
250.24(A)(2)

Transformers which supply the service and are located outside the building, at least one additional grounding connection shall be installed from the grounded service conductor to a grounding electrode, either at the transformer or elsewhere outside the building. **(See Figure 2-9)**

Circuit and System Grounding

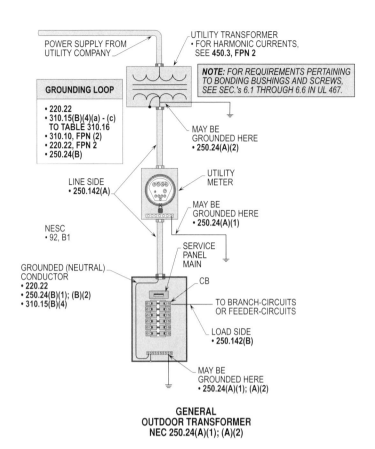

Figure 2-9. This illustration shows AC grounded systems shall be grounded at each service and the outside supply transformer.

DUAL FED SERVICES
250.24(A)(3)

A single grounding electrode connection to the tie point of the grounded circuit conductor from each power source shall be permitted to be installed where services are dual fed (double ended) in a common enclosure or grouped together in service enclosures and employing a secondary tie. **(See Figure 2-10)**

MAIN BONDING JUMPER AS WIRE OR BUSBAR
250.24(A)(4)

Where the main bonding jumper (wire or busbar) is installed from the neutral bar or bus to the equipment grounding terminal bar or bus in the service equipment, the grounding electrode conductor shall be permitted to be connected to the equipment grounding terminal bar or bus to which the main bonding jumper is connected.

LOAD SIDE GROUNDING CONNECTIONS
250.24(A)(5)

Unless otherwise permitted, a grounding connection shall not be permitted to be made to any grounded circuit conductor on the load side of the service disconnecting means. **(See Figure 2-11)**

Grounding Note: See **250.30(B)** for separately derived system connections, **250.32** for connections at separate buildings or stuctures and **250.142** for use of the grounded service conductor for grounding equipment.

Quick Calc

Sizing copper busbar to provide 250 amp rating.

Sizing
360.6
1/4" x 1" x 1000 = 250 A

Solution: The copper 1/4 in. x 1 in. busbar is equal to a 4/0 AWG THWN copper wire (conductor).

Figure 2-10. This illustration shows the requirements for grounding and bonding dual fed services from a service-entrance supply.

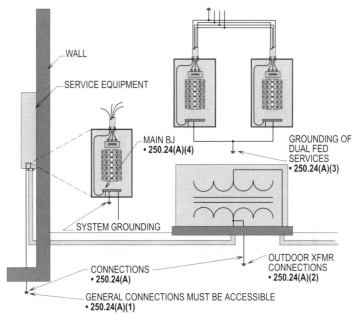

DUAL FED SERVICES
NEC 250.24(A)(3)

Figure 2-11. This illustration shows the grounded circuit conductor shall not be permitted to have a grounding connection installed on the load side of the service disconnecting means.

Grounding Tip 32: All equipment grounding conductors and equipment bonding jumpers on the load side shall be sized from **Table 250.122** based on the rating of the OCPD ahead of the circuit conductors.

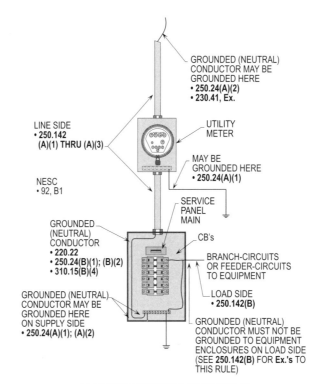

LOAD SIDE GROUNDING CONNECTIONS
NEC 250.24(A)(5)

CALCULATING FAULT CURRENT

By dividing the length of wire between supply and load by 1000 and multiplying by the resistance, the amount of fault current that will flow may be calculated. The resistance of each length is added together and divided into the voltage-to-ground to derive the fault current at the location of such short. **(See Figure 2-12)**

Circuit and System Grounding

FINDING TOTAL RESISTANCE
TABLE 8, CH. 9

Feeder-circuit conductors (HOT)
Step 1: Finding known values
Table 8, Ch. 9
100'
4 AWG cu. = .308 R

Step 2: Calculating resistance

$$R = \frac{100'}{1000'} \times .308 \, R$$

$$R = .0308 \, (\sqrt{})$$

Branch-circuit conductors (HOT)
Step 1: Finding known values
Table 8, Ch. 9
100'
12 AWG cu. = 1.93 R

Step 2: Calculating resistance

$$R = \frac{100'}{1000'} \times 1.93 \, R$$

$$R = .193 \, R \, (\sqrt{})$$

THE CURRENT FLOW THROUGH EARTH, IF EARTH WAS USED AS SOLE GROUND PATH

$$I = E \div R$$
$$I = 120 \, V \div 10R = 12 \, A$$

NOTE: THIS AMOUNT OF CURRENT WILL NOT NORMALLY FLOW, IT WILL USUALLY BE MUCH LESS.

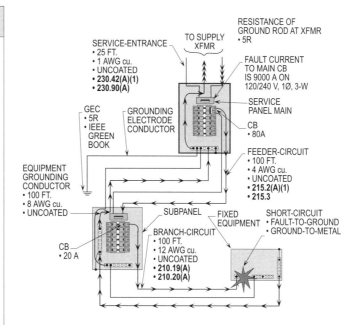

Figure 2-12. This illustration shows the procedures for calculating the fault current at the point of the fault.

Grounding Tip 33: When applying the rule of thumb method, the grounded (neutral) conductor can be computed by multiplying the size of the main OCPD for the service by 189. (See Quick Calc on page 2-13 and Figure 2-15 in this book)

POINT OF FAULT AT EQUIPMENT

Branch-circuit conductor (EGC)
Step 1: Finding known values
Table 8, Ch. 9
100'
12 AWG cu. = 1.93 R

Step 2: Calculating resistance
$$R = \frac{100'}{1000'} \times 1.93 \, R$$
$$R = .193 \, R \, (\sqrt{})$$

Feeder-circuit conductors (EGC)
Step 1: Finding known values
Table 8, Ch. 9
100'
8 AWG cu. = .778 R

Step 2: Calculating resistance
$$R = \frac{100'}{1000'} \times .778 \, R$$
$$R = .0778 \, R \, (\sqrt{})$$

Total resistance
Step 1: Finding total resistance ($\sqrt{}$)
Branch-circuit = .193 R (HOT) + .193 R (EGC)
Branch-circuit = .386 R
Feeder-circuit = .0308 (HOT) + .0778 (EGC)
Feeder-circuit = .1086 R

Step 2: Calculating total resistance
Branch-circuit = .386 R
Feeder-circuit = .1086 R
Total = .4946

Calculating fault-current

$$I = E \div R$$

$$I = 120 \div .50 \, R$$

$$I = 240 \, A$$

Fault-current is equal to about 240 amps when .4946 R is rounded up to .50 R.

NOTE: ALL WIRES ARE STRANDED

CALCULATING FAULT CURRENT
NEC 110.9
NEC 110.10

GROUNDED CONDUCTOR BROUGHT TO SERVICE EQUIPMENT
250.24(B)

The grounded (neutral) conductors shall be run to each service disconnecting means and shall be bonded to each disconnecting means enclosure for AC systems which operate at less than 1000 volts and are grounded at any point. **(See Figure 2-13)**

The grounded (neutral) conductor(s) shall be installed in accordance with the following conditions of use:

- Routing
- Parallel conductors
- High impedance

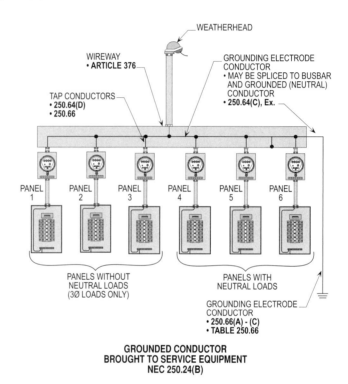

Figure 2-13. The grounded (neutral) conductor shall be run to each service disconnecting means, even if it's not used to supply single-phase loads with neutral connections.

Grounding Tip 34: A good example of this rule is where six disconnects are installed at the service equipment and three are used with grounded (neutral) conductor connections and three are not. The three without grounded (neutral) conductor connections are still required to have the grounded (neutral) conductor run to each disconnect to provide a low-impedance path for the fault current to return over and trip the cut-out fuses on the primary side of the power transformer.

GROUNDED CONDUCTORS BROUGHT TO ASSEMBLY LISTED FOR USED AS SERVICE EQUIPMENT
250.24(B), Ex.

Grounding Tip 35: Where the neutral is run to a subpanel in the same facility, the neutral shall be isolated and shall not be bonded to the metal of the subpanels enclosure.

Where more than one service disconnecting means is located in an assembly which is listed for use as service equipment, the grounded (neutral) conductor shall be permitted to be run to the assembly and the grounded (neutral) conductor shall be bonded to the assembly enclosure.

ROUTING
250.24(B)(1)

The grounded (neutral) conductor shall be installed with the ungrounded (phase) conductors and shall not be permitted to be smaller than the required grounding electrode conductor per **Table 250.66**, but shall not be required to be larger than the largest ungrounded service (phase) conductor. The grounded (neutral) conductor shall not be permitted to be smaller than 12 1/2 percent of the area of the largest service-entrance (phase) conductor larger than 1100 KCMIL copper or 1750 KCMIL aluminum. **(See Figure 2-14)**

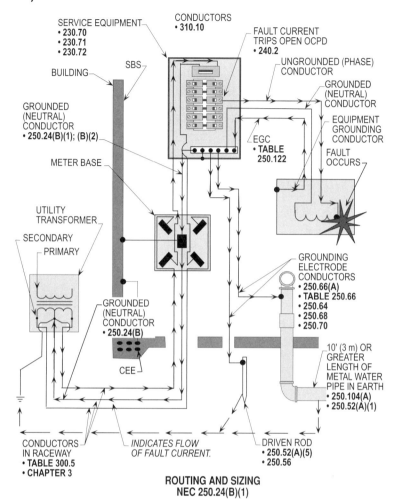

Figure 2-14. The grounded (neutral) conductor (fault-current return path) shall be installed with the ungrounded (phase) conductors and run to each service if the secondary of the utility's power transformer is grounded.

It is estimated that the amount of ground-fault current that will flow in the system is approximately 5 to 10 percent in the ground and 90 to 95 percent on the grounded (neutral) conductor between the supply transformer and service equipment. The rating of the grounded (neutral) conductor shall be computed at least 12 1/2 percent of the largest ungrounded (phase) conductor. However, the equipment grounding conductor should be calculated at not less than 25 percent of the largest ungrounded (phase) conductor to ensure grounding conductors provide safe and dependable fault-ground paths. For an OCPD to clear a circuit safely, a fault current of at least 6 to 10 times its rating shall be available.

Note that fault-current will from the point of fault through the grounded (phase or neutral) conductor to the supply transformer and then will return through the phase which was faulted to ground and open the OCPD. The overcurrent protection will rapidly trip a faulted phase that has a fault current of 6 to 10 times its rating.

For example: What is the minimum and maximum fault current required to clear a 150 amp OCPD?

Step 1: Calculating percentage
250.24(B)
150 A x 6 = 900 A
150 A x 10 = 1500 A

Solution: **The minimum fault current is 900 amps and the maximum fault current is 1500 amps.**

See Figure 2-15 for a detailed illustration when sizing the minimum and maximum fault current needed to clear and OCPD.

Figure 2-15. This illustration shows an exercise problem for sizing the minimum and maximum available fault current to clear an OCPD.

ILLUSTRATION	FORMULA REQUIREMENTS	EXERCISE PROBLEMS
SERVICE CONDUCTORS • PHASES = 292 A • NEUTRAL = 95 A OCPD • 300A MBJ — GEC — GES NEC 250.24(B)(1) TABLE 250.66 **KNOWN VALUES** PHASES = 292 A NEUTRAL = 95 A	PHASES (THWN) • 292 A = 350 KCMIL cu. • OCPD 300 A **GEC - TABLE 250.66** 2 AWG OR SMALLER = 8 1 AWG OR 1/0 AWG = 6 2/0 AWG OR 3/0 AWG = 4 OVER 3/0 AWG THRU 350 KCMIL = 2 OVER 350 THRU 600 KCMIL = 1/0 OVER 600 THRU 1100 KCMIL = 2/0 OVER 1100 KCMIL = 3/0 **QUICK CALC. USE 189 MULITPLIER** MAIN OCPD x 189 = SIZE GROUNDED (NEUTRAL) CONDUCTOR **Step 1:** Size of OCPD OCPD = 300 A **Step 2:** Sizing grounded (neutral) conductor Table 8, Ch. 9 300 A x 187 = 56,100 CM 56,100 CM requires 2 AWG cu. conductor	**Step 1:** Calculating percentage 300 A x ____ A [A] = ____ A [B] ____ A [C] x 10 = ____ A [D] **Solution:** The minimum fault current is ____ amps [E] and the maximum fault current is ____ amps [F].

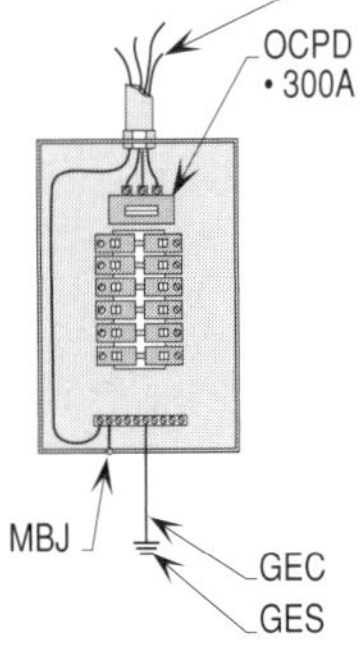

Grounding Tip 36: Where unscheduled outages are desired, high-impedance grounding is recommended. The fault current due to one phase going to ground can be limited as low as 4 amps or less or up to 50 amps. Note that the electrical system will continue to operate until such fault is found and cleared.

Grounded (neutral) conductors shall be installed to provide an effective path for fault currents to travel over where phase-to-ground faults occur in the electrical system. The grounded (neutral) conductors shall be sized as large as the grounding electrode conductor per **250.66** and **Table 250.66**. The grounded (neutral) conductor shall be sized at least 12 1/2 percent of the area of the largest ungrounded (phase) conductor where the service conductors are installed larger than 1100 KCMIL copper or 1750 KCMIL aluminum.

Grounding Note: Two-phase or three-phase conductors shall not be permitted to be run to the service without installing a grounded (neutral) conductor when the utility company's secondary is grounded. An example of this rule is where all the service loads are three-phase, 480 volt and a step-down separately derived system is installed to supply single-phase loads of 120/208 volt.

For example: What size copper grounded (neutral) conductor is required based on the service-entrance conductors being rated 250 KCMIL THWN-THHN copper?

Step 1: Finding the grounded (neutral) conductor
250.142(A)(1); 250.24(B)(1); Table 250.66
250 KCMIL requires 2 AWG cu.

Solution: The size grounded (neutral) conductor is 2 AWG copper.

See Figure 2-16 for a detailed illustration when sizing the grounded (neutral) conductor based on the size of the service-entrance conductors.

ILLUSTRATION	FORMULA REQUIREMENTS	EXERCISE PROBLEM
SERVICE CONDUCTORS • PHASES = 292 A • NEUTRAL = 95 A OCPD • 300A MBJ GEC GES NEC 250.24(B)(1) TABLE 250.66 **KNOWN VALUES** PHASES = 292 A NEUTRAL = 95 A	PHASES (THWN) • 292 A = 350 KCMIL cu. NEUTRAL • 95 A = 3 cu. **APPLYING TABLE 250.66** 2 AWG OR SMALLER = 8 1 AWG OR 1/0 AWG = 6 2/0 AWG OR 3/0 AWG = 4 OVER 3/0 AWG THRU 350 KCMIL = 2 OVER 350 THRU 600 KCMIL = 1/0 OVER 600 THRU 1100 KCMIL = 2/0 OVER 1100 KCMIL = 3/0 **Note 1:** See Figure 2-22(a) on page 2-20 in this book for an illustration when sizing the main bonding jumper. **Note 2:** See Figure 5-22(a) on page 5-20 in this book for an illustration when sizing the equipment bonding jumper.	**Step 1:** Finding the grounded (neutral) conductor 350 KCMIL THWN requires ▨▨▨ cu. [A] **Solution:** The size grounded (neutral) conductor required is ▨▨▨ AWG copper [B]. **Note 3:** The size grounding electrode conductor, main bonding jumper and equipment bonding jumper are also required to be 2 AWG copper per **Table 250.66**, **250.28(D)** and **250.102(C)**.

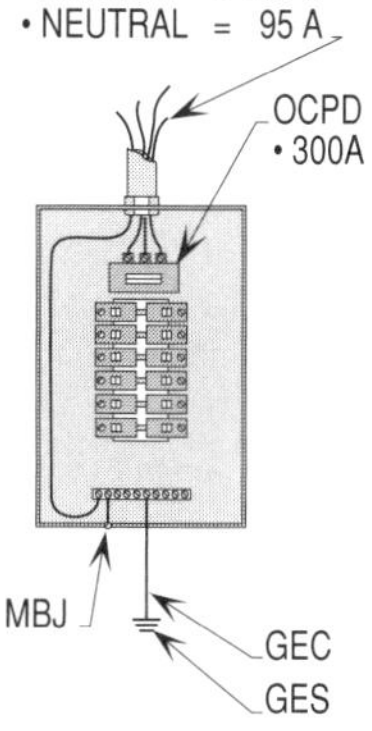

Figure 2-16. This illustration shows an exercise problem for sizing the grounded (neutral) conductor to be used as a normal current-carrying conductor plus utilized as an equipment grounding conductor during a ground-fault condition.

PARALLEL CONDUCTORS
250.24(B)(2)

The grounded (neutral) conductor shall be based on the total circular mil area of the service-entrance conductors when installed in parallel. Grounded (neutral) conductors which are installed in two or more raceways shall be based on the size of the ungrounded service-entrance (phase) conductor in the raceway, but not smaller than 1/0 AWG.

Grounding Note: See **310.4** for grounded (neutral) conductors which are connected in parallel.

> **For example:** In a paralleled three-phase, four-wire service with 4 - 600 KCMIL THWN copper conductors per phase, what size grounded (neutral) conductor is required in each conduit run?
>
> **Step 1:** Finding total KCMIL
> **250.24(B)(1)**
> 600 KCMIL x 4 = 2400 KCMIL
>
> **Step 2:** Finding KCMIL to size grounded (neutral) conductor
> **250.24(B)(2)**
> 2400 KCMIL x .125 = 300 KCMIL
>
> **Step 3:** Dividing KCMIL in each conduit run
> **250.24(B)(2)**
> 300 KCMIL ÷ 4 = 75 KCMIL
>
> **Step 4:** Finding CM rating
> **Table 8, Ch. 9**
> 75 KCMIL x 1000 = 75,000 CM
>
> **Step 5:** Sizing grounded (neutral) conductor for each conduit run
> **250.24(B)(2); Table 8, Ch. 9; 310.4**
>
> **Solution:** **At least 1/0 THWN copper grounded (neutral) conductors are required per 250.24(B)(2) and 310.4.**

Grounding Tip 37: When grounded (neutral) conductors are connected in parallel, see **250.24(B)(2)** and **310.4**.

See **Figure 2-17(a)** for a detailed illustration when sizing the grounded (neutral) conductor to be used in a paralleled installation.

Figure 2-17(a). This illustration shows an exercise problem for sizing the grounded (neutral) conductor to be used in a parallel hook-up.

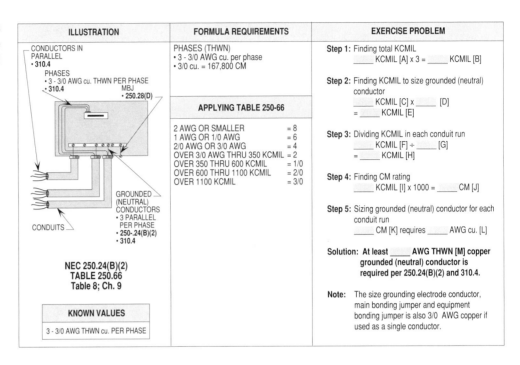

Grounding Tip 38: Before selecting the size of the grounded (neutral) conductor in a parallel hook-up, review **250.24(B)(1)** and **(B)(2)** and **310.4** very carefully.

For example: In a paralleled three-phase, four-wire service with 3 - 700 KCMIL THWN copper conductors per phase, what size grounded (neutral) conductor is required in each run?

Step 1: Finding total KCMIL
250.24(B)(1)
700 KCMIL x 3 = 2100 KCMIL

Step 2: Finding KCMIL to size grounded (neutral) conductor
250.24(B)(2)
2100 KCMIL x .125 = 262.5 KCMIL

Step 3: Dividing KCMIL in each conduit run
250.24(B)(2)
262.5 KCMIL ÷ 3 = 87.5 KCMIL

Step 4: Finding CM rating
Table 8, Ch. 9
87.5 KCMIL x 1000 = 87,500 CM

Step 5: Sizing grounded (neutral) conductor for each conduit run
250.24(B)(2); Table 8, Ch. 9; 310.4
87,500 CM requires 1/0 AWG cu.

Solution: **At least 1/0 AWG THWN copper grounded (neutral) conductors are required per 250.24(B)(2) and 310.4.**

See **Figure 2-17(b)** for a detailed illustration when sizing the grounded (neutral) conductor for a paralleled installation.

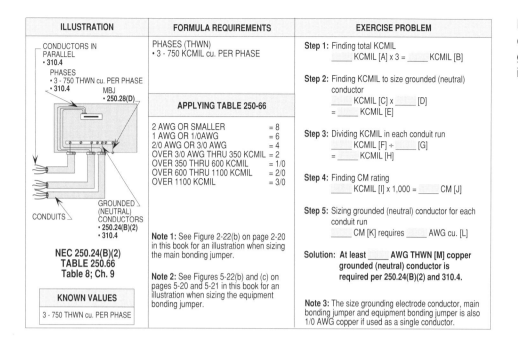

Figure 2-17(b). This illustration sows an exercise problem for sizing the copper grounded (neutral) conductor to be used in a parallel hook-up.

HIGH-IMPEDANCE
250.24(B)(3)

The grounded (neutral) conductor on a high-impedance grounded neutral system shall be grounded per **250.36**.

GROUNDING ELECTRODE CONDUCTOR
250.24(C)

A grounding electrode conductor shall be used to connect the equipment conductors, the service equipment enclosures, and where the system is grounded, the grounded service conductor to the grounding electrode or electrodes. **(See Figure 2-18)**

Where the three-phase, three-wire critical load is relatively large compared with loads that require a grounded (neutral) circuit conductor, a high-impedance grounded service power supply is sometimes is used. This arrangement requires an on-site transformer for loads that require a grounded (neutral) circuit conductor.

Note that high-impedance grounded neutral system connections shall be made per **250.36**. High-impedance grounded systems should not be used unless they are equipped with ground-fault indicators or alarms, or both, and qualified persons are available to quickly locate and remove ground-faults. If ground-faults are not promptly removed, the service reliability will be reduced.

Grounding Tip 39: A High-Impedance Grounding Unit can be sized to allow 4 to 5 amps to flow when a ground-fault occurs to one of the ungrounded (phase) conductors. This current can be used to sound an alarm without interrupting the service supply from the utility transformer.

UNGROUNDED SYSTEM GROUNDING CONNECTIONS
250.24(D)

A premises wiring system that is supplied by an AC service supply that is ungrounded, shall have at the service, a grounding electrode conductor connected to the grounding electrode system. The grounding electrode conductor shall be connected to a metal enclosure of the service conductors at any accessible point from the load end of the service drop or service lateral to the service disconnecting means.

Grounding Tip 40: When the electrical system supply is ungrounded, the electrical equipment down stream from the service equipment shall be grounded with an equipment grounding means that connects to a grounding point in the service equipment enclosure.

Figure 2-18. This illustration shows the procedure for using a grounding electrode conductor to connect equipment grounding conductors, the service equipment enclosures, and where the system is grounded, the grounded service conductor to the grounding electrode or electrodes.

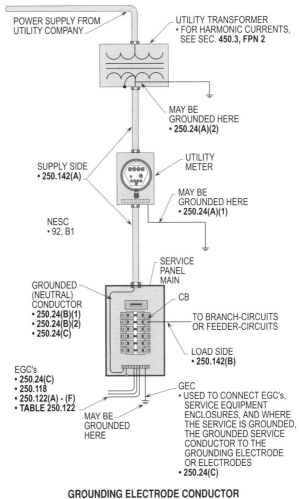

On an ungrounded system, none of the circuit conductors of the system are intentionally grounded. Note that the bonding together of all conductive enclosures and equipment in each circuit with equipment grounding conductors shall be done. These equipment grounding conductors shall be run with or enclose the circuit conductors, and they shall provide a permanent, low-impedance path for ground-fault currents to be detected. **(See Figure 2-19)**

Quick Calc

- 300 A main OCPD x 189
- 56,700 CM
- Table 8, Ch. 9
- 56,700 CM requires 2 AWG cu.

Solution: The size grounded (neutral) conductor required is 2 AWG copper.

CONDUCTOR TO BE GROUNDED - AC SYSTEMS 250.26

The grounded (neutral) conductor whether a grounded phase or neutral is usually installed with a white or gray insulation or otherwise identified. The following systems shall be grounded and have a grounded (phase) conductor or grounded (neutral) conductor: **(See Figure 2-20)**

- 120 volt, two-wire, single-phase - one neutral conductor
- 120/240 volt, three-wire, single-phase - the neutral conductor
- 120/208 volt, four-wire, three-phase wye - the common conductor
- 277/480 volt, four-wire, three-phase wye - the common conductor
- 240 volt, three-wire, three-phase delta - the common conductor
- 480 volt, three-wire, three-phase delta - the common conductor
- 600 volt, three-wire, three-phase delta - the common conductor
- 480 volt corner grounded delta - one phase conductor

Circuit and System Grounding

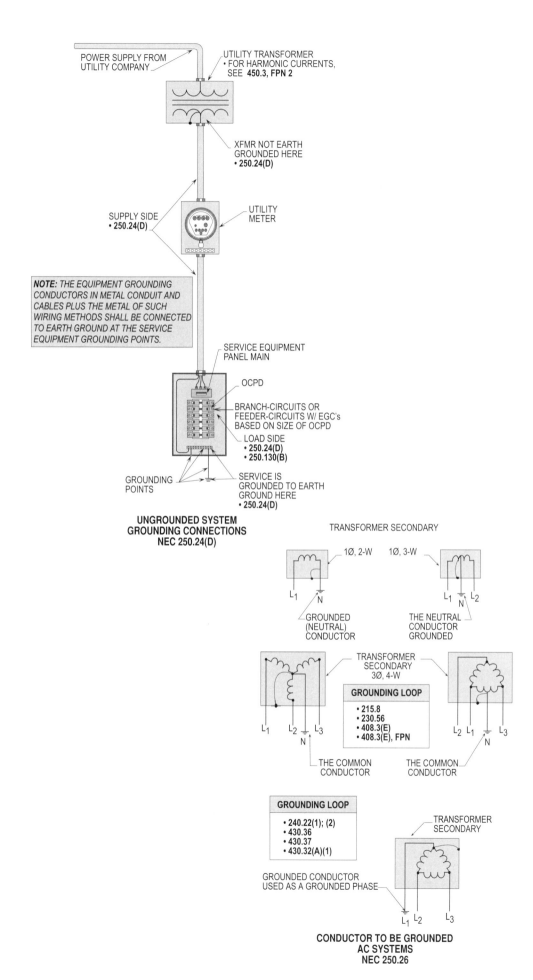

Figure 2-19. This illustration shows the procedures for installing an ungrounded service supply.

Grounding Tip 41: Ungrounded systems are recommended when a facility cannot afford an unscheduled outage of power. When one phase of the system accidentally goes to ground, no outage occurs, for the result is the same if this phase was intentionally connected to earth ground. Note that there can be a big problem if a second phase unintentionally goes to ground before the first fault is cleared.

Figure 2-20. The grounded (neutral) conductor, whether a grounded phase or neutral is usually installed with a white or gray insulation or otherwise identified.

MAIN BONDING JUMPER
250.28

An unspliced main bonding jumper (for grounded systems) shall be installed to connect the equipment grounding conductors and the service disconnecting enclosure to the grounded (neutral) conductor of the system within the enclosure for each service disconnect. The following conditions of use shall be complied with when installing main bonding jumper:

- Material
- Construction
- Attachment
- Size

MAIN BONDING JUMPER - MATERIAL
250.28(A)

Unspliced main bonding jumpers shall be of copper or other corrosion-resistant material. Main bonding jumpers shall be installed as a wire, bus, screw or similar suitable conductor. **(See Figure 2-21)**

Figure 2-21. This illustration shows the items permitted to be used as a main bonding jumper.

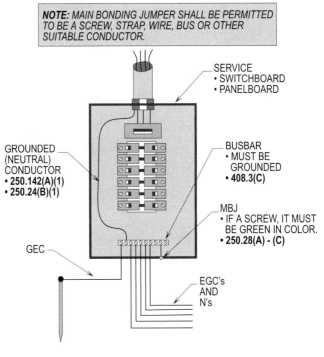

MAIN BONDING JUMPER
NEC 250.28

Grounding Tip 42: If the main bonding jumper is a screw, it shall be identified with a green color so that engineers, electricians and inspectors can easily verify that the grounded busbar is bonded and grounded to the metal case of the enclosure housing the service equipment components.

MAIN BONDING JUMPER - CONSTRUCTION
250.28(B)

When a screw is used as the main bonding jumper, it shall be identified with a green colored finish which is visible with the screw installed. **(See Figure 2-21)**

MAIN BONDING JUMPER - ATTACHMENT
250.28(C)

Section **250.8** specifies the manner in which the main and equipment bonding jumpers shall be attached for electrical circuits and equipment. Section **250.70** specifies the manner by which grounding electrodes shall be attached together after sizing and selecting such grounding conductors. **(See Figure 2-21)**

MAIN BONDING JUMPER - SIZE
250.28(D)

The size of the service main bonding jumper and the service equipment bonding jumper shall be based on the size of the service-entrance conductors and shall be sized per **Table 250.66**. These bonding jumers shall be sized at least 12 1/2 percent of the area of the largest ungrounded (phase) conductor where the service conductors are installed larger than 1100 KCMIL copper or 1750 KCMIL aluminum.

For example: What size copper main bonding jumper is required based on the service-entrance conductors being rated at 250 KCMIL THWN-THHN copper?

    Step 1:   Finding the main bonding jumper
                 250.28(D); Table 250.66
                 250 KCMIL requires 2 AWG cu.

    Solution: **The size of the main bonding jumper is required to be 2 AWG copper.**

See **Figure 2-22(a)** for an exercise problem when sizing the main bonding jumper.

For example: In a paralleled three-phase, four-wire service with 3 - 700 KCMIL THWN copper conductors per phase, what size main bonding jumper is required to ground the busbar to the service equipment enclosure?

    Step 1:   Finding total KCMIL
                 250.28(D)
                 700 KCMIL x 3 = 2100 KCMIL

    Step 2:   Finding KCMIL to size MBJ
                 250.28(D)
                 2100 KCMIL x .125 = 262.5 KCMIL

    Step 3:   Sizing MBJ
                 250.28(D); Table 8, Ch. 9
                 262.5 KCMIL requires 300 KCMIL

    Solution: **The size of the main bonding jumper is required to be 300 KCMIL copper.**

See **Figure 2-22(b)** for an exercise problem when sizing the main bonding jumper where the ungrounded (phase) conductors are in parallel.

Grounding Tip 43: The main bonding jumper shall be at least equivalent to the grounding electrode conductor and the grounded (neutral) conductor sized per **Table 250.66** based on the largest ungrounded (phase) conductor supplying the service equipment.

Grounding Tip 44: The main bonding jumper must be the same size as the grounded (neutral) conductor per **250.24(B)(1).**

GROUNDING SEPARATELY DERIVED AC SYSTEMS
250.30

Low-voltage and high-voltage feeder-circuits are sometimes installed from floor-to-floor in a high-rise building with transformers installed on each floor to reduce the voltage to 120/240 or 120/208 volts for general use lighting and receptacle loads in large building applications. Such grounding, since the 1978 NEC, may be installed either at the transformer or at the load served which is connected and supplied from the secondary side per **240.21(C)(2), (C)(3) and 240.92(B)(1) thru (B)(3). (See Figure 2-23)**

Figure 2-22(a). The size of the main bonding jumper and the service equipment bonding jumper shall be based on the size of the service-entrance conductors and shall be sized per **Table 250.66**.

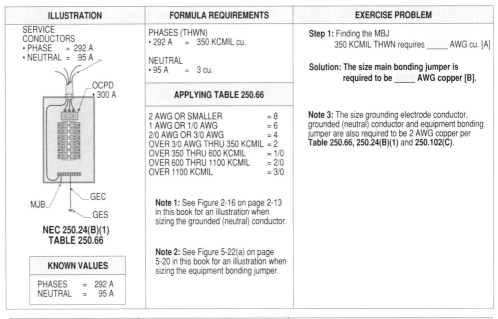

Figure 2-22(b). This illustration shows the main bonding jumper being sized for a parallel hook-up of the ungrounded (phase) conductors.

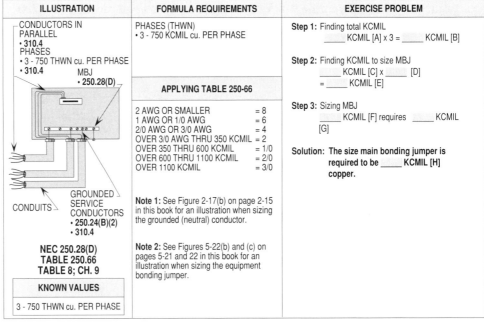

Figure 2-23. The bonding jumper and grounding electrode conductor is designed and installed based on the derived phase conductors supplying the panel, switch, or other equipment connected from the secondary side of the transformer.

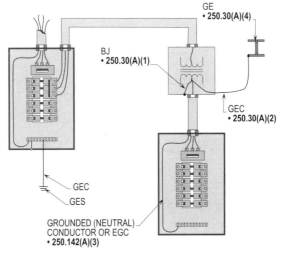

GROUNDED SYSTEMS
250.30(A)

A separately derived system that is grounded shall comply with the following:

- Bonding jumper
- Grounding electrode conductor
- Grounding electrode conductor taps
- Grounding electrode
- Equipment bonding jumper size
- Grounded conductor

Grounding Tip 45: If a separately derived system is installed for supplying computer equipment, an isolated k-rated transformer should be used.

BONDING JUMPER
250.30(A), Ex.

The grounding connection requirements for high-impedance grounded neutral systems shall not be required to comply with **250.30(A)(1)** and **(A)(2)** but shall be made per **250.36** and **250.186**.

BONDING JUMPER
250.30(A)(1)

The bonding jumper shall be designed and installed based on the derived ungrounded (phase) conductors supplying the panel, switch or other equipment connected from the secondary side of the transformer and sized per **250.28(D)** and **250.102(C)**. The bonding jumper shall be sized per **250.66** and **Table 250.66** from the ungrounded (phase) conductors up to 1100 KCMIL for copper and 1750 KCMIL for aluminum. The bonding jumper shall be sized at least 12 1/2 percent (.125) of the area of the largest ungrounded (phase) conductor where the service conductors are installed larger than 1100 KCMIL copper or 1750 KCMIL aluminum. The bonding jumper shall be installed and connected at any point on the separately derived system from the source to the first system disconnecting means or overcurrent protection device. If the ungrounded (phase) conductors are larger than 1100 KCMIL for copper and 1750 KCMIL for aluminum, the bonding jumper will normally be larger than the grounding electrode conductor.

Grounding Tip 46: When the size of the grounding electrode conductor from **Table 250.66** has been selected, the grounded (neutral) conductor and main bonding jumper shall be equal to the size grounding electrode conductor selected, if the ungrounded (phase) conductors are rated 1100 KCMIL or less.

For example: What size copper bonding jumper is required to bond and ground the secondary of a separately derived system having 4 - 3/0 AWG THWN copper conductors connected to its secondary?

    Step 1: Finding BJ
            250.30(A)(1); Table 250.66
            3/0 AWG THWN cu. requires 4 AWG cu.

    Solution: **The size of the bonding jumper is required to be 4 AWG copper.**

Grounding Tip 47: The bonding jumper used to bond the secondary side of a separately derived system shall be at least the same size as the grounded (neutral) conductor and grounding electrode conductor.

See **Figure 2-24(a)** for an exercise problem when sizing the copper bonding jumper to bond and ground the secondary of a separately derived system.

Figure 2-24(a). The size of the bonding jumper for a separately derived system shall be based on the size of the secondary conductors and selected per **Table 250.66**.

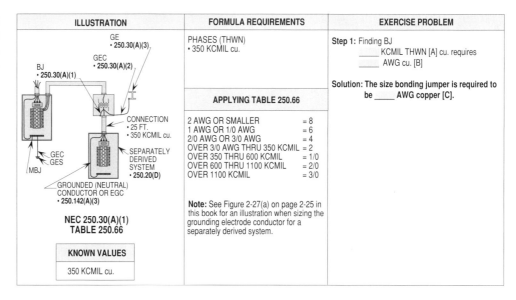

Grounding Tip 48: The 300 KCMIL bonding jumper is used to transfer the maximum amount of fault current that the system may encounter.

For example: In a paralleled three-phase, four-wire service with 4 - 600 KCMIL THWN copper conductors per phase, what size bonding jumper is required to bond and ground the separately derived system?

Step 1: Finding total KCMIL for BJ
250.30(A)(1); 250.28(D)
600 KCMIL x 4 = 2400 KCMIL

Step 2: Finding KCMIL for BJ
250.30(A)(1); 250.28(D)
2400 KCMIL x .125 = 300 KCMIL

Step 3: Sizing BJ
250.30(A)(1); 250.28(D); Table 8, Ch. 9
300 KCMIL requires 300 KCMIL

Solution: **The size of the bonding jumper is required to be 300 KCMIL copper.**

See Figure 2-24(b) for an exercise problem when sizing the copper bonding jumper to bond and ground the secondary of a separately derived system (parallel hook-up)

Figure 2-24(b). This illustration shows the procedure for sizing the bonding jumper for a separately derived system with a parallel hook-up from its secondary to the panelboard

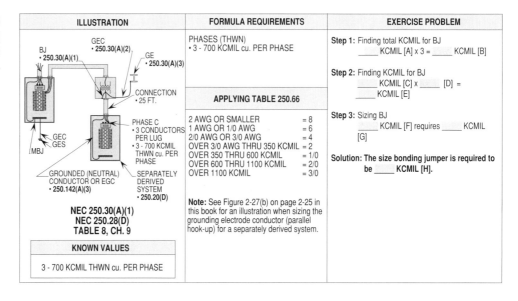

ADDITIONAL BONDING JUMPER
250.30(A)(1), Ex. 1

An additional bonding jumper connection (more than one) shall be permitted to be made only where doing so will not create a parallel path for the grounded circuit conductor. the additional bonding jumper shall not be permitted to be smaller than the other bonding jumper but shall shall not be required to be larger than the ungrounded (phase) conductor(s). **(See Figure 2-25)**

This section states that no parallel may be formed when installing an additional bonding jumper at equipment supplied by a separately derived system. The basic rule requies the bonding jumper and the grounding electrode conductor connection to the grounded system conductor to be made at the same point. There is an Ex. to this rule, but only for cases, where there won't be a parallel path, such as a nonmetallic conduit (PVC) run back to the transformers source and so located to not share conductive contact with common structural elements or equipment grounding conductors. The purpose of this rule is to keep the grounded (neutral) conductor current confined and flowing over insulated electrical circuit conductors.

Grounding Tip 49: To apply **Ex. 1 to 250.30(A)(1)**, the wiring method between the secondary of the transformer and the load supplied shall be of the nonmetallic type and not capable of providing a parallel path for current flow.

Figure 2-25. This illustration shows the grounding and bonding necessary when an additional bonding jumper is used.

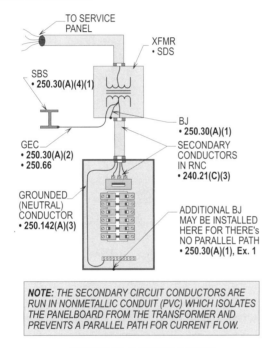

ADDITIONAL BONDING JUMPER
NEC 250.30(A)(1), Ex. 1

BONDING JUMPER - CLASS 1, CLASS 2 OR CLASS 3 CIRCUITS
250.30(A)(1), Ex. 2

The bonding jumper shall be sized not smaller than the derived ungrounded (phase) conductors and shall not be smaller than 14 AWG copper or 12 AWG aluminum for systems which supplies a Class 1, Class 2 or Class 3 circuit, and is derived from a transformer which is rated not more than 1000 volt-amperes. **(See Figure 2-26)**

Grounding Tip 50: A Class 1 circuit operates at 30 volts or less with an output of 1000 VA or less per **725.21**.

GROUNDING ELECTRODE CONDUCTOR
250.30(A)(2)

The grounding electrode conductor shall be installed in accordance with **250.30(A)(2)(a)** and **(b)** based on the following type of systems:

- Single separately derived systems
- Multiple separately derived systems

Figure 2-26. This illustration shows the bonding requirements for a small transformer using the equipment grounding conductor instead of installing a grounding electrode conductor and connecting it to a grounding electrode.

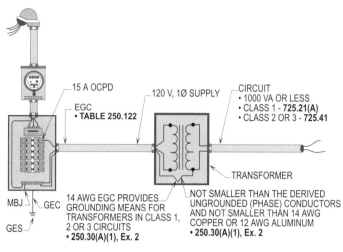

SINGLE SEPARATELY DERIVED SYSTEMS
250.30(A)(2)(a)

Grounding Tip 51: Because there is no OCPD ahead of the supply conductors, the grounding electrode conductor shall be sized from the largest transformer secondary conductor run from the transformer to the load served. In other words, it shall be sized as if it were part of the service supply conductors.

The grounding electrode conductor shall be designed and installed based on the derived ungrounded (phase) conductors supplying the panel, switch or other equipment connected from the secondary of the transformer and shall be sized per **Table 250.66**.

The grounding electrode conductor shall be installed and connected at any point on the separately derived system from the source to the first system disconnecting means or overcurrent protection device. When the KCMIL rating is greater than 1100 for copper and 1750 for aluminum, the grounding electrode conductor will usually be smaller than the bonding jumper.

A common continuous grounding electrode conductor shall be permitted to be extended from the grounding electrode system and run through the building and the connection made at an accessible location near the separately derived system required to be grounded per **250.30(A)(3)**.

> **For example:** What size copper grounding electrode conductor is required to bond and ground the secondary of a separately derived system having 4 - 3/0 AWG THWN copper conductors connected to its secondary?
>
> **Step 1:** Finding size GEC
> **250.30(A)(2)(a); Table 250.66**
> 3/0 AWG THWN cu. requires 4 AWG cu.
>
> **Solution:** **The size of the grounding electrode conductor is required to be 4 AWG copper.**

See Figure 2-27(a) for an exercise problem when sizing the copper grounding electrode conductor.

Circuit and System Grounding

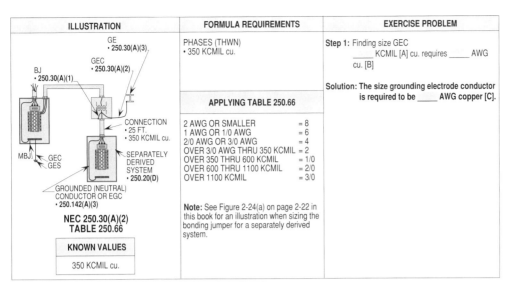

Figure 2-27(a). This illustration shows an exercise problem for sizing the grounding electrode conductor based on the size of the ungrounded (phase) conductors.

Grounding Tip 52: The largest grounding electrode conductor based on **Table 250.66** is 3/0 AWG and it does not matter if the KCMIL rating of the transformers secondary exceeds 1100 KCMIL.

For example: In a paralleled three-phase, four-wire service with 4 - 600 KCMIL THWN copper conductors per phase, what size grounding electrode conductor is required to bond and ground the separately derived system?

Step 1: Finding total KCMIL for GEC
250.30(A)(2)(a); 250.28(D)
600 KCMIL x 4 = 2400 KCMIL

Step 2: Sizing GEC
250.30(A)(2)(a); Table 250.66
2400 KCMIL requires 3/0 AWG cu.

Solution: **The size of the grounding electrode conductor is required to be 3/0 AWG copper.**

See Figure 2-27(b) for an exercise problem when sizing the copper grounding electrode conductor.

Grounding Tip 53: The largest grounding electrode conductor required by **Table 250.66** is 3/0 AWG copper and it does not matter if the KCMIL rating of the secondary conductors exceeds 1100 KCMIL. However, the bonding jumper shall be sized at 12 1/2 percent for three parallel runs of 700 KCMIL. This calculation (700 KCMIL x 3 x .125 = 262.5 KCMIL) requires a 300 KCMIL copper per **Table 8 to Ch. 9**.

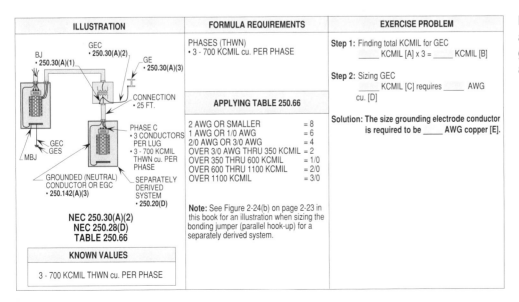

Figure 2-27(b). This illustration shows an exercise problem for sizing the grounding electrode conductor in a parallel hook-up.

MULTIPLE SEPARATELY DERIVED SYSTEMS
250.30(A)(2)(b)

A properly sized continuous common grounding electrode conductor shall be permitted to ground all separately derived systems on each floor of a high-rise building. The continuous common grounding electrode conductor shall be sized in accordance with **250.66** based on the total area of the derived ungrounded (phase) conductors smaller than 3/0 AWG copper or 250 KCMIL aluminum. **(See Figure 2-28)**

Figure 2-28. This illustration shows a properly sized continuous common grounding electrode conductor to serve as grounding means on each floor of a high-rise building.

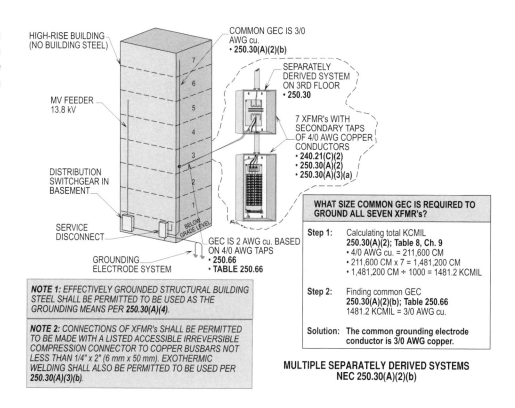

Grounding Tip 54: If it were not for this Ex. to 250.30(A)(2), a low-voltage transformer of 24 volts would have to be grounded with at least a 8 AWG copper grounding electrode conductor per **Table 250.66**. Note that there would not be enough room in the enclosure of the transformer to terminate a 8 AWG copper grounding electrode conductor.

GROUNDING ELECTRODE CONDUCTOR - CLASS 1, CLASS 2 OR CLASS 3 CIRCUITS
250.30(A)(2)(a), Ex.

A Class 1, Class 2 or Class 3 remote-control or signaling transformer that is rated 1,000 VA or less and has a grounded secondary conductor bonded to the metal case of the transformer, no grounding electrode conductor shall be required to be installed. However, the metal transformer case shall be properly grounded by a grounded metal raceway or equipment grounding conductor that supplies its primary or by means of an equipment grounding conductor that connects the case back to the grounding electrode for the primary system.

At least a 14 AWG copper conductor shall be used to bond and ground the transformer secondary to the transformer frame. This leaves the supply raceway, equipment grounding conductor or both to the transformer to provide the grounding return path back to the common service ground.

Circuit and System Grounding

GROUNDING ELECTRODE CONDUCTOR TAPS
250.30(A)(3)

Taps shall be permitted to be connected to the from a separately derived system to a continuous common grounding electrode conductor. Each tap conductor shall be connected to the grounded conductor of the separately derived system to the continuous common grounding electrode conductor. The following shall be considered when designing and installing grounding electrode conductor taps:

- Tap conductor size
- Connections
- Installation
- Bonding

TAP CONDUCTOR SIZE
250.30(A)(3)(a)

Each tap conductor from the separately derived system to the continuous common grounding electrode conductor shall be sized per **250.66** based on the derived ungrounded (phase) conductors of the separately derived system that it supplies. **(See Figure 2-29)**

Quick Calc

Sizing main bonding jumper based on OCPD

- 300 A main OCPD x 189
- 56,700 CM
  Table 8, Ch. 9
- 56,700 CM requires 2 AWG cu.

Solution: The size main bonding jumper required is 2 AWG copper.

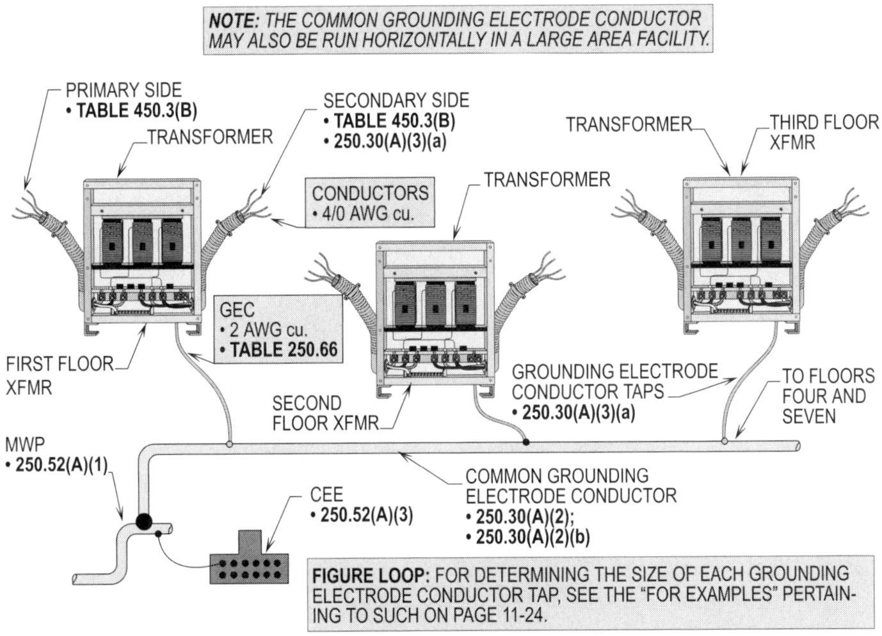

Figure 2-29. This illustration shows that each tap conductor from the separately derived system to the continuous common grounding electrode conductor shall be sized per **250.66** based on the derived ungrounded (phase) conductors of the separately derived system that it supplies.

TAP CONDUCTOR SIZE
NEC 250.30(A)(3)(a)

CONNECTIONS
250.30(A)(3)(b)

Connections of the grounding electrode conductor from the separately derived system to the continuous common grounding electrode conductor shall be made at an accessible location by an irreversible compression connector listed for the purpose, listed connections to copper busbars not less than 1/4 in. x 2 in. (6 mm x 50 mm) or by the exothermic welding process.

2-29

INSTALLATION
250.30(A)(3)(c)

The taps to each separately derived system and the continuous common grounding electrode conductor shall comply with **250.64(A), (B), (C)** and **(E)** which covers the installation requirements for grounding electrode conductors.

BONDING
250.30(A)(3)(d)

If the structural steel or interior metal piping is available and the continuous common grounding electrode conductor is also installed, these electrodes shall be bonded together as near as practicable to the location of the separately derived system.

GROUNDING ELECTRODE
250.30(A)(4)

The grounding electrode conductor shall be as near as possible and preferably in the same area as the grounding electrode conductor connection to the system. From the following choices, one shall be selected and installed in the order as they are listed:

- Nearest effectively grounded structural building steel
- Nearest effectively grounded metal water pipe with 5 ft. (1.5 m) from the point of entrance into the building
- Other electrodes as specified in **250.52** where the electrodes specified in **250.30(A)(4)(1) and (A)(4)(2)** are not available

Grounding Note: Metal water pipe located in the area shall be bonded to the grounded conductor per **250.104(A)(4)**.

See Figure 2-30 for a detailed illustration pertaining to the installation of the grounding electrode conductor.

Figure 2-30. The grounding electrode conductor shall be as near as possible and preferably in the same area as the grounding electrode conductor connec-

Grounding Tip 55: Where a separately derived system is grounded to a concrete-encased electrode, the grounding electrode conductor shall be at least a 4 AWG copper per **250.66(B)** and **250.52(A)(3)**. If its grounded to a ground ring, the grounding electrode conductor shall be at least a 2 AWG copper or the size of the ring conductor per **250.66(C)** and **250.52(A)(4)**.

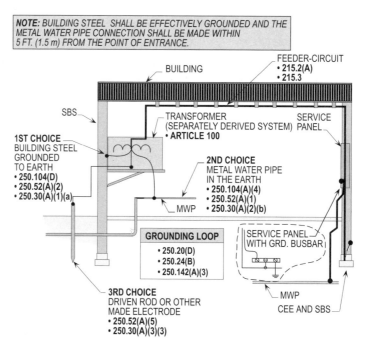

GROUNDING ELECTRODE
NEC 250.30(A)(4)

The grounding electrode conductor shall not be required to be installed larger than 3/0 AWG for copper or 250 KCMIL for aluminum when connecting to the nearest building steel or nearest metal water pipe system. The grounding electrode conductor shall not be required to be installed larger than 6 AWG copper or 4 AWG aluminum when connecting to a driven rod or other made electrodes.

GROUNDING ELECTRODE - INTERIOR METAL WATER PIPE
250.30(A)(4)(2), Ex.

Where only qualified persons will service metal water pipes located in commercial and industrial buildings, such water pipes shall be permitted by this exception to be used as the only grounding electrode conductor. However, the entire length of the piping that is subject to ground currents shall be exposed. Check with the AHJ to verify if he or she consider the 4 in. (100 mm) concrete between floors of a high-rise building which conceals such piping to be exposed. Note that the AHJ may or may not consider the piping exposed. **(See Figure 2-31)**

Quick Calc

Service conductors are 250 KCMIL cu. (sizing main bonding jumper)

Table 250.66
- 250 KCMIL cu. requires
- GEC to be 2 AWG cu.
- Ground conductor to be 2 AWG cu.
- MBJ to be 2 AWG cu.

Solution: Main bonding jumper must be at least 2 AWG copper.

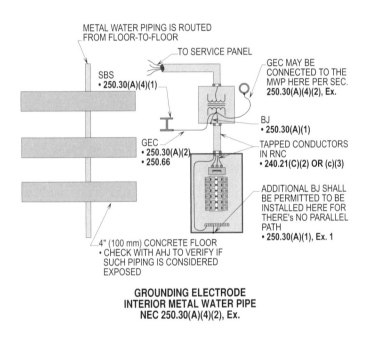

Figure 2-31. If the metal water pipe is exposed and qualified personnel service the installation, the connection of the grounding electrode to the water pipe shall not be required to be run within 5 ft. (1.5 m) of where it enters the building.

Grounding Tip 56: Under certain conditions, the AHJ may consider the metal water piping run through the concrete floor to be exposed for the sake of interpreting **250.30(A)(4)(2), Ex.** See **250.52(A)(1), Ex.** for requirements that allow this type of grounding technique.

GROUNDING ELECTRODE - USED IN SERVICE EQUIPMENT
250.30(A)(4)(3), Ex.

Where a separately derived system originates in listed equipment suitable for use as service equipment, the grounding electrode used for the service of feeder shall be permitted to be used as the grounding electrode for the separately derived system provided the grounding electrode conductor from the service or feeder to the grounding electrode is of sufficient size for the separately derived system. To apply this rule, the equipment ground bus internal to the service equipment shall not be permitted to be smaller than the required grounding electrode conductor. If this is the case, the grounding electrode connection for the separately derived system shall be permitted to be made to the bus. **(See Figure 2-32)**

Figure 2-32. This illustration shows that a separately derived system shall be permitted to be grounded to the grounding bus supplied by a feeder-circuit.

Grounding Tip 57: When applying this Ex. to 250.30(A)(4)(3), the grounding electrode conductor from the transformer does not have to be run and connected to a grounding electrode. It can be connected in the service equipment to the same electrode system.

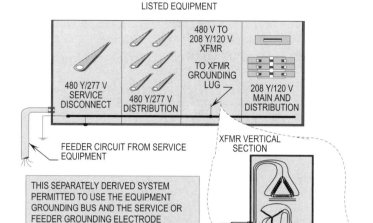

EQUIPMENT BONDING JUMPER SIZE
250.30(A)(5)

Where the equipment bonding jumper is run with the derived ungrounded (phase) conductors from the source of the separately derived system to the first disconnecting means, the equipment bonding jumper shall be sized per **250.28(A)** thru **(D)** based on the largest ungrounded (phase) conductor derived from the source of the separately derived system.

GROUNDED CONDUCTOR
250.30(A)(6)

Where the equipment bonding jumper is located at the first disconnecting means and not at the separately derived system, a grounded (neutral) conductor shall be installed to carry excessive current should a ground-fault condition occur. The grounded (neutral) conductor shall be routed with the ungrounded (phase) conductors and shall not be smaller than the required grounding electrode conductor specified in **Table 250.66**, But shall not be required to be larger than the largest ungrounded (phase) conductor. In addition, for ungrounded (phase) conductors larger than 1100 KCMIL copper or 1750 KCMIL aluminum, the grounded (neutral) conductor shall not be smaller than 12 1/2 percent of the area of the largest ungrounded (phase) conductor. The grounded (neutral) conductor of a three-phase, three-wire delta system shall be the same size as the ungrounded (phase) conductors.

> **For example:** What size grounded (neutral) conductor is required from the secondary side of a separately derived system feeding a panelboard with a 250 KCMIL ungrounded (phase) conductors?
>
> **Step1:** Calculating grounded (neutral) conductor
> **250.30(A)(6)(a); Table 250.66**
> 250 KCMIL cu. requires 2 AWG cu.
>
> **Solution:** The size grounded (neutral) conductor is 2 AWG copper.

Circuit and System Grounding

Where the derived ungrounded (phase) conductors are installed in parallel, the size of the grounded (neutral) conductor shall be based on the total circular mil area of the parallel conductors. Where installed in two or more raceways, the size of the grounded (neutral) conductor in each raceway shall be based on the size of the ungrounded (phase) conductors in the raceway but not smaller than 1/0 AWG copper.

For example: What size grounded (neutral) conductor is required from the secondary side of a separately derived system feeding a power panelboard with 3 - 3/0 AWG ungrounded (phase) conductors?

Step 1: Calculating grouned (neutral) conductor
250.30(A)(6)(b); Table 250.66
3/0 AWG cu. requires 4 AWG cu.

Step 2: Finding size per **250.30(A)(6)(b)**
Minimum size requires 1/0 AWG cu.

Solution: The size grounded (neutral) conductor is 1/0 AWG copper.

UNGROUNDED SEPARATELY DERIVED AC SYSTEMS
250.30(B)

In an ungrounded system there is no grounded (phase or neutral) conductor to be grounded. In this case, the equipment grounding conductor from the separately derived system is connected to the grounding electrode conductor. As noted, the equipment grounding conductor could be a grounding conductor, or if the equipment grounding conductor on the premises is conduit, the conduit or any other form of metal-clad cable wiring could be the equipment grounding conductor. **(See Figure 2-33)**

Grounding Note: A grounding electrode shall be required even though no conductor from the supply is grounded and the equipment grounding conductor scheme shall be connected to the grounding electrode conductor that is connected to the grounding electrode system.

Grounding Tip 58: The same grounding procedure for ungrounded systems are required as for grounded systems. The only difference is there is not a grounded (neutral) conductor present.

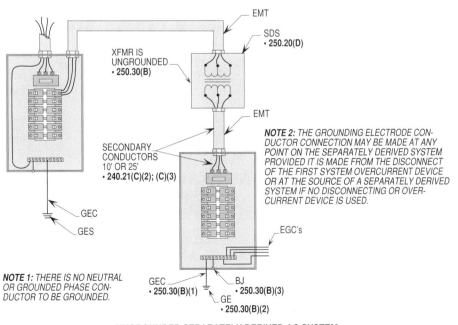

Figure 2-33. This illustration shows a separately derived system that is operated ungrounded.

GROUNDING ELECTRODE CONDUCTOR
250.30(B)(1)

Grounding Tip 59: For information of protection against corrosion of grounding electrode conductors connected to grounding electrodes, see Sec. 5 in UL 467.

Section **250.66** covers the sizing of grounding electrode conductors. The grounding electrode conductor for the derived phase conductors shall be used to connect the grounded (neutral) conductor of the derived system to the grounding electrode.

This grounding connection shall be permitted to be made at any point on the separately derived system provided it is made from the disconnect of the first system overcurrent device. It shall also be permitted to be made at the source of a separately derived system if no disconnecting or overcurrent device is used. This requirement indicates that it could be made from the secondary of the transformer from which the separately derived system originates.

Refer to Figures 2-24(a) and (b) for a detailed illustration when sizing the grounding electrode conductor when there is a grounded (neutral) conductor present.

GROUNDING ELECTRODE
250.30(B)(2)

Grounding Tip 60: For more detailed information pertaining to types of grounding electrodes, see **Sec. 9 in UL 467**.

Grounding Tip 61: If any or all of the grounding electrodes in 250.52(A)(1) thru (A)(7) are available, they shall be bonded and grounded together to form a grounding electrode system.

The grounding electrode shall be as close as practical, and is preferred in the same area as the grounding conductor to the system. Where grounding electrodes are used, they should be:

- The nearest effectively grounded metal building or structure.
- The nearest water pipe that is effectively grounded by sufficient metal water pipe buried in the earth. (within 5 ft. (1.5 m) from the point of entrance)
- Made electrodes as specified in **250.52(A)(4) thru (A)(7)**, provided that the electrodes that were specified in the above are not available.

Refer to Figure 2-30 for a detailed illustration when selecting the grounding electrode for a separately derived system.

TWO OR MORE BUILDINGS OR STRUCTURES SUPPLIED FROM A COMMON SERVICE
250.32

Grounding Tip 62: It is highly recommended that ungrounded systems should not be used unless they are equipped with ground-fault indicators or alarms or both. Qualified persons should be available to quickly locate and remove ground-faults per **250.21, FPN**.

These requirements apply to two or more buildings or structures supplied from one service. Basically, the grounded system of each building or structures shall have a grounding electrode that is also connected to the metal enclosure of the disconnecting means which is actually defined as a feeder-circuit. However, a grounding electrode shall be provided at each building or sturcture and connected to the disconnecting means as well as the grounded circuit conductor of the AC supply on the supply side of the building or structure disconnecting means. **(See Figure 2-34)**

Circuit and System Grounding

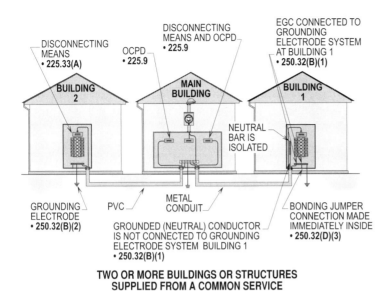

Figure 2-34. This illustration shows the used of the grounded (neutral) conductor and the equipment grounding conductor in a feeder-circuit supplying another building or structure.

Grounding Tip 63: When terminating the grounded (neutral) conductor and equipment grounding conductor at another structure or building, comply with the requirements in **408.20**.

GROUNDING ELECTRODE
250.32(A)

Where two or more buildings or structures are supplied from one service by a feeder or branch-circuit, any of the grounding electrodes that are listed in **250.52(A)(1)** thru **(A)(6)** shall be bonded together to complete the grouding electrode system if one or all are available. **(See Figure 2-35)**

Grounding Tip 64: For detailed information pertaining to rod electrodes, see Sec. 9.2 in UL 497.

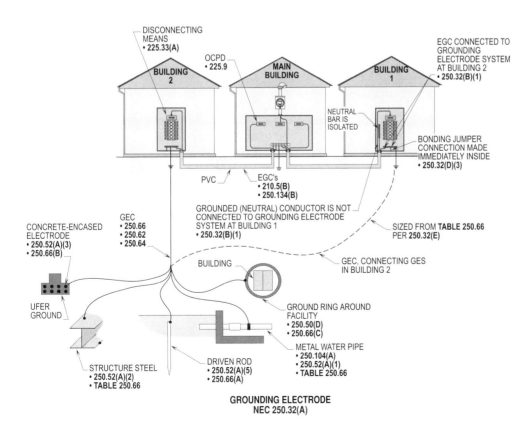

Figure 2-35. If any of the electrodes that are listed in **250.52(A)(1) through (A)(7)** are available, they shall be bonded together to complete the grounding electrode system.

Grounding Tip 65: The OCPD's in the main building can be used to protect the feeder-circuit conductors and panelboards in buildings 1 and 2 per **225.9** and **240.4(A)** thru **(G)**.

GROUNDING ELECTRODE - ONE BRANCH-CIRCUIT
250.32(A), Ex.

A grounding electrode at a separate building or structure shall not be required to be installed where only one branch-circuit serves the building or structure and the branch-circuit includes an equipment grounding conductor for grounding noncurrent-carrying parts of all equipment.

For further information, see **250.32(D)** for applying the requirements when installing the disconnecting means per **225.31, Ex.'s 1** and **2**. **(See Figure 2-36)**

Figure 2-36. A grounding electrode shall not be required to be installed at a separate bulding or structure where only one branch-circuit is run and the branch-circuit includes an equipment grounding conductor for grounding noncurrent-carrying parts of all equipment.

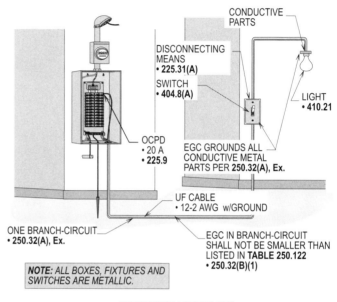

TWO OR MORE BUILDINGS OR STRUCTURES - GROUNDED SYSTEMS
250.32(B)(1); (B)(2)

Grounding Tip 66: The OCPD ahead of the feeder-circuit can be used to protect the circuit conductors and the equipment served per **225.9**. For example, a 225 amp OCPD in the main building can protect 4/0 AWG copper conductors and a 225 amp panelboard for short-circuits, ground-faults and overload conditions.

Where one or more buildings are supplied from a common AC grounded service, each panelboard at each building structure shall be separately grounded. The grounded (neutral) conductor or equipment grounding conductor run from the service panel of a building to a panel in a separate building or structure shall be at least the size specified in **Table 250.122**. The supply to each building or structure shall be disconnected by one to six disconnecting means per **225.33(A)** or one of the exceptions shall be permitted to applied under certain conditions. OCPD's shall comply with the rules of **Article 240** per **225.9**. See **240.3(A)** thru **(G)** for such rules and regulations. **(See Figure 2-37)**

Grounding Note: The metal of a panelboard enclosure shall be grounded in the manner specified in **Article 250**, or **408.3(C)** and **408.20**. An approved terminal bar for equipment grounding conductors shall be provided and secured inside of the enclosure for the attachment of all feeder and branch-circuit equipment grounding conductors when the panelboard is used with nonmetallic raceways, cable wiring or where separate grounding conductors are provided. The terminal bar for the equipment grounding conductors shall be bonded to the enclosure or panelboard frame (if it is metal), or else connected to the grounding conductor that is run with the conductors which supplies the panelboard. Grounded (neutral) conductors shall not be permitted to be connected to a neutral bar, unless it is identified for such use. In addition, is shall be located at the connection between the grounded (neutral) conductor and the grounding electrode.

Circuit and System Grounding

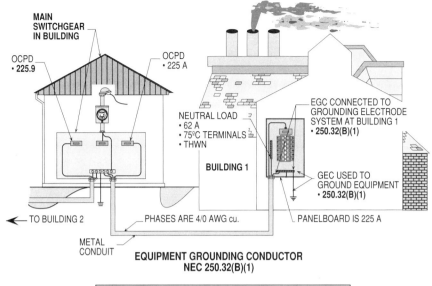

Figure 2-37. Two or more buildings shall be permitted to be supplied by a feeder-circuit and the grounded (neutral) conductor or equipment grounding conductor shall be permitted to be used as a grounding means to ground the subpanel or switchgear. Grounded (neutral) conductor is isolated in panelboard in building 2.

EQUIPMENT GROUNDING CONDUCTOR
250.32(B)(1)

Where a building or structure is supplied from a service in another building by more than one branch-circuit, a grounding electrode shall be installed at the additional building(s) or structure(s) being served. The equipment grounding conductor run to the other building or structure shall be sized per **Table 250.122**.

Where livestock is housed, the equipment grounding conductor shall be insulated or covered where routed with the feeder or branch-circuit and bonded to the metal case of the enclosure. Note that the neutral bus and grounded (neutral) conductor is isolated from the case. See **250.142(B)** and **384.20** for rules pertaining to this type of installation. **(See Figure 2-38)**

For example: A 150 amp (75°C terminals) panelboard in building 1 is supplied by 3 - 1/0 AWG THWN copper conductors and the neutral load is 48 amps. The feeder-circuit is protected by a 150 amp OCPD (75°C terminals). What size neutral and equipment grounding conductor is required using THWN copper conductors in the feeder-circuit?

Step 1: Finding neutral in feeder
215.2(A); Table 310.16
48 A load requires 8 AWG cu.

Step 2: Finding EGC in feeder
250.122(A); Table 250.122
150 A OCPD requires 6 AWG cu.

Soution: **The size of the neutral is 8 AWG copper and the equipment grounding conductor is 6 AWG copper.**

Grounding Tip 67: The disconnecting means for the facility supplied by the feeder-circuit can be located in the main building. With this type of installation, there can be more than six mains in the building supplied by the feeder-circuit per **250.33(A), 250.32, Ex. 1** and **250.32(D)**.

Grounding Tip 68: When the grounded (neutral) conductor is used as a current-carrying conductor plus an equipment grounding conductor for grounding equipment, it must be sized per **220.22** and **Table 250.122** to ensure its large enough to be used as both a neutral and equipment grounding conductor combined. The neutral in the subpanel to the separate building in Figure 2-36 only has to be sized based on the neutral loads per **220.22** and **310.15(B)(4)(a)** thru **(c)**. The equipment grounding conductor is sized per **Table 250.122** based on the OCPD ahead of the feeder-circuit conductors.

See Figure 2-39 for an exercise problem when sizing the grounded (neutral) conductor and equipment grounding conductor from a building or structure (grounded system) to another building.

Grounding Note: An individual equipment grounding conductor shall be permitted to be installed to provide an effective return path for the fault current to travel over instead of using a grounded (neutral) conductor for such use. This particular installation is used to prevent the grounded (neutral) conductor and equipment grounding conductor from joining together at both ends, which if done, provides a parallel path for stray currents to travel over and this is not desirable for such an installation and causes problems for certain types of equipment.

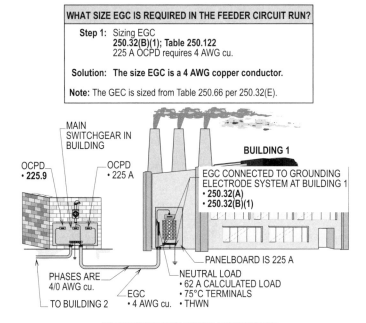

Figure 2-38. This type of installation is used to prevent the grounded (neutral) conductor and equipment grounding conductor from joining together at both ends, which if done, provides a parallel path for stray currents to travel over.

Figure 2-39. This illustration shows an exercise problem for the sizing the grounded (neutral) conductor and equipment grounding conductor from a building or structure to another building.

Grounding Tip 69: The panelboard in building 1 is wired and used as a subpanel which isolates the grounded (neutral) conductor and bonds the equipment grounding conductor to the metal of the enclosure as required per **408.20**. (See Grounding Tip 69 on page 2-37)

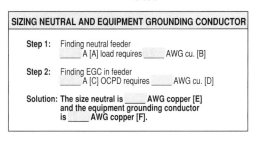

GROUNDED CONDUCTOR
250.32(B)(2)

If a grounded (neutral) conductor, without an equipment grounding conductor, is routed in the circuit to a separate building, it shall be equal to the size required per **Table 250.122** to ensure the capacity to clear a ground-fault condition. See **250.142(A)(2)** for permission to use the grounded (neutral) conductor as a current-carrying conductor and equipment grounding conductor in a feeder-circuit. **(See Figure 2-40)**

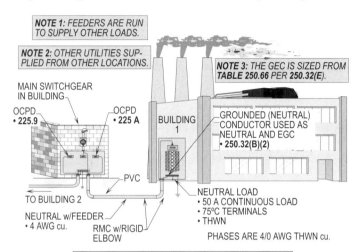

Figure 2-40. Where the grounded (neutral) conductor is used as an equipment grounding conductor plus a neutral, the raceway system that contains the ungrounded (phase) conductors shall be of the nonmetallic type.

Grounding Tip 70: Where the grounded (neutral) conductor is used as an equipment grounding conductor plus a neutral, the raceway system that contains the ungrounded (phase) conductors must be of the nonmetallic type. For example, rigid down at each end with rigid metal conduit and an ELL which converts rigid metal conduit to rigid nonmetallic conduit (PVC). In other words, nonmetallic (PVC) is run in the ditch and connects the two facilities together with rigid metal conduit stubbed up at each building and this type of installation prevents paralleled paths between the conduits and neutral.

For example: A 225 amp (75°C terminals) panelboard in building 1 is supplied by 3 - 4/0 AWG THWN copper conductors and the neutral load is 50 amps. The feeder-circuit is protected by a 225 amp OCPD (75°C terminals). What size neutral is required to service as an equipment grounding conductor?

- **Step 1:** Calculating neutal load
  220.22; 225.3(B); 310.15(B)(4)(c)
  50 A x 125% = 62.5 A

- **Step 2:** Sizing neutral
  Table 310.16
  62.5 A requires 6 AWG cu.

- **Step 3:** Sizing neutral to be used as EGC
  250.32(B)(2); Table 250.122
  225 A OCPD requires 4 AWG cu.

- **Soution:** **The grounded (neutral) conductor is required to be 4 AWG copper to service as an EGC.**

See **Figure 2-41** for an exercise problem when sizing the grounded (neutral) conductor to be used as a grounded (neutral) conductor and equipment grounding conductor from a building or structure (grounded system) to another building.

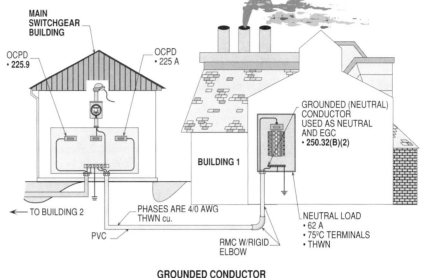

Figure 2-41. This illustration shows an exercise problem for the sizing the grounded (neutral) conductor to be used as an grounded (neutral) conductor and equipment grounding conductor from a building or structure to another building.

Grounding Tip 71: The panelboard in building 1 is wired and used as a panelboard would be for service equipment. For permission to use a panel-board in this manner, see **250.142(B)** and **250.32(B)(2)**. (See Note 3 in Figure 2-38 on page 2-33)

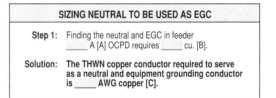

Figure 2-42. A grounding electrode for an ungrounded system shall be connected only to the subpanel or switchgear enclosure when installed at one or more buildings. The equipment grounding conductor shall be sized per **Table 250.122** based on the OCPD ahead of the feeder-circuit conductors.

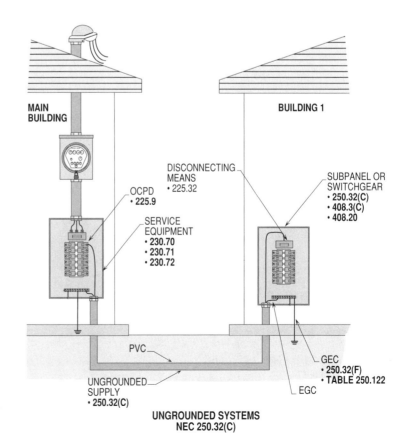

UNGROUNDED SYSTEMS
250.32(C)

A grounding electrode for an ungrounded system shall be connected only to the service equipment enclosure where installed at one or more buildings. The feeder or branch-circuit shall be grounded to an enclosure at the other building being served from the service of the main building. All equipment grounding conductors at the separate building are connected to a grounding bus which is bonded to the enclosure. Such enclosure is bonded and grounded to the grounding electrode conductor and grounding electrode system. **(See Figure 2-42)**

DISCONNECTING MEANS LOCATED IN SEPARATE BUILDING OR STRUCTURE ON SAME PREMISES
250.32(D)

Where one or more disconnecting means supply one or more additional buildings or structures under single management, and where these disconnecting means are located remote from those buildings or structures per the requirements of **225.31, Ex.'s 1 and 2**, all of the following conditions shall be complied with:

- The connection of the grounded (neutral) conductor to the grounding electrode at a separate building or structure shall not be made.

- An equipment grounding conductor for grounding any noncurrent-carrying equipment, interior metal piping systems and building and structural metal frames is run with the circuit conductors to a separate building or structure and bonded to the existing grounding electrode systems.

Grounding Note: Where there are no existing electrodes, a grounding electrode in **250.52(A)(1)** thru **(A)(7)** shall be sized and installed if such building or structure is supplied by more than one branch-circuit.

- Bonding the equipment grounding conductor to the grounding electrode at a sepa-rate building or structure shall be made in a junction box, panelboard or similar enclosure located immediately inside or outside the separate building or structure.

Grounding Note: An equipment grounding conductor shall be run with feeder-circuit conductors and the grounded (neutral) conductor shall not be bonded to the enclosure or equipment grounding bus and the equipment grounding bus shall be connected to a new or existing grounding electrode system at the second building. For safety, all non-current-carrying metal parts of equipment, building steel and interior metal piping systems shall be connected to the grounding electrode system.

See Figure 2-43 for a detailed illustration when applying these requirements.

Grounding Tip 72: If there are more than six disconnecting means in building 1 in Figure 2-43, there shall be a disconnecting means installed in the main building that complies with **225.32, Ex. 1** and **250.32(D)**.

GROUNDING ELECTRODE CONDUCTOR
250.32(E)

The size of the grounding electrode conductor to the grounding electrode(s) shall not be smaller than given in **Table 250.66** based on the largest ungrounded (phase) conductor.

For example: What size grounding electrode conductorA 225 amp (75°C terminals) panelboard in building 1 is supplied by 3 - 4/0 AWG THWN copper conductors and the neutral load is 50 amps. The feeder-circuit is protected by a 225 amp OCPD (75°C terminals). What size neutral is required to service as an equipment grounding conductor?

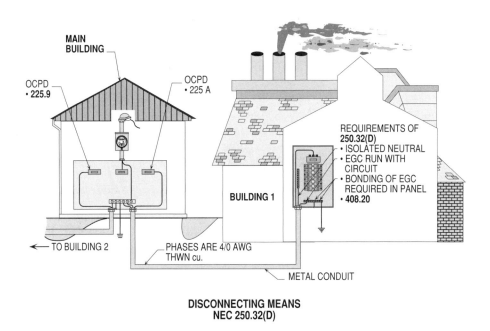

Figure 2-43. If the disconnecting means for building 1 is located in the main building, an equipment grounding conductor shall be routed with feeder-circuit conductors. The equipment grounding conductor shall be bonded to panel enclosure and the grounded (neutral) conductor shall be isolated from panelboard enclosure.

PORTABLE AND VEHICLE-MOUNTED GENERATORS
250.34

Portable is defined as equipment that is easily carried from one location to another. Mobile equipment is capable of being moved, as on wheels or rollers, such as vechicle-mounted or placed on a trailer. Under certain conditions of use, the frame of a portable generator shall not be required to be connected to ground such as to a ground rod, water pipe, structural steel, etc.

PORTABLE GENERATORS
250.34(A)

Grounding Tip 73: It is not necessary to use a GFCI on a 20 or 15 amp circuit where the generator is grounded per **250.34(A)** and used as required **527.6(a), Ex. 1**.

The frame of a portable generator shall not be required to be grounded if it supplies only the equipment on the generator or cord-and-plug-connected equipment connected to receptacles mounted on the generator, provided all the following conditions are complied with:

- An equipment grounding conductor is installed to bond the receptacles to the frame of the generator.

- The equipment grounding conductor in the cord is installed to bond the exposed noncurrent-carrying metal parts of the equipment to the frame of the generator.

See Figure 2-44 for a detailed illustration when applying these requirements.

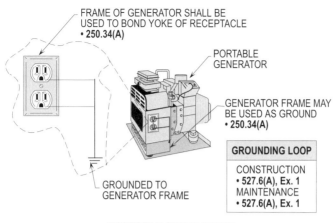

Figure 2-44. The frame of a portable generator shall not be required to be grounded if it supplies only the equipment on the generator or cord-and-plug connected equipment to receptacles mounted on the generator.

VEHICLE-MOUNTED GENERATORS
250.34(B)

The frame of the vehicle shall be permitted to ground a circuit supplied by a generator on the vehicle provided:

- The generator frame is bonded to the vehicle frame.

- The generator supplies only equipment mounted to the vehicle or if the generator supplies cord-and-plug connected equipment through receptacles mounted on the vehicle or both equipment located on the vehicle and cord-and-plug connected equipment through receptacles mounted on the vehicle or on the generator.

- Exposed metal parts of the equipment served are bonded to the generator frame either direct or through the receptacles.

See Figure 2-45 for a detailed illustration when applying these requirements.

The rules apply only for circuits that require grounding accoridng to **250.20**. Two-wire DC or two-wire AC circuits of less than 50 volts need not be grounded. All three-wire circuits shall be grounded. DC circuits of 51 to 300 volts shall be grounded. AC circuits of 50 to 150 volts shall be grounded.

Grounding Note: Vehicle-mounted generators that provide a grounded (neutral) conductor and are installed as separately derived systems supplying equipment and receptacles on the vehicle shall have the grounded (neutral) conductor bonded to the generator frame and vehicle frame. The noncurrent-carrying parts of the equipment shall be bonded to the generator frame.

GROUNDED CONDUCTOR BONDING
250.34(C)

A system conductor that is required to be grounded per **250.26** shall be bonded to the generator frame where the generator is a component of a separately derived system. Portable and vehicle-mounted generators installed as separately derived systems and that provide a grounded (neutral) conductor, such as three-phase, four-wire wye or single-phase, 240/120 volt volt or three-phase, four-wire delta connected, shall have the grounded (neutral) conductor bonded to the generator frame. **(See Figure 2-46)**

Grounding Note: For grounding portable generators supplying fixed wiring systems, see **250.20(D)** for additional requirements.

Grounding Tip 74: The following systems are not required to be grounded:

- Less than 50 V, 2-wire DC
- Less than 50 V, 2-wire AC

The following systems are required to be grounded:

- All 3-wire DC
- 51-300 volts DC
- 51-150 volts AC

Grounding Tip 75: For more information pertaining to generator installations, See **Article 445** in the NEC.

Figure 2-45. The frame of a generator does not have to be grounded if it supplies only the equipment supplied from the generator or cord-and-plug connected equipment connected to receptacles mounted and bonded to the frame of the vehicle.

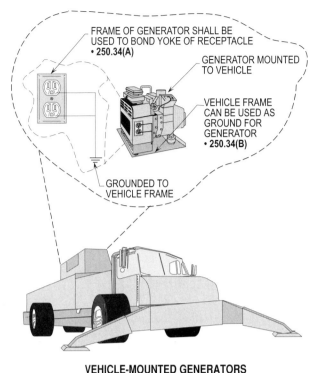

Grounding Tip 76: If the portable generator is connected to a driven rod, GFCI-protected circuits must be provided to supply 125 volt, single-phase power to 15, 20 and 30 amp receptacle outlets.

Figure 2-46. This illustration shows a portable generator with its grounded (neutral) connection properly bonded to the frame.

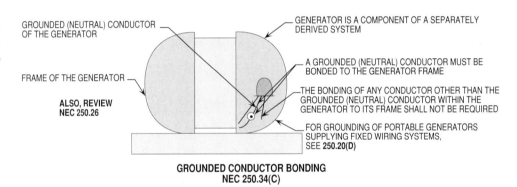

HIGH-IMPEDANCE GROUNDED NEUTRAL SYSTEMS 250.36

Grounding Tip 77: In a high-impedance grounding system, the ground-fault current return path for the service equipment is completed through the high-impedance grounding unit. The unit limits the line-to-ground fault current to a magnitude that can be tolerated until such fault is located and removed from the system. Note that ground-faults that are not promptly removed reduces the reliability of the service.

High-impedance grounded neutral systems in which a grounding impedance, usually a resistor, limits the ground-fault current to a low value shall be permitted for three-phase AC systems of 480 volts to 1000 volts where all of the following conditions are complied with:

- The conditions of maintenance and supervision ensure that only qualified persons will service the installation
- Continuity of power is required
- Ground detectors are installed on the system
- Line-to-neutral loads are not served

The grounded (neutral) conductor may be grounded through an impedance coil placed in the grounding wire. When an impedance coil is used, no direct ground is permitted. The impedance neutral shall be identified and insulated. Insulation shall be equal to that of the phase wires. However, equipment grounding conductors may be insulated or bare.

The requirements of **250.36(A)** thru **(F)** below are to be complied with when high-impedance grounded neutral systems are used.

Grounding Note: When a ground-fault is detected, it must be cleared as soon as possible. A system with a ground-fault must not be allowed to operate for long periods of time for it is not safe to do so.

GROUNDING IMPEDANCE LOCATION
250.36(A)

When high-impedance grounding is used it shall be located between the grounding source grounded (neutral) conductor and the grounding electrode. Where a grounded (neutral) conductor does not exist, the high-impedance shall be installed from a grounding transformer to the grounding electrode.

NEUTRAL CONDUCTOR
250.36(B)

The grounded (neutral) conductor from its connection point to the grounding impedance must be fully insulated and must have an ampacity that is not less than the maximum current rating of the grounding impedance. The grounded (neutral) conductor may not, however, be smaller than 8 AWG copper or 6 AWG aluminum.

SYSTEM NEUTRAL CONNECTION
250.36(C)

The grounded (neutral) conductor shall only be connected through the grounding impedance to the grounding electrode.

Grounding Note: When a circuit is closed, there is a value of charging current that sometimes exceeds the overcurrent protection of that system. Therefore, the impedance usually is selected based on the ground-fault current slightly greater than or equal to this capacity charging current. Impedance grounding also aids in limiting transient overvoltages to safe values. For more information, refer to **ANSI/IEEE Standard 142,** which gives recommendations for Industrial Commercial Power Systems. The design of impedance grounding in reality becomes an engineering and design problem.

NEUTRAL CONDUCTOR ROUTING
250.36(D)

The impedance grounding system from the neutral point of a transformer or generator may be installed in a separate raceway. Thus, the grounded (neutral) conductor is not required to be run in the same raceways as the ungrounded (phase) conductors to the disconnecting means or overcurrent device. The impedance grounding unit limits the ground current so it may be treated as a solidly grounded neutral.

Grounding Tip 78: When using reactance grounding, the fault current should be at least 60 percent of, but not more than, the three-phase fault current. This is particularly important if generators are connected since such machines are generally braced only for the expected three-phase fault current. However, when using low-resistance grounding, the fault current will normally be in the range from 10 percent to 60 percent of the three-phase fault current. For high-resistance grounding, the fault current is intended to be limited to a smaller fraction of the three-phase fault current, generally less than 50 amps.

Grounding Tip 79: The grounded (neutral) conductor appears to be a grounding electrode conductor where it connects to the High-Impedance unit **(See Figure 2-47)**. However, its sized from the amperage rating of the High-Impedance unit and not from **Table 250.66** based on the size of the ungrounded (phase) conductors.

Grounding Tip 80: This requirement of not routing the grounded (neutral) conductor with the ungrounded (phase) conductors is very useful in adding a high-impedance grounding unit to an existing ungrounded electrical supply system.

EQUIPMENT BONDING JUMPER
250.36(E)

An equipment bonding jumper that is connected between equipment grounding conductors and the grounding impedance must be an unspliced conductor and must be run from the first disconnecting means to the grounded side of the impedance.

GROUNDING ELECTRODE CONDUCTOR LOCATION
250.36(F)

Grounding Tip 81: If practical, always select a grounding electrode having a resistance of 1 to 5 ohms.

The grounding conductor is to be attached at any point from the grounded side of a grounding impedance to the equipment grounding connection which may be at the service equipment or disconnecting means of the first system. This might sound like paralleling grounding conductors, but the impedance grounding conductor originates from the grounded (neutral) conductor and the equipment grounding conductor referred to here connects from the grounding electrode to the equipment disconnecting means of the first system.

See Figure 2-47 for a detailed illustration when applying these requirements.

See Figure 2-47 for a detailed illustration when applying these requirements.

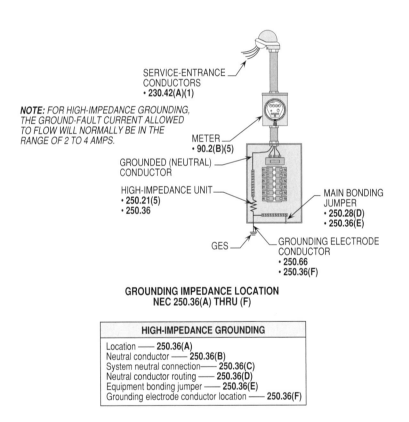

GROUNDING IMPEDANCE LOCATION
NEC 250.36(A) THRU (F)

HIGH-IMPEDANCE GROUNDING
Location —— **250.36(A)**
Neutral conductor —— **250.36(B)**
System neutral connection —— **250.36(C)**
Neutral conductor routing —— **250.36(D)**
Equipment bonding jumper —— **250.36(E)**
Grounding electrode conductor location —— **250.36(F)**

Name _____   Date _____

Chapter 2
Circuit and System Grounding

	Section	Answer

1. AC circuits of less than 50 volts shall be grounded where a transformer installed to supply low-voltage systems receives it's supply from a transformer exceeding 150 volts-to-ground. _____   T   F

2. AC systems of 1 kV and over shall not be required to be grounded if supplying mobile or portable equipment. _____   T   F

3. Circuits installed for industrial electric furnaces shall be grounded for AC systems of 50 volts to 1000 volts. _____   T   F

4. Circuits for electric cranes operating over combustible fibers in Class III locations shall be grounded. _____   T   F

5. Where a transformer supplying the service is located outside the building, at least one additional grounding connection shall be made from the grounded service conductor to a grounding electrode. _____   T   F

6. A single grounding electrode connection to the point of the grounded circuit conductor from each power source shall be permitted to be installed where services are dual fed in a common enclosure and employing a secondary tie. _____   T   F

7. The grounded (neutral) conductor shall be routed with the ungrounded (phase) conductors and shall be no smaller than the required grounding electrode conductor. _____   T   F

8. Grounded (neutral) conductors which are installed in two or more raceways shall be based on the size of the ungrounded service entrance conductors in parallel but not smaller than 4 AWG. _____   T   F

9. An unspliced main bonding jumper (for grounded systems) shall be installed to connect the equipment grounding conductors and the service disconnecting enclosure to the grounded (neutral) conductor of the system within the enclosure for each service disconnect. _____   T   F

10. An additional bonding jumper connection (more than one) shall be permitted to be made for a separately derived system where doing so will not create a parallel path for the grounded circuit conductor(s). _____   T   F

11. The grounding electrode for a separately derived system shall be as near as possible and preferably in the same area as the grounding electrode conductor connection to the system. _____   T   F

12. An effectively grounded metal water pipe within 6 ft. (1.8 m) from the point of entrance into the building shall be permitted to be used as a grounding electrode for an ungrounded system for a separately derived system. _____   T   F

Section	Answer		
_____	T	F	**13.** A grounding electrode at a separate building or structure shall be required to be installed where only one branch-circuit serves the building or structure and the branch-circuit includes an equipment grounding conductor.
_____	T	F	**14.** Where one or more buildings are supplied from a common AC grounded service, each panelboard at each building structure shall be separately grounded.
_____	T	F	**15.** A grounded (neutral) conductor shall be permitted to be used as an equipment grounding conductor plus a grounded (neutral) conductor to a separate building or structure where the raceway system is of the nonmetallic type.
_____	T	F	**16.** A grounding electrode for an ungrounded system shall be connected only to the service equipment enclosure where installed at one or more buildings.
_____	T	F	**17.** The connections of the grounded (neutral) conductor to the grounding electrode at a separate building or structure shall be made where one or more disconnecting means supply one or more additional buildings or structures under single management.
_____	T	F	**18.** Vehicle-mounted generators installed as separately derived systems and that provide a grounded (neutral) conductor shall not be required to have the neutral conductor bonded to the generator frame.
_____	T	F	**19.** When high-impedance grounding is used it shall be located between the grounding source grounded (neutral) conductor and the grounding electrode.
_____	T	F	**20.** An equipment bonding jumper that is connected between the equipment grounding conductor and the grounding impedance shall be a spliced conductor and shall run from the first disconnecting means to the grounded side of the impedance.
_____	_____		**21.** AC circuits of less than 50 volts shall be grounded where a transformer is installed to receive its supply from a transformer with an _____ system.
_____	_____		**22.** AC circuits of less than _____ volts shall be grounded where low-voltage overhead conductors are installed outside and not inside.
_____	_____		**23.** AC circuits of 50 volts to 1000 volts shall be grounded where the maximum voltage-to-ground on the ungrounded (phase) conductors does not exceed _____ volts.
_____	_____		**24.** AC circuits of 50 volts to 1000 volts shall be grounded where the system is nominally rated as 240/120 volt, three-phase, four-wire delta connected in which the _____ of one phase is used as a circuit conductor with a phase conductor having a higher voltage-to-ground than the other two.
_____	_____		**25.** AC systems of _____ kV and over shall be grounded if supplying mobile or portable equipment.

	Section	Answer

26. AC systems of 50 volts to 1000 volts shall not be required to be grounded where separately derived systems are installed with primaries not exceeding _____ volts and used exclusively for secondary control circuits where conditions of maintenance and supervision ensure that only qualified persons will service the installation.

27. Circuits for electric cranes operating over combustible fibers in Class _____ locations shall not be required to be grounded.

28. Connections of the grounding electrode conductor from the separately derived system to the continuous common grounding electrode conductor shall be made at an accessible location by an _____ compression connector, listed for the purpose.

29. A grounding connection shall not be permitted to be made to any _____ circuit conductor on the load side of the service disconnecting means (general rule).

30. The grounded (neutral) conductors shall be run to each service disconnecting means and shall be bonded to each disconnecting means enclosure for AC systems which operate at less than _____ volts and are grounded at any point.

31. The grounded (neutral) conductor shall not be smaller than _____ percent of the area of the largest service-entrance (phase) conductors larger than 1100 KCMIL copper or 1750 KCMIL aluminum.

32. Grounded (neutral) conductors which are installed in two or more raceways shall be based on the size of the ungrounded service-entrance (phase) conductors in the raceway but not smaller than _____ AWG.

33. Where more than one service disconnecting means is located in an assembly which is _____ for use as service equipment, the grounded (neutral) conductor shall be permitted to be run to the assembly and the grounded (neutral) conductor shall be bonded to the assembly enclosure.

34. On an _____ system, none of the circuit conductors of the system are intentionally grounded.

35. Unspliced main bonding jumpers shall be of _____ or other corrosion-resistant material.

36. A continuous common grounding electrode conductor for multiple separately derived systems of a high-rise building shall not be required to be sized greater than _____ AWG.

37. When a screw is used as the main bonding jumper, it shall be identified with a _____ colored finish which is visible with the screw installed.

38. The bonding jumper for a separately derived system which supplies Class 1, Class 2 or Class 3 circuits shall be sized no smaller than _____ AWG copper.

2-49

Section	Answer

_____ _____ 39. The grounding electrode conductor for a separately derived system shall not have to be installed larger than _____ AWG copper when connecting to the nearest building steel or nearest metal water pipe system.

_____ _____ 40. If the metal water pipe is exposed and qualified personnel service the installation for a separately derived system, the connection of the grounding electrode conductor to the metal water pipe shall not be required to be run within _____ ft. of where it enters the building.

_____ _____ 41. What is the minimum and maximum fault current required to clear a 200 amp OCPD?

_____ _____ 42. What size copper grounded (neutral) conductor is required based on the service-entrance conductors being rated at 300 KCMIL THWN copper?

_____ _____ 43. What size copper grounded (neutral) conductor is required based on the service-entrance conductors being rated at 400 KCMIL THWN copper?

_____ _____ 44. In a paralleled three-phase, four-wire service with 4 - 4/0 AWG THWN copper conductors per phase, what size grounded (neutral) conductor is required in each conduit run?

_____ _____ 45. In a paralleled three-phase, four-wire service with 4 - 250 KCMIL THWN copper conductors per phase, what size grounded (neutral) conductor is required in each conduit run?

_____ _____ 46. In a paralleled three-phase, four-wire service with 3 - 600 KCMIL THWN copper conductors per phase, what size grounded (neutral) conductor is required in each conduit run?

_____ _____ 47. In a paralleled three-phase, four-wire service with 4 - 700 KCMIL THWN copper conductors per phase, what size grounded (neutral) conductor is required in each conduit run?

_____ _____ 48. What size copper main bonding jumper is required based on the service-entrance conductors being rated at 300 KCMIL THWN copper?

_____ _____ 49. What size copper main bonding jumper is required based on the service-entrance conductors being rated at 400 KCMIL THWN copper?

_____ _____ 50. In a paralleled three-phase, four-wire service with 4 - 250 KCMIL THWN copper conductors per phase, what size main bonding jumper is required to ground the busbar to the service equipment enclosure?

_____ _____ 51. In a paralleled three-phase, four-wire service with 3 - 600 KCMIL THWN copper conductors per phase, what size main bonding jumper is required to ground the busbar to the service equipment enclosure?

_____ _____ 52. In a paralleled three-phase, four-wire service with 4 - 700 KCMIL THWN copper conductors per phase, what size main bonding jumper is required to ground the busbar to the service equipment enclosure?

	Section	Answer

53. What size copper bonding jumper is required to bond and ground the secondary of a separately derived system having 4 - 4/0 AWG THWN copper conductors connected to its secondary?

54. What size copper bonding jumper is required to bond and ground the secondary of a separately derived system having 4 - 300 KCMIL THWN copper conductors connected to its secondary?

55. In a paralleled three-phase, four-wire service with 4 - 500 KCMIL THWN copper conductors per phase, what size bonding jumper is required to bond and ground the separately derived system.

56. In a paralleled three-phase, four-wire service with 4 - 750 KCMIL THWN copper conductors per phase, what size bonding jumper is required to bond and ground the separately derived system.

57. What size grounding electrode conductor is required to bond and ground the secondary of a separately derived system having 4 - 4/0 AWG THWN copper conductors connected to its secondary?

58. What size grounding electrode conductor is required to bond and ground the secondary of a separately derived system having 4 - 300 KCMIL THWN copper conductors connected to its secondary?

59. In a paralleled three-phase, four-wire service with 4 - 500 KCMIL THWN copper conductors per phase, what size grounding electrode conductor is required to bond and ground the separately derived system.

60. In a paralleled three-phase, four-wire service with 4 - 750 KCMIL THWN copper conductors per phase, what size grounding electrode conductor is required to bond and ground the separately derived system.

61. A 175 amp (75°C terminals) panelboard in building 1 is supplied by three 2/0 AWG THWN copper conductors and the neutral load is 50 amps. In building 1 (grounded system), the feeder is protected by a 175 amp OCPD (75°C terminals). What size grounded (neutral) conductor and equipment grounding conductor is required using THWN copper conductors in the feeder-circuit.

62. A 200 amp (75°C terminals) panelboard in building 1 is supplied by three 3/0 AWG THWN copper conductors and the neutral load is 58 amps. In building 1 (grounded system), the feeder is protected by a 200 amp OCPD (75°C terminals). What size grounded (neutral) conductor and equipment grounding conductor is required using THWN copper conductors in the feeder-circuit.

63. A 100 amp OCPD and feeder-circuit from building 1 (grounded system) supplies a 100 amp panelboard in building 2. What size THWN copper feeder-circuit is required to serve as a grounded (neutral) conductor and equipment grounding conductor to supply the panelboard in building 2?

Section   Answer

_____   _____   **64.** A 200 amp OCPD and feeder-circuit from building 1 (grounded system) supplies a 200 amp panelboard in building 2. What size THWN copper feeder-circuit is required to serve as a grounded (neutral) conductor and equipment grounding conductor to supply the panelboard in building 2?

_____   _____   **65.** A 300 amp OCPD and feeder-circuit from building 1 (grounded system) supplies a 300 amp panelboard in building 2. What size THWN copper feeder-circuit is required to serve as a grounded (neutral) conductor and equipment grounding conductor to supply the panelboard in building 2?

_____   _____   **66.** A 100 amp OCPD and feeder-circuit from building 1 (ungrounded system) supplies a 100 amp panelboard in building 2. What size equipment grounding conductor is required using THWN copper conductors in the feeder-circuit?

_____   _____   **67.** A 200 amp OCPD and feeder-circuit from building 1 (ungrounded system) supplies a 200 amp panelboard in building 2. What size equipment grounding conductor is required using THWN copper conductors in the feeder-circuit?

_____   _____   **68.** A 300 amp OCPD and feeder-circuit from building 1 (ungrounded system) supplies a 300 amp panelboard in building 2. What size equipment grounding conductor is required using THWN copper conductors in the feeder-circuit?

_____   _____   **69.** A 200 OCPD and feeder-circuit from building 1 (ungrounded system) supplies a 200 amp panelboard in building 2. What size grounding conductor is required using THWN copper conductors in the feeder-circuit?

_____   _____   **70.** A 300 OCPD and feeder-circuit from building 1 (ungrounded system) supplies a 300 amp panelboard in building 2. What size grounding conductor is required using THWN copper conductors in the feeder-circuit?

3

GROUNDING ELECTRODE SYSTEM AND GROUNDING ELECTRODE CONDUCTOR

Grounding electrodes and the grounding electrode conductors that connect the electrodes to the system grounded (neutral) conductor are not intended to carry ground-fault currents that are due to ground-faults in equipment, raceways or other conductor enclosures. In solidly grounded systems, the ground-fault current flows through the equipment grounding conductors from a ground-fault located anywhere in the system to the main bonding jumper connecting the equipment grounding conductors and the system grounded (neutral) conductor to the grounded busbar in the service equipment.

In solidly grounded service-supplied systems, the ground-fault current return path is completed through the main bonding jumper in the service equipment and the grounded (neutral) conductor to the supply transformer. Such fault-current should open the primary OCPD (usually cut-out fuses) located on the supply of the transformer.

GROUNDING ELECTRODE SYSTEM
250.50

If available on the premises at each building or structure served, the following electrodes shall be bonded together to form the grounding electrode system:

- Metal underground water pipe
- Metal frame of the building or structure
- Concrete-encased electrode
- Ground ring
- Rod and pipe electrodes
- Plate electrodes
- Other local metal underground system or structures

Grounding Tip 82: If any of the electrodes that are listed in **250.52(A)(1)** thru **(A)(7)** are available, they shall be bonded together to complete the grounding electrode system. The method in which the grounding electrode system is connected (bonded) depends upon whether the installation in new, existing or has been remodeled.

Figure 1-3(a). This illustration shows the types of grounding electrodes that shall be bonded together to form the grounding electrode system if available.

Grounding Tip 83: If all or less than all of the electrodes in **250.52(A)(1)** thru **(A)(7)** are available, they shall be bonded and grounded together to form one complete grounding electrode system. For example, when a service is enlarged and some of these electrodes are not accessible without damaging the structure in some way, then they shall not be considered accessible.

Figure 1-3(b). This illustration shows the types of grounding electrodes that shall be bonded together to form the grounding electrode system if available.

If one of the above electrodes are not available, one or more the following electrodes shall be installed and used:

- Ground ring
- Rod and pipe electrodes
- Plate electrodes
- Other local metal underground systems or structures

See Figure 3-1(a) and (b) for a detailed illustration when applying these requirements.

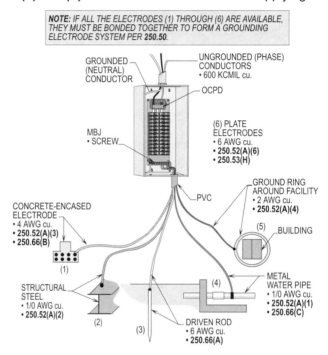

GROUNDING ELECTRODE SYSTEM
NEC 250.50
NEC 250.52(A)(1) THRU (A)(6)

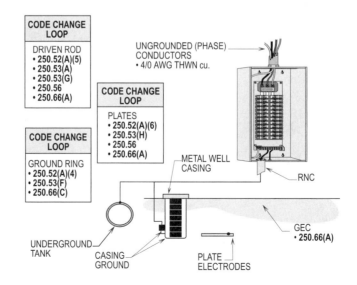

OTHER LOCAL METAL UNDERGROUND SYSTEMS OR STRUCTURES
NEC 250.52(A)(7)

GROUNDING ELECTRODES
250.52

The following types of electrodes shall be permitted for grounding:

- Metal underground water pipe
- Metal frame of the building or structure
- Concrete-encased electrode
- Ground ring
- Rod and pipe electrodes
- Plate electrodes
- Other local metal underground system or structures

METAL UNDERGROUND WATER PIPE
250.52(A)(1)

Metal water pipe with lengths of at least 10 ft. (3 m) long in the earth shall be connected to the grounded (neutral) bar in the service equipment enclosure. The grounded (neutral) bar shall be bonded to the service equipment enclosure, the grounded (neutral) conductor, the neutral conductors and the equipment grounding conductors by a bonding jumper sized per **Table 250.66**. Metal water piping shall be electrically continuous or made electrically continuous by bonding around insulated joints or sections.

The grounding connection to the metal water pipe shall be made at a point no more than 5 ft. (1.52 m) where the piping enters a building. It is recommended that the grounding conductor from the supplementary driven ground rod be terminated to the common grounded bar in the service equipment panel. **(See Figure 3-2)**

Grounding Tip 84: For requirements pertaining to a metal water pipe or copper tubing where such is not considered an electrode, see **250.104(A)**.

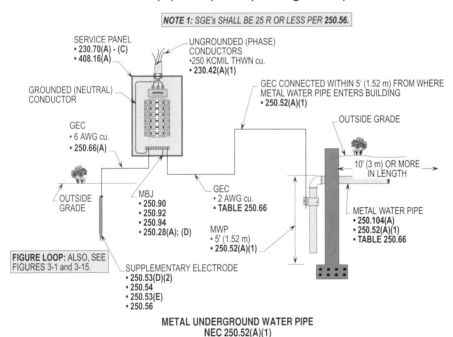

Figure 3-2. In most all installations (existing), the grounding electrode conductor connects the service equipment grounded (neutral) conductor to a copper tubing or metal water piping system or a driven rod or all other electrodes, if available.

INDUSTRIAL AND COMMERCIAL BUILDINGS
250.52(A)(1), Ex.

The grounding electrode conductor shall be permitted to be installed further than 5 ft. (1.52 m) in industrial and commercial buildings where conditions of maintenance and supervision ensure that only qualified personnel will service the installation and the entire length of the interior metal water pipe that is being used for the conductor that is

Grounding Tip 85: The **Ex. to 250.52(A)(1)** does not apply to single-family dwelling units. The connection to the metal water pipe shall always be accessible and located on the piping within 5 ft. (1.52 m) where it enters the dwelling where such piping is classified as an electrode per **250.52(A)(1)**.

exposed. Under these conditions, such metal water pipes shall be permitted to be used as grounding electrode conductors to ensure continuity. **(See Figure 3-3)**

Figure 3-3. In industrial and commercial buildings, where the metal water pipe is exposed and maintained by qualified personnel, the grounding electrode conductor shall be permitted to be connected at any accessible location.

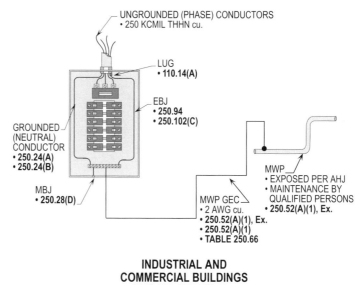

INDUSTRIAL AND COMMERCIAL BUILDINGS
NEC 250.52(A)(1), Ex.

METAL FRAME OF THE BUILDING OR STRUCTURE 250.52(A)(2)

Grounding Tip 86: Building structural steel that is exposed and forms the structural frame work of the facility shall be bonded and grounded into the grounding electrode system per **250.52(A)(2)** and **250.104(C)**.

In grounding installation, using structural building steel, such electrode shall be connected to the service equipment grounding conductor (grounded (neutral) conductor) by a grounding electrode conductor sized per **Table 250.66** based on the size of the service conductors. **(See Figure 3-4)**

Figure 3-4. This illustration shows structural steel used as a grounding electrode.

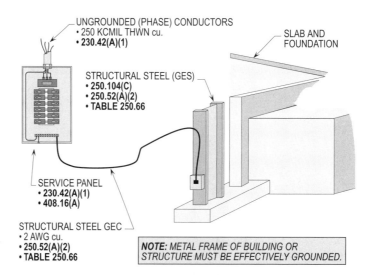

Grounding Tip 87: When using structural building steel as a grounding electrode, always review **250.52(A)(2)** and **250.104(C)** very carefully.

METAL FRAME OF THE BUILDING OR STRUCTURE
NEC 250.52(A)(2)

CONCRETE-ENCASED ELECTRODE
250.52(A)(3)

Concrete-encased electrodes consist of 1/2 in. x 20 ft. (13 mm x 6 m) lengths of reinforcing rebar located in the foundation. The 20 ft. (6 m) length of rebar shall be permitted to be in one continuous piece or many pieces spliced together to form a 20 ft. (6 m) or more continuous length. The metal reinforcing rebars shall be of the conductive type. A concrete-encased electrode shall also be permitted to be a 4 AWG bare copper conductor at least 20 ft. (6 m) long that is installed in the footing of the foundation. The 20 ft. (6 m) long 4 AWG conductor shall be located in at least 2 in. (50 mm) of concrete near the bottom of the foundation or footing. The reinforcing rebars or 4 AWG bare copper conductor installed properly usually provides a resistance of about 3 to 5 ohms resistance. **(See Figure 3-5)**

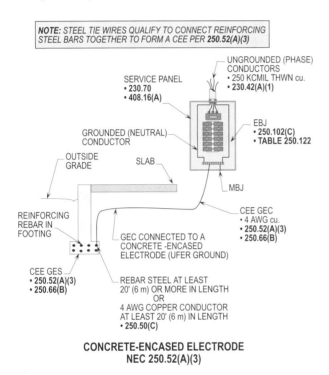

Figure 3-5. Concrete-encased electrodes consist of 1/2 in. x 20 ft. (13 mm x 6 m) lengths of reinforcing rebar located in the foundation. A concrete-encased electrode shall be permitted to be a 4 AWG bare copper conductor at least 20 ft. (6 m) long that shall be installed in the footing of the foundation.

Grounding Tip 88: There have been reports that concrete-encased electrodes have provided resistance readings of a little above zero ohms.

GROUND RING
250.52(A)(4)

A bare copper conductor not smaller than 2 AWG and at least 20 ft. (6 m) shall be permitted to be installed in the earth to form a ground ring grounding system that encircles the building or structure. **(See Figure 3-6)**

Grounding Note: A more reliable and dependable ground may be obtained by burying the conductor about 3 ft. (900 mm) from the building and periodically driving ground rods and attaching them to the bare conductor by approved connecting methods.

Grounding Tip 89: The grounding electrode conductor to a ground ring is usually sized as large as the conductor used to form the ring per **250.66(C)** and **250.52(A)(4)**.

ROD AND PIPE ELECTRODES
250.52(A)(5)

A driven electrode used to ground the service shall be permitted to be a 1/2 in. x 8 ft. (13 mm x 2.5 m) copper (brass) rod or a 3/4 x 10 ft. (21 x 3 m) galvanized pipe. A copper rod is usually the type of driven elctrode that is installed. Driven rods are utilized to connect the grounded (neutral) bar terminal to earth ground. Where the water pipe system in the ground is nonmetallic (PVC) and converts to metal pipe or copper tubing above ground,

Grounding Tip 90: The resistance to ground on driven rods can be measured using an earth ground type megger.

a driven rod shall be permitted to be used to bond and ground the metal water pipe system and connect the electrical system to earth ground. **(See Figure 3-7)**

The metal water pipe system shall be bonded into the grounding electrode system, whether or not the piping in the ground is nonmetallic (PVC) or metal. The following are acceptable made electrodes which shall be permitted to be used:

- 1/2 in. x 8 ft. (13 mm x 2.5 m) copper rod
- 3/4 x 10 ft. (21 x 3 m) metal pipe or conduit
- 5/8 in. x 8 ft. (15.87 mm x 2.5 m) rebar
- 5/8 in. (16 mm) or larger stainless steel rod or 1/2 in. x 8 ft. (13 mm x 2.5 m) if listed for such use

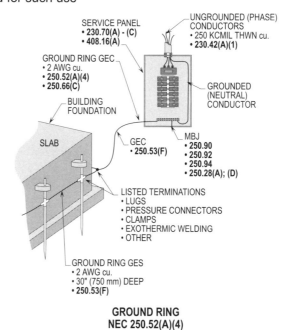

Figure 3-6. A bare copper conductor not smaller than 2 AWG and at least 20 ft. (6 m) shall be permitted to be installed in the earth to form a ground ring grounding system that encircles the building or structure.

Grounding Tip 91: The ground ring around the building shall be connected together at each end to form a complete circle around the foundation.

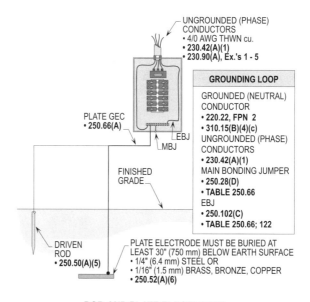

Figure 3-7. Rod and plate electrodes shall be installed when none of the electrodes are present in **250.52(A)(1)** thru **(A)(6)** are available.

Quick Calc
Rule of Thumb Method

GEC run 150 ft.

Table 8 to Ch. 9
- 2 AWG cu. = 66,360 CM
- 66,360 CM x 150% = 99,540 CM
- 99,540 CM requires 1/0 AWG cu.

Solution: The 2 AWG copper grounding electrode conductor must be in-creased to 1/0 AWG copper.

Note: The grounding electrode conductor in the Quick Calc. is routed 150 ft. from the service to the grounding electrode. (Refer to grounding tip 111 on page 3-19 in this book)

PLATE ELECTRODES
250.52(A)(6)

Plate electrodes shall be a minimum of 2 sq. ft. (0.186 m sq. ft.), and if made of iron or steel they shall be at least 1/4 in. (6.4 mm) thick. If made of a nonferrous metal such as copper, they shall be 0.06 in. (1.5 mm) in thickness. **(See Figure 3-7)**

OTHER LOCAL METAL UNDERGROUND SYSTEMS OR STRUCTURES
250.52(A)(7)

Other local metal underground systems and structures such as piping systems and underground tanks shall be permitted to be installed as an grounding electrode.

Grounding Tip 92: A metal casing used with a water pump to draw water from a well and deliver it to the facility shall be permitted to be used as a grounding electrode conductor.

GROUNDING ELECTRODE SYSTEM INSTALLATION
250.53

To complete the grounding electrode system, the following grounding electrodes shall have specific installation requirements applied:

- Rod, pipe and plate electrodes
- Electrode spacing
- Bonding jumper
- Metal underground water pipe
- Supplemental electrode bonding connection size
- Ground ring
- Rod and pipe electrodes
- Plate electrodes

ROD AND PIPE ELECTRODES
250.53(A)

Rod, pipe and plate electrodes shall be embedded below permanent moisture level, where practicable. Rod, pipe and plate electrodes shall be from nonconductive coatings such as paint and enamel.

ELECTRODE SPACING
250.53(B)

Where more than one rod, pipe or plate electrode is used, each electrode of one grounding system (including that used for air terminals) shall not be less than 6 ft. (1.83 m) from any other electrode of another grounding system. Two or more grounding electrodes that are effectively bonded together shall be considered a single grounding electrode system.

BONDING JUMPER
250.52(C)

Where bonding jumper(s) are used to connect the grounding electrodes together to form the grounding electrode system, the bonding jumper(s) shall be installed in accordance with **250.64(A), (B)** and **(E)**, they shall be sized in accordance with **250.66** and shall be connected in accordance with **250.70**.

METAL UNDERGROUND WATER PIPE
250.52(D)

Grounding Tip 93: To fully understand the requirements of not interrupting the metal water piping system, See **250.52(A)(1)** and **250.68(B)**.

The continuity of the grounding path for the bonding connection to the interior metal water piping shall not be permitted to rely on water meters, filtering devices or similar equipment. **(See Figure 3-8)**

Figure 3-8. This illustration makes it clear that continuity of the grounding path for the bonding connection to the interior metal water piping shall not be permitted to rely on water meters, filtering devices or similar equipment.

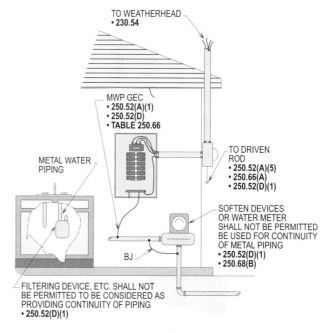

METAL UNDERGROUND WATER PIPE
NEC 250.53(D)

Grounding Note: Many rural areas doe not have water meters but may have water filtering or softening equipment that could be removed. Such removal would break or impair the grounding path or bonding connection of the metal interior water piping. Proper bonding and joining together of the metal piping can be accomplished by using correctly sized bonding jumpers.

Grounding Tip 94: The supplementary grounding electrode used to supplement the metal water piping or copper tubing shall have a resistance measurement of 25 ohms or less per **250.53(D)(2)** and **250.56**.

The copper tubing or metal water pipe in a new installation may be installed in the earth or above grade. Either installation requires the copper or metal water pipe to be supplemented by an additional electrode. The additional grounding electrode to supplement the metal water pipe shall be permitted to be any of the electrodes listed in **250.52(A)(2)** thru **(A)(6)**. **(See Figure 3-9)**

SUPPLEMENTAL ELECTRODE BONDING CONNECTION SIZE
250.52(E)

Where a rod, pipe or plate electrode is used as the supplemental electrode, that portion of the bonding jumper that is the sole connection to the supplemental grounding electrode shall not be required to be larger than 6 AWG copper wire or 4 AWG aluminum wire.

GROUND RING
250.52(F)

Where a ground ring is used as the grounding electrode, the ground ring shall be buried not less than 30 in. (750 mm) below the earth's surface.

Grounding Electrode System and Grounding Electrode Conductor

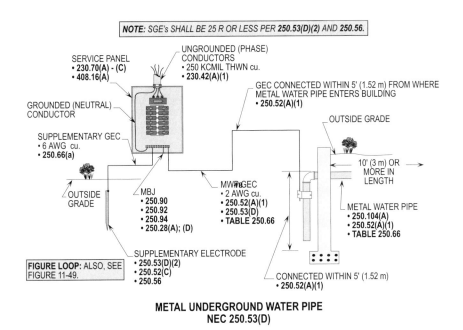

Figure 3-9. In most all installation (existing), the grounding electrode conductor connects the service equipment grounded (neutral) conductor to a copper tubing or a metal water pipe system which is supplemented by a driven rod or other electrodes (shall be permitted for this purpose), if available.

ROD AND PIPE ELECTRODES
250.52(G)

Rod and pipe electrodes shall be installed where there is at least 8 ft. (2.44 m) of length in contact with the soil. Where a rock bottom is encountered, the electrode shall be permitted to be driven at an angle not to exceed 45 degrees from the vertical or shall be permitted to be buried at least 30 in. (750 mm) deep. **(See Figure 3-10)**

Grounding Tip 95: If a driven rod is exposed above grade so that the connection of the grounding electrode conductor to the rod can be seen, a 10 ft. (3 m) rod shall used and not an 8 ft. (2.5 m) rod per **250.52(A)(5)** and **250.53(G)**.

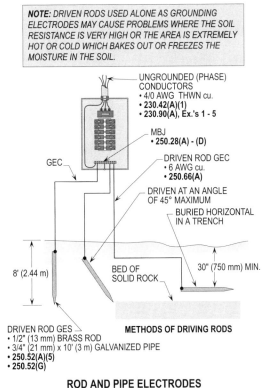

Figure 3-10. This illustration shows the methods in which rod and pipe electrodes are to be installed.

Grounding Tip 96: The Note in Figure 3-10 points to the fact that baked-out or frozen soil will cause the resistance of the soil to increase. This increased resistance can cause the voltage-to-ground to fluctuate and create unwanted noise problems for computers.

3-9

PLATE ELECTRODES
250.52(H)

Where a plate electrode is used as the grounding electrode, the plate electrode shall be buried not less than 30 in. (750 mm) below the earth's surface.

SUPPLEMENTARY GROUNDING
250.54

Grounding Tip 97: It is an NEC violation to ground a motor to a driven rod and use the earth as an equipment grounding conductor. The rod can be driven and used for lightning protection, but for safety, the supply circuit shall be provided with an equipment grounding conductor per **250.4(D)** and **250.54**.

Supplementary grounding electrodes shall be permitted to be installed to augment the equipment grounding conductors specified in **250.118** and shall not be required to comply with the electrode bonding requirements of **250.50** or **250.53(C)** or the resistance requirements of **250.56**, but the earth shall not be permitted to be used as the sole equipment grounding conductor. **(See Figure 3-11)**

Grounding Note: This for instance, requires an equipment grounding conductor to be used, because the earth resistance in almost every case is too high to properly cause OCPD's to operate when a ground-fault occurs.

Figure 3-11. Supplementary grounding electrodes (SGE) shall be permitted to be used to supplement the equipment grounding conductor but are never to be used as the sole grounding means.

Grounding Tip 98: The metal enclosure of the motor shall be permitted to be grounded to the structural building steel or driven rod for the purpose of providing protection from lightning strikes or static electricity.

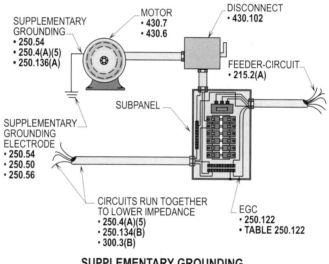

SUPPLEMENTARY GROUNDING
250.54

RESISTANCE OF ROD, PIPE AND PLATE ELECTRODES
250.56

Grounding Tip 99: The NEC only requires a 6 ft. (1.8 m) spacing between 2 - 8 ft. (2.5 m) driven rods per **250.56**. It is recommended in the IEEE Green Book to space them at least 16 ft. apart for best results. In other words, double the length of one rod by 2 (8' x 2 = 16' apart).

Made electrodes shall have a resistance to ground of 25 ohms or less, wherever practicable. When the resistance is greater than 25 ohms, two or more electrodes shall be permitted to be connected in parallel or extended to a greater length. Note that a made electrode or supplementary grounding electrode that measures more than 25 ohms shall be augmented by one additional electrode of a type permitted by **250.52(A)(2)** thru **(A)(7)**.

Continuous metal water piping systems usually have a ground resistance of less than 3 ohms. Metal frames of buildings normally have a good ground and usually have a resistance of less than 25 ohms. As pointed out in **250.52(A)(2)**, the metal frame of a building, if effectively grounded shall be permitted to be used as the ground. Local metallic water systems and well casings make good grounding electrodes in most all types of installations. **(See Figure 3-12(a), (b) and (c))**

Grounding Electrode System and Grounding Electrode Conductor

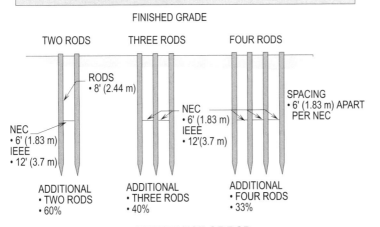

Figure 3-12(a). This illustration shows the procedure for calculating the amps to a driven rod, based on 25 ohms.

Figure 3-12(b). Rod, pipe and plate electrodes shall have a resistance to ground of 25 ohms or less, wherever practicable. When the resistance is greater than 25 ohms, two or more electrodes shall be permitted to be connected in parallel or extended to a greater length

3-11

Figure 3-12(c). If a resistance of 25 ohms is not obtained, a rod, pipe or plate electrode or supplementary grounding electrode shall be augmented by one additional electrode per **250.52(A)(2)** thru **(A)(7)**.

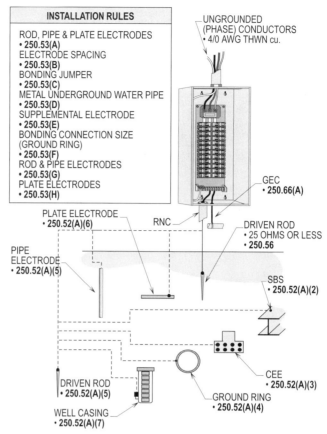

**RESISTANCE OF ROD, PIPE AND PLATE ELECTRODES
NEC 250.56**

SOIL TREATMENT

When made electrodes are used, grounding may be greatly improved by the use of chemicals, such as magnesium sulfate, copper sulphate or rock salt. A trench, a basin or a doughnut-type hole system may be dug around the ground rod into which the chemicals are put. Rain and snow will dissolve the chemicals and allow them to penetrate the soil and saturate the rod providing lower resistance. **(See Figure 3-13)**

COMMON GROUNDING ELECTRODE
250.58

Grounding Tip 100: The grounded (neutral) conductor on the supply side of the service can be a grounded (phase) conductor (corner grounded) as well as a neutral per **250.26(4)** and **240.22(1)** and **(2)**.

Where an AC wiring system has a grounded circuit conductor (maybe a neutral), the same electrode (metal water pipe, driven ground rod, etc.) that grounds the neutral (if present) must also be used to ground equipment. The 120/240, 120/208 and 277/480 volt systems have a neutral. The neutral is the grounded circuit conductor which is grounded at the service equipment enclosure. The same electrode system that grounds the neutral also grounds equipment on the premises.

Review carefully **250.24** and **250.32** for they deal with these specific grounding techniques used at the service equipment and at other structures.

Grounding Note: Where two or more grounding electrodes are effectively bonded together they are considered a single grounding electrode system when applying this requirement. **(See Figure 3-14)**

Grounding Electrode System and Grounding Electrode Conductor

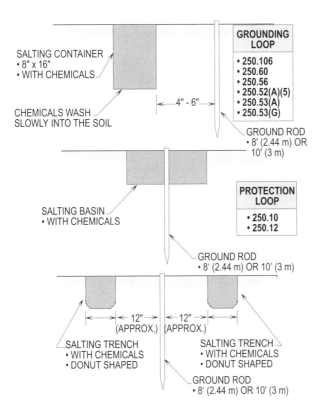

Figure 3-13. When made electrodes are used, grounding may be greatly improved by the use of chemicals, such as magnesium sulfate, copper sulphate or rock salt.

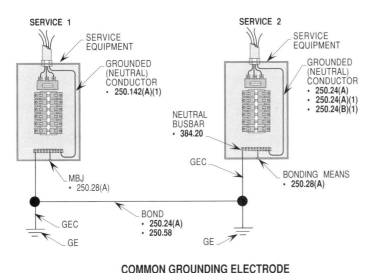

Figure 3-14. Grounding electrodes which are effectively bonded together for two or more services are considered a single grounding electrode system which will bond the metal enclosures together to create an equipotential plane. (See **547.2** for equipotential plane defined)

USE OF AIR TERMINALS
250.60

Lightning down conductors and made electrodes which are connected to lightning rods for grounding shall not be permitted to be used in place of made electrodes for grounding electrical wiring systems and equipment. However, they shall be permitted to be bonded together to limit the difference of potential that might appear between them. See **250.50** and **250.106**. **(See Figure 3-15)**

Grounding Tip 101: It is most important for safety to connect the grounding electrode for lightning protection down conductors into the facility's grounding electrode system per **250.52(A)(1)** thru **(A)(7)**, **250.60** and **250.106**.

Figure 3-15. This illustration shows the requirements for installing and grounding lightning protection systems.

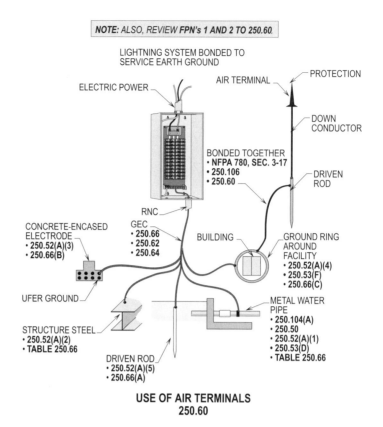

See **250.106**, **800.40(D)**, **810.21(J)** and **820.40(D)** for required bonding and grounding of other interconnected systems. **(See Figure 3-16)**

Grounding Note: Difference of potential between two different grounding systems on the same building, if tied together, will reduce the potential difference between them.

For example, lightning rods and the electrical systems grounding electrodes are bonded together for this reason.

Figure 3-16. This illustration shows the requirements for installing and grounding lightning protection systems.

Grounding Tip 102: For specific requirements pertaining to lightning down conductors, see pages 13-5 through 13-10 in Chapter 13 of this book.

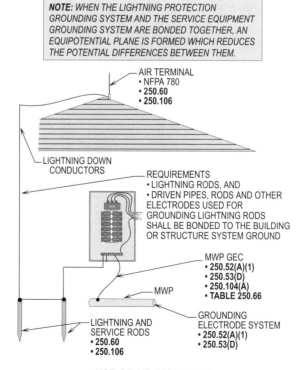

GROUNDING ELECTRODE CONDUCTOR MATERIAL
250.62

The grounding electrode conductor shall be of copper, aluminum or copper clad aluminum and shall be installed as solid or stranded, insulated, covered or bare. The grounding electrode conductor selected shall be resistant to any corrosive condition existing at the installation or be suitably protected against corrosion. **(See Figure 3-17)**

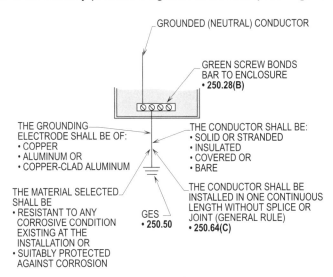

Figure 3-17. This illustration shows the materials that the grounding electrode may be made of and be used to ground the service equipment to a grounding electrode.

GROUNDING ELECTRODE CONDUCTOR INSTALLATION
250.64

The following wiring methods shall be permitted to be utilized for bonding, grounding and enclosing grounding electrode conductors:

- Aluminum or copper-clad aluminum conductors
- Securing and protection from physical damage
- Continuous
- Grounding electrode conductor taps
- Enclosures for grounding electrode conductors
- To electrode(s)

See Figure 3-18 for a detailed illustration when applying these requirements.

Grounding Tip 103: Raceways, cables, wireways, gutters and other protective enclosures shall be permitted to be used to protect grounding electrode conductors.

ALUMINUM OR COPPER-CLAD ALUMINUM CONDUCTORS
250.64(A)

Bare or aluminum or copper-clad aluminum conductors shall not be permitted to be used in the following locations:

- Where in direct contact with masonry or the earth
- Where subject to corrosive conditions

Aluminum or copper-clad aluminum grounding conductors shall not be permitted to be installed within 18 in. (450 mm) of the earth where such coductors are installed outside.

3-15

Figure 3-18. These enclosures shall be permitted to enclose grounding electrode conductors and also serve as a grounding and bonding means.

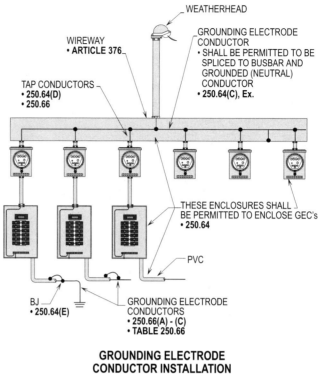

GROUNDING ELECTRODE
CONDUCTOR INSTALLATION
NEC 250.64

SECURING AND PROTECTION FROM PHYSICAL DAMAGE
250.64(B)

The following requirements shall be used to select and install the grounding electrode conductor that is used to connect the service enclosure to the grounding electrode system:

- Grounding electrode conductors 4 AWG or larger shall be securely fastened to the surface. Grounding electrode conductors not exposed to severe physical damage shall not require additional protection.

- A 6 AWG grounding electrode conductor shall be permitted to be run along the building surface construction without metal covering if not exposed to physical damage. Rigid metal conduit, intermediate metal conduit, rigid nonmetallic conduit, electrical metallic tubing or cable armor shall be permitted to be used where the grounding electrode conductor is exposed to physical damage.

- Rigid metal conduit, intermediate metal conduit, rigid nonmetallic conduit, electrical metallic tubing or cable armor shall be permitted to be used for installing a grounding electrode conductor smaller than 6 AWG. In most cases, a 8 AWG grounding electrode conductor is required to be installed per **Table 250.66** due to no OCPD ahead of service conductors.

See Figure 3-19 for a detailed illustration pertaining to these requirements.

CONTINUOUS
250.64(C)

The grounding electrode conductor shall be installed in one continuous length and shall not be permitted to be spliced from the service equipment to the grounding electrode, unless the splice is by means of irreversible compression-type connectors listed for the purpose or by the exothermic welding process. **(See Figure 3-20)**

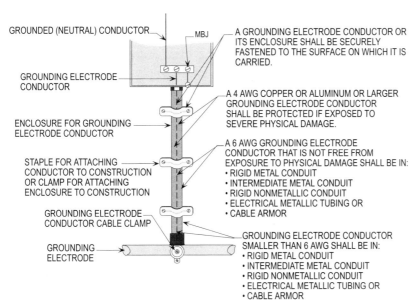

Figure 3-19. This illustration shows the requirements for installing and protecting the grounding electrode conductor with specific wiring methods.

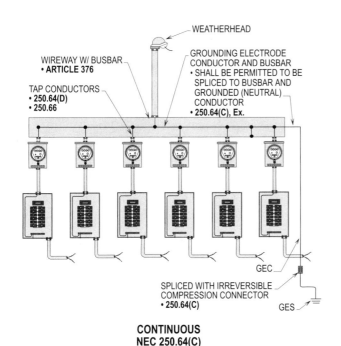

Figure 3-20. A grounding electrode conductor shall be permitted to be spliced in busbars located in meters, gutter, enclosures, wireways or with irreversible compression connectors.

Grounding Tip 104: The requirements of **250.64(C)** were added to the NEC to allow a grounding electrode conductor to be spliced if it's damage during construction stages or a service panel was moved in a remodel of a facility.

SPLICES IN BUSBARS
250.64(C), Ex.

A grounding electrode conductor shall be permitted to be spliced in busbars if they are located as meter enclosures, etc. **(See Figure 3-21)**

Grounding Tip 105: The spliced grounded neutral or grounded (neutral) conductor shall be sized according to the provisions in **250.66, 250.24(B)(1)** and **Table 250.66**. Note that the tapped grounding electrode conductors shall be sized in the same manner.

GROUNDING ELECTRODE CONDUCTOR TAPS
250.64(D)

A grounding electrode conductor shall be permitted to be connect to taps in a service consisting of more than a single enclosure. The tap conductors shall be connected to the grounding electrode conductor in such a manner that the grounding electrode conductor remains without a splice or joint. The tapped conductor shall be designed and connected to a grounding electrode conductor per **Table 250.66** based on the largest ungrounded (phase) conductors. The tapped grounding conductor is then connected to an enclosure. **(See Figure 3-21)**

Figure 3-21. A grounding electrode conductor shall be permitted to be connected to taps in a service consisting of more than a single enclosure. The tap conductors shall be connected to the grounding electrode conductor in such a manner that the grounding electrode conductor remains without a splice or joint.

SPLICES IN BUSBARS
NEC 250.64(C), Ex.

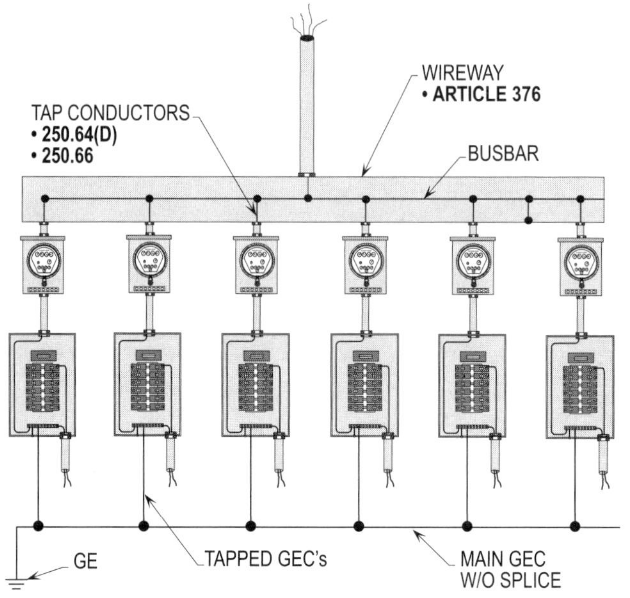

Grounding Tip 106: Section **250.64(C)** permits the grounding electrode conductor to be spliced by an irreversible compression-type connector or by exothermic welding. **(Refer to Figure 3-2)**

ENCLOSURES FOR GROUNDING ELECTRODE CONDUCTORS 250.64(E)

Grounding Tip 107: The fittings permitted to be used as a bonding means to satisfy this requirement in **250.64(E)** are found in **250.92** and **300.15(A)**.

If a metal enclosure is used over the grounding electrode, it shall be electrically continuous from its origin to the grounding electrode. It shall be securely fastened to the grounding clamp so that there is electrical continuity from where it originates to the grounding electrode. If a metal enclosure is used for physical protection for the grounding electrode conductor and it is not continuous from the enclosure to the grounding electrode, it shall be bonded at both ends. **(See Figure 3-22)**

Figure 3-22. This illustration shows requirements for bonding a raceway containing the grounding electrode conductor.

ENCLOSURES FOR GROUNDING
ELECTRODE CONDUCTORS
NEC 250.64(E)

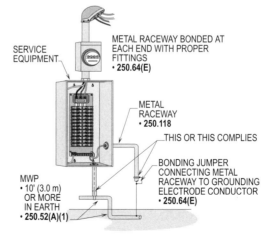

Grounding Tip 108: By bonding the ends of the raceway to the grounding electrode conductor, the metal of the raceway will not act as the core of a "choke coil" and restrict the flow of current through the grounding electrode conductor.

TO ELECTRODE(S)
250.64(F)

A grounding electrode conductors shall be permitted to be run to any convenient grounding electrode that is available in the grounding electrode system or to one or more grounding electrode(s) individually. The grounding electrode conductor shall be sized for the largest grounding electrode conductor required among all the electrodes connected to it.

SIZE OF ALTERNATING-CURRENT GROUNDING ELECTRODE CONDUCTOR
TABLE 250.66

The following grounding electrodes shall be grounded by conductors sized and selected as listed in **Table 250.66** and **250.66(A)** thru **(C)**.

- Connections to metal water pipe
- Connections to structural steel
- Connections to rod, pipe or plate electrodes
- Connections to concrete-encased electrodes
- Connections to ground rings

Grounding Tip 109: Table 250.66 is used (only) to size the grounding electrode conductor to ground the service equipment to a metal water pipe or to structural steel per 250.52(A)(1), 250.104(A), 250.52(A)(2) and 250.104(C). To size the grounding electrode conductors of other electrodes, see 250.66(A) thru (C) and 250.52(A)(3) thru (A)(7).

CONNECTIONS TO METAL WATER PIPE
250.66; TABLE 250.66; 250.104(A); 250.52(A)(1); 250.53(D)

The procedure for selecting the grounding electrode conductor to ground the service to a metal water pipe shall be determined by the size of the service-entrance conductors. **(See Figure 3-2)**

Grounding Tip 110: Because there is no OCPD ahead of the ungrounded (phase) conductors, the grounding electrode conductors run to a metal water pipe shall be sized from **Table 250.66** based on the largest ungrounded (phase) conductors.

For example: What size copper GEC is required to ground a service to a metal water pipe supplied by 250 KCMIL copper conductors?

    Step 1: Finding size GEC
           Table 250.66
           250 KCMIL cu. = 2 AWG cu.

    Solution: The size grounding electrode conductor is required to be 2 AWG copper.

CONNECTIONS TO STRUCTURAL BUILDING STEEL
250.66; TABLE 250.66; 250.104(C); 250.52(A)(2)

The procedure for selecting the grounding electrode conductor to ground the service to structual steel shall be determined by the size of the service-entrance conductors. **(See Figure 3-4)**

For example: What size copper grounding electrode conductor is required to ground a service to structural steel supplied by 250 KCMIL copper conductors?

    Step 1: Finding the GEC
           Table 250.66
           250 KCMIL cu. requires 2 AWG cu.

    Solution: The size grounding electrode conductor is required to be 2 AWG copper.

Grounding Tip 111: If the grounding electrode conductor is run greater than a 100 ft. and connected to a grounding electrode, it shall be increased 25 percent in size for each 25 ft. exceeding 100 ft. For example, if a 2 AWG copper grounding electrode conductor is run 150 ft., it shall be upsized to 99,540 CM. (66,360 CM x 150% = 99,540 CM)

CONNECTIONS TO ROD, PIPE OR PLATE ELECTRODES
250.66(A)

Grounding Tip 112: A 6 AWG copper grounding electrode conductor will carry about 4.8 amps if the driven rod has a resistance of 25 ohms with a voltage of 120 volts-to-ground. (See Quick Calc below)

The service equipment shall be permitted to be grounded with a rod, pipe or plate electrode where there are no other electrodes available in **250.52(A)(1)** thru **(A)(7)**. A driven rod or supplementary grounding electorde with a resistance of 25 ohms or less is considered low enough to allow the grounded system to operate safely and function properly. If the driven rod is used as a supplementary grounding electrode to the metal water pipe system, it should be connected to the grounded terminal bar in the service equipment panelboard. For further information, see **250.52(A)(4)** thru **(A)(7)** and **250.53(A), (G)** and **(H)** for the application of this rule when using one of the made electrodes listed there. The grounding electrode conductor shall not be required to be larger than 6 AWG copper or 4 AWG aluminum where connected to made electrodes such as driven rods. **(See Figure 3-10)**

> **For example:** What is the current flow in a 6 AWG copper grounding electrode conductor connecting the common grounded terminal bar in the service equipment to a driven rod? (The supply voltage is 120/208 volt, three-phase system)
>
> **Step 1:** Finding amperage
> **250.56**
> I = 120 V ÷ 25R
> I = 4.8 A
>
> **Solution:** The normal current flow is about 4.8 amps.

CONNECTIONS TO CONCRETE-ENCASED ELECTRODES
250.66(B)

Grounding Tip 113: Rebar that is coated, taped or insulated shall not be permitted to be used as a concrete-encased electrode. Stress-cables shall not be permitted to be used as a concrete-encased electrode. If a concrete-encased electrode is desired, at least a 4 AWG copper conductor shall be installed per **250.52(A)(3)**.

Grounding Tip 114: Concrete in contact with earth will always draw some moisture which causes the grounding resistance to be very low. With the grounding electrodes available to be bonded together with the rebar in the concrete, resistance readings have been measured just a little above zero ohms.

Lengths of rebar 1/2 in. (13 mm) in diameter and at least 20 ft. (6 m) long in length shall be permitted to be used as a grounding electrode to ground the service equipment. The size of the grounding electrode conductor shall be a 4 AWG or larger copper conductor per **250.66(B)** and **250.52(A)(3)**. The rebar system shall be permitted to be one length that is 1/2 in. x 20 ft. (13 mm x 6 m) long or a number of such that are spliced together to form a 20 ft. (6 m) length. This grounding method is known in the electrical industry as the UFER ground.

A concrete-encased electrode shall also be permitted to be a 4 AWG copper conductor instead of 1/2 in. x 20 ft. (13 mm x 6 m) rebar. The 4 AWG copper conductor shall be located within 2 in. (50 mm) of concrete located within or near the bottom of the foundation.

The size of the grounding electrode shall be at least a 4 AWG copper per **250.66(B)**. The 4 AWG copper grounding electrode conductor is selected based on the 4 AWG copper grounding electrode installed in the foundation. **(See Figure 3-5)**

CONNECTIONS TO GROUND RINGS
250.66(D)

Grounding Tip 115: The ground ring shall be installed at least 30 in. (750 mm) in the earth soil to be considered a ground ring electrode.

A bare copper conductor not smaller than 2 AWG and at least 20 ft. (6 m) long shall be installed to create a ground ring that will encircle a building or structure. This ring shall be permitted to be utilized as the main grounding electrode if necessary or just bonded in as one per **250.50** and **250.52(A)(4)**. **(See Figure 3-6)**

GROUNDING ELECTRODE CONDUCTOR AND BONDING JUMPER CONNECTION TO GROUNDING ELECTRODES 250.68

The following methods shall be adhered to when connecting the grounding electrode conductor to grounding electrodes:
- Accessibility
- Effective grounding path

ACCESSIBILITY
250.68(A)

The 1987 NEC made it very clear that installers and inspectors shall work together to see that grounding clamps and fittings used to connect the grounding electrode conductor or bonding jumpers to water pipes or building steel are accessible. All installations built before the adoption of the 1978 NEC did not necessarily require the grounding clamps or fittings to be accessible. This revision in the 1978 NEC came about due to the grounding clamps or fittings being removed or damaged during construction and was never fixed or replaced. A missing ground clamp could cause the grounded (neutral) conductor to float and proper voltage regulation would not be accomplished. **(See Figure 3-23)**

Grounding Tip 116: Only the grounding connection of the grounding electrode conductor to a metal water pipe or building steel shall be required to be accessible after the installation per **250.68(A)**. However, this is not required for driven rods, concrete-encased electrodes or ground rings per **250.68(A), Ex.**.

Figure 3-23. The 1987 NEC made it very clear that installers and inspectors shall work together to see that grounding clamps and fittings used to connect the grounding electrode conductor to metal water pipes or building steel are accessible. All installations built before the adoption of the 1978 NEC did not necessarily require the grounding clamps or fittings to be accessible.

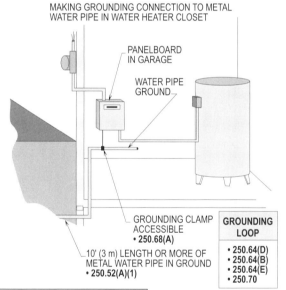

MINIMUM SIZE WATER-PIPE ELECTRODES AND GROUNDING ELECTRODE CONDUCTORS	
SIZE OF GEC's	SIZE OF WATER PIPE, IN INCHES
8 AWG	1/2" (16)
6 AWG	3/4" (21)
4 AWG	1" (27)
2 AWG	1 1/4" (35)
1/0 AWG	1 1/2" (41)
2/0 AWG	2" (53)
3/0 AWG OR 4/0 AWG	2 1/2" (63)
THE SIZE OF THE METAL WATER PIPES USED TO CONNECT THE GEC's ARE RECOMMENDED SIZES.	

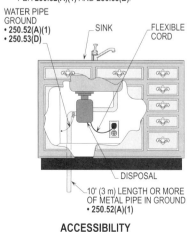

ACCESSIBILITY
NEC 250.68(A)

ACCESSIBILITY NOT REQUIRED
250.68(A), Ex.

Grounding clamps, fittings or other approved methods that are connect the grounding electrode conducor to concrete-encased electrodes, ground rings, plates or driven rods are not required to be accessible. However, they shall be approved (listed) for direct burial.

EFFECTIVE GROUNDING PATH
250.68(B)

The connection of the grounding electrodes or bonding jumpers to the grounding electrode shall be in a manner that ensures a permanent and effective grounding path. When metal piping systems are used as a grounding electrode, effective bonding conductors of sufficient length shall be installed around insulated joints and sections and around any equipment that is likely to be disconnected for repairs or replacement. **(See Figure 3-24)**

Grounding Tip 117: Grounded effectively is defined as intentionally connected to earth through a ground connection or connections of sufficiently low-impedance and having sufficient current-carrying capacity to prevent the buildup of voltages that may result in undue hazards to connected equipment or to persons.

Figure 3-24. This illustration shows water meters and other devices shall not be permitted to be used for continuity of the metal water piping.

Grounding Tip 118: Table 250.66 is only used to size grounding electrode conductors to ground service equipment to earth ground when there is no OCPD ahead of the ungrounded (phase) conductors.

Grounding Tip 119: The size of the grounding electrode conductors in **Table 250.66** shall be determined by the size of the ungrounded (phase) conductors which are based on 100 ft. lengths or less.

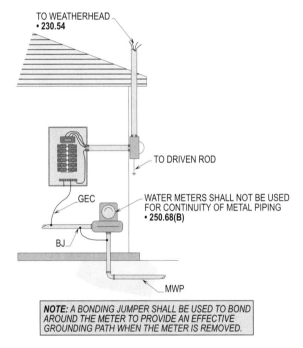

**EFFECTIVE GROUNDING PATH
NEC 250.68(B)**

Grounding Electrode System and Grounding Electrode Conductor

Table 250.66. Grounding Electrode Conductor for Alternating-current Systems

Size of Largest Ungrounded Service-Entrance Conductor or Equivalent Area for Parallel Conductors (AWG / KCMIL)		Size of Grounding Electrode Conductor (AWG / KCMIL)	
Copper	Aluminum or Copper-clad Aluminum	Aluminum or Copper-clad Copper	Aluminum
2 AWG or smaller	1/0 AWG or smaller	8 AWG	6 AWG
1 AWG or 1/0 AWG	2/0 AWG or 3/0 AWG	6 AWG	4 AWG
2/0 AWG or 3/0 AWG	4/0 AWG or 250 KCMIL	4 AWG	2 AWG
Over 3/0 AWG through 350 KCMIL	Over 250 KCMIL through 500 KCMIL	2 AWG	1/0 AWG
Over 350 KCMIL through 600 KCMIL	Over 500 KCMIL through 900 KCMIL	1/0 AWG	3/0 AWG
Over 600 KCMIL through 1100 KCMIL	Over 900 KCMIL through 1750 KCMIL	2/0 AWG	4/0 AWG
Over 1100 KCMIL	Over 1750 KCMIL	3/0 AWG	250 KCMIL

A 2 AWG copper grounding electrode conductor is required to ground a service supplied by 250 KCMIL conductors.

METHODS OF GROUNDING AND BONDING CONDUCTOR CONNECTIONS TO ELECTRODES 250.70

When connecting grounding or bonding conductors to grounding fittings by suitable lugs, use pressure connectors, clamps or other listed means including exothermic welding. Soldering shall never be used. The ground clamps shall be of a material that is compatible for both the grounding electrode and the grounding electrode conductor. No more than one conductor shall be connected to an electrode unless the connector is listed for the purpose per **110.14(A)**. Grounded fittings used to connect the grounding conductor shall be pemitted to be any of the following:

- Exothermic welding
- Listed lugs
- Listed pressure connectors
- Listed clamps

Direct burial grounding clamps shall be utilized on grounding electrodes in the earth such as pipes, rods, rebar, encased electrodes, etc. **(See Figure 3-25)**

Grounding Tip 120: Grounding clamps and fittings that are used to connect grounding electrode conductors in the earth soil shall be listed for direct burial per **110.3(B)** and **250.70**.

Figure 3-25. When connecting grounding conductors to grounding fittings by suitable lugs; use pressure connectors, clamps or other listed means, including exothermic welding.

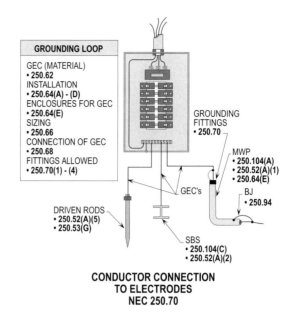

Name _____   Date _____

Chapter 3
Grounding Electrode System and Grounding Electrode Conductor

	Section	Answer

1. The grounding electrode system shall be permitted to consist of more than one electrode. _____   T   F

2. A plate electrode that measures more than 25 ohms shall be augmented by two additional electrodes. _____   T   F

3. The grounding electrode conductor shall be permitted to be installed further than 5 ft. (1.52 m) in industrial and commercial buildings where conditions of maintenance and supervision ensure that only qualified personnel will service the installation and the entire length of the interior metal water pipe that is being used for the conductor is exposed. _____   T   F

4. Metal water piping shall be electrically continuous or made electrically continuous by bonding around insulated joints or sections. _____   T   F

5. Metal water piping used as a grounding electrode shall not be required to be supplemented by an additional electrode. _____   T   F

6. Metal reinforcing bars for concrete-encased electrodes shall not be required to be of the conductive type to be used as grounding electrodes. _____   T   F

7. Underground metal gas piping systems shall not be permitted to be installed and used as a grounding electrode to earth ground the service equipment. _____   T   F

8. Metal underground systems and structures such as piping systems and underground tanks shall not be permitted to be installed as an grounding electrode. _____   T   F

9. Where a rock bottom is encountered, a driven electrode shall be permitted to be driven at an angle not to exceed 45 degrees from the vertical. _____   T   F

10. The earth shall be permitted to be used as the sole equipment grounding conductor. _____   T   F

11. An insulated aluminum grounding electrode conductor shall be permitted to be installed where in direct contact with masonry or the earth. _____   T   F

12. A grounding electrode conductor shall be permitted to be spliced in busbars if they are located as a meter enclosure, etc. _____   T   F

13. Grounding clamps and fittings used to connect the grounding electrode conductor to metal water pipes or building steel shall not be required to be accessible. _____   T   F

Section	Answer		
_____	T	F	**14.** The connection of the grounding electrode conductor to the grounding electrode shall be in a manner that ensures a permanent and effective grounding path.
_____	T	F	**15.** Soldering shall be permitted for connection of the grounding electrode conductor to the grounding electrode.
_____	_____		**16.** Metal water piping with lengths at least _____ ft. long in the earth shall be connected to the grounded neutral bar in the service equipment enclosure.
_____	_____		**17.** The grounding connection to the metal water pipe shall be made at a point no more than _____ ft. where the piping enters the building.
_____	_____		**18.** Concrete-encased electrodes consists of 1/2 in. x _____ ft. lengths of reinforcing rebar located in the foundation.
_____	_____		**19.** A concrete-encased electrode shall be permitted to consist of a _____ AWG copper conductor at least 20 ft. long that is installed in the footing of the foundation.
_____	_____		**20.** A bare copper conductor not smaller than _____ AWG and at least 20 ft. long in length at least 2 1/2 ft. below the earth shall be permitted to be used to create a ground ring grounding system for encircling the building or structure.
_____	_____		**21.** A driven electrode used to ground the service shall be permitted to be a 1/2 in. x _____ ft. copper (brass) rod.
_____	_____		**22.** A driven electrode shall be installed where there is at least _____ ft. in contact with the soil.
_____	_____		**23.** Where a rock bottom in encountered, a driven electrode shall be permitted to be buried in a trench at least _____ in. deep.
_____	_____		**24.** Plate electrodes shall be a minimum of 2 sq. ft. and if made of iron or steel they shall be at least _____ in. thick.
_____	_____		**25.** Made electrodes shall have a resistance to ground of _____ ohms or less, wherever practicable.
_____	_____		**26.** Aluminum or copper-clad aluminum grounding conductors shall not be installed within _____ in. of the earth where such conductors are installed outside.
_____	_____		**27.** Grounding electrode conductors _____ AWG or larger shall be securely fastened to the surface.
_____	_____		**28.** A _____ AWG grounding electrode conductor shall be permitted to be run along the building surface construction without metal covering if not exposed to physical damage.

3-26

                                                                                      Section   Answer

29. Rigid metal conduit is a type of wiring method that shall be permitted to be   _____   _____
    used for installing a grounding electrode conductor smaller than _____
    AWG.

30. Plate electrodes shall not be installed less than _____ in. below the surface   _____   _____
    of the earth.

31. What size copper grounding electrode conductor is required to ground a   _____   _____
    service to a metal water pipe supplied by 4/0 AWG THWN copper conduc-
    tors?

32. What size copper grounding electrode conductor is required to ground a   _____   _____
    service to a metal water pipe supplied by 300 KCMIL THWN copper con-
    ductors?

33. What size copper grounding electrode conductor is required to ground a   _____   _____
    service to building steel supplied by 4/0 AWG THWN copper conductors?

34. What size copper grounding electrode conductor is required to ground a   _____   _____
    service to building steel supplied by 300 KCMIL THWN copper conductors?

35. What size copper grounding electrode conductor is required to connect the   _____   _____
    common grounded terminal bar in the service equipment to a driven rod?
    The size of the service-entrance conductors are 4/0 AWG THWN copper
    con-ductors.

36. What size copper grounding electrode conductor is required to connect the   _____   _____
    common grounded terminal bar in the service equipment to a driven rod?
    The size of the service-entrance conductors are 300 KCMIL THWN copper
    conductors.

37. What size grounding electrode conductor is required to connect the com-   _____   _____
    mon grounded terminal bar in the service equipment to a concrete-encased
    electrode? The size of the service-entrance conductors are 4/0 AWG THWN
    copper conductors.

38. What size grounding electrode conductor is required to connect the com-   _____   _____
    mon grounded terminal bar in the service equipment to a concrete-encased
    electrode? The size of the service-entrance conductors are 300 KCMIL
    THWN copper conductors.

39. What size grounding electrode conductor is required to ground 4/0 AWG   _____   _____
    THWN copper conductors to a ground ring?

40. What size grounding electrode conductor is required to ground 300 KCMIL   _____   _____
    THWN copper conductors to a ground ring?

4

ENCLOSURE, RACEWAY AND SERVICE CABLE GROUNDING

Equipment grounding is the bonding together of all conductive enclosures for conductors and equipment in each circuit with an equipment grounding means. These equipment grounding conductors along with metal conduits, metal clad of cables, etc. shall be run with or enclose the circuit conductors in such a manner as to provide a permanent, low-impedance conductive path for ground-fault currents. In solidly grounded systems, the equipment grounding conductors are bonded to the grounded (phase or neutral) conductor and to the system grounding means by a main bonding jumper located in the service equipment or subpanel supplying power to the loads of the facility.

SERVICE RACEWAYS AND ENCLOSURES
250.80

All metal service enclosures and service raceways shall be grounded. However, a metal conduit elbow in a nonmetallic conduit run shall not be required to be grounded so long as it is no less than 18 in. (450 mm) below grade.

Metal raceways would include, electrical metallic tubing and the metal sheath or armor of service cables. Grounding can be accomplished by bonding the conduit, electrical metallic tubing, metal sheath or armor to the grounded service-entrance cabinet, or by connection to a grounded neutral. Bonding shall be in accordance with **250.94**. **(See Figure 4-1)**

Figure 4-1. Metal enclosures and raceways are required to be bonded and grounded with connectors, couplings and threaded fittings and hubs to ensure that electrical continuity is maintained from point to point.

Grounding Tip 121: A potential difference with a reference to ground can be dangerous to life and the associated current flow can generate enough heat anywhere in the ground-current path to cause a fire to occur.

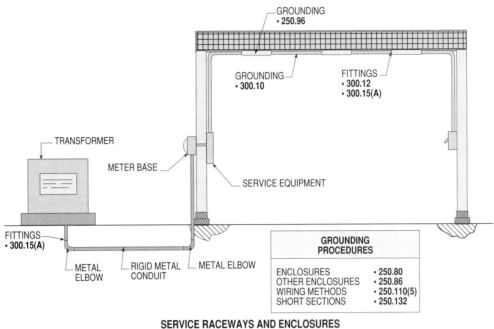

NOT REQUIRED TO BE GROUNDED
250.80, Ex.

A metal elbow that is installed in an underground installation of rigid nonmetallic conduit and is isolated from possible contact by a minimum cover of 18 in. (450 mm) to any part of the elbow shall not be required to be grounded. **(See Figure 4-2)**

Figure 4-2. This illustration shows an underground run of conduit where the metal elbow in the earth is not required to be grounded.

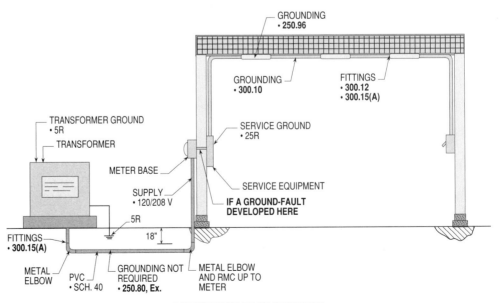

UNDERGROUND SERVICE CABLE OR CONDUIT
250.84

A continuous underground system provides grounding for the conduit or metal sheath of the underground services (laterals) connected to the underground system. Therefore, in this case the conduit or metal sheath of the service (lateral) need not be grounded at the building. This exception applies for laterals connected to "continuous underground systems," and does not apply for a lateral run underground to a pole and overhead conductors.

The grounded (neutral) conductor and service enclosure shall be grounded, the same as for overhead services to ensure safe operation of the electrical system.

UNDERGROUND SERVICE CABLE
250.84(A)

The sheath or armor of a continuous underground metal-sheathed service cable system that is mechanically connected to the underground system shall not be required to be grounded at the building. The sheath or armor shall be permitted to be insulated from the interior conduit or piping.

UNDERGROUND SERVICE CONDUIT CONTAINING CABLE
250.84(B)

An underground service conduit that contains a metal-sheathed cable bonded to the underground service shall not be required to be grounded at the buildings. The sheath or armor shall be permitted to be insulated from the interior conduit or piping. **(See Figure 4-3)**

Grounding Tip 122: Consider a 120/208 volt, three-phase, four-wire system with a transformers neutral connected to earth ground which has a resistance of 5 ohms and the conduit enclosing the circuit conductors is earth grounded at the service equipment with a resistance of 25 ohms, what is the fault current if phase C developed a fault to the conduit?

- GFC = 120 V ÷ 25R + 5R
- GFC = 4 A

Solution: The ground-fault current is 4 amps.

Grounding Tip 123: The reason the rigid metal conduit and metal elbows cannot be connected together by a bonding jumper run on the outside of the rigid nonmetallic conduit is, Sec. 250-102(e) only permits a bonding jumper to be routed in lengths not greater than 6 ft. on the outside of a raceway.

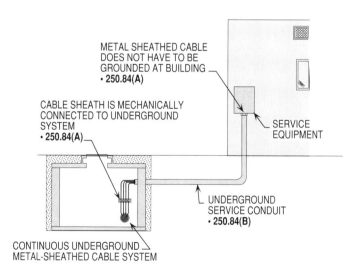

Figure 4-3. Under certain conditions of use, the metal-sheathed cable in an underground conduit system does not have to be grounded at the building.

OTHER CONDUCTOR ENCLOSURES AND RACEWAYS
250.86

Conduit, electrical metallic tubing, metal raceways and BX sheath, wherever used, shall be grounded, except as allowed per **250.112(I)**. Properly grounded metal provides safety for personnel by establishing a condition of essentially equal potential, covering the entire area of the electrical system for all machine frames and enclosures including all nonelectrical structures. Without equipment grounding, accidental contact of the electric circuit to a noncurrent-carrying metallic body raises the potential of the metallic object above ground potential and creates the threat of electrical shock and fire hazards.

NOT REQUIRED TO BE GROUNDED
250.86, Ex.'s 1 thru 3

Grounding Tip 124: How short is considered short? Well most AHJ's consider short as being a 10 ft. or less length of conduit. Note that this is based on a standard length of manufactured conduit.

An exception to this rule shall be permitted if a short piece of conduit, electrical metallic tubing or AC cable is added to a nonmetallic wiring system, such as Romex. If the conduit, electrical metallic tubing or AC cable is less than 25 ft. (7.5 m) in length and free from probable contact with ground, metal lath, piping or conductive material, it need not be grounded. If 25 ft. (7.5 m) or more in length, it shall always be grounded. If within reach of ground or a grounded object, it shall be grounded regardless of length. Short sections of enclosures used to protect cable assemblies shall not be required to be grounded.

A metal conduit elbow in a rigid nonmetallic conduit run shall not be required to be grounded so long as it is not less than 18 in. (450 mm) below grade or beneath a 2 in. (50 mm) thick concrete slab. **(See Figure 4-4)**

Figure 4-4. This illustration shows metal conduits and enclosures that do not have to be grounded.

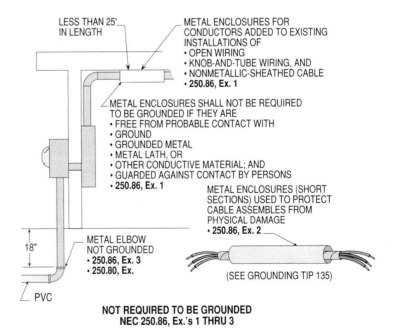

Name _____   Date _____

Chapter 4
Enclosure, Raceway and Service Cable Grounding

	Section	Answer

1. All metal service enclosures and service raceways shall be grounded. _____ T F

2. A metal elbow that is installed in an underground installation of rigid non-metallic conduit and is isolated from possible contact by a minimum cover of 12 in. to any part of the elbow shall not be required to be grounded. _____ T F

3. The sheath or armor of a continuous underground metal-sheathed service cable system that is mechanically connected to the underground system shall not be required to be grounded at the building. _____ T F

4. An underground service conduit that contains a metal-sheathed cable bonded to the underground service shall be grounded at the building. _____ T F

5. Short sections of metal enclosures or raceways used to provide supports or protection of cable assemblies from physical damage shall be grounded. _____ T F

5

BONDING

When performed properly, bonding connects the equipment frames or enclosures to the ground bus. The ground bus provides a protective ground network throughout the area served by the electrical system to establish a uniform potential and is connected to the grounding electrodes. The grounding electrodes are embedded in the earth for the purpose of dissipating the currents conducted to them by the grounding electrode conductors.

BONDING
250.90

Bonding is the permanent joining of metallic parts to form an electrically conductive path that ensures electrical continuity and the capacity to conduct safely any current likely to be imposed. Such bonding jumpers shall be sized to handle these larger available fault currents when necessary to prevent damaging components and equipment.

SERVICES
250.92

The following bonding requirements shall be considered for services:

- Bonding of services
- Method of bonding at the service

BONDING OF SERVICES
250.92(A)

The following noncurrent-carrying metal parts of service equipment shall be bonded together:

- Service raceways, cable trays, cable bus framework, auxiliary gutters or service cable armor or sheath except as permitted per **250.84**

- All service enclosures containing service conductors, including meter fittings, meter base enclosures or CT cans, interposed on the service raceway or armor

- Any metallic raceway or armor enclosing a grounding electrode conductor per 250.64(B). Bonding shall apply at each end and to all intervening raceways, boxes and enclosures between the service equipment and the grounding electrode.

Service-entrance raceways, meter base enclosures, service equipment enclosures and all other metal enclosures of the service equipment shall be bonded together to ensure a continuous metal-to-metal grounding system. **(See Figure 5-1)**

Figure 5-1. All metal raceways, metal-clad of cables, meter base and metal equipment enclosures shall be required to be bonded together for safety.

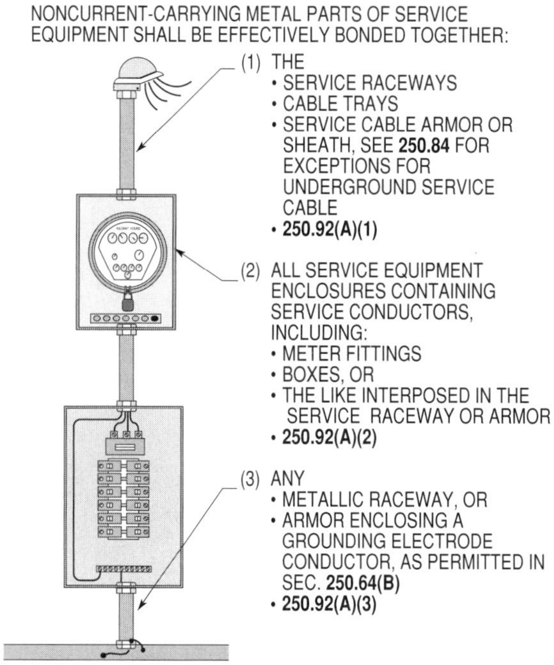

Grounding Tip 125: For information pertaining to the procedures of performance testing, See Chapters 11 through 14 in UL 467.

NONCURRENT-CARRYING METAL PARTS OF SERVICE EQUIPMENT SHALL BE EFFECTIVELY BONDED TOGETHER:

(1) THE
- SERVICE RACEWAYS
- CABLE TRAYS
- SERVICE CABLE ARMOR OR SHEATH, SEE **250.84** FOR EXCEPTIONS FOR UNDERGROUND SERVICE CABLE
- **250.92(A)(1)**

(2) ALL SERVICE EQUIPMENT ENCLOSURES CONTAINING SERVICE CONDUCTORS, INCLUDING:
- METER FITTINGS
- BOXES, OR
- THE LIKE INTERPOSED IN THE SERVICE RACEWAY OR ARMOR
- **250.92(A)(2)**

(3) ANY
- METALLIC RACEWAY, OR
- ARMOR ENCLOSING A GROUNDING ELECTRODE CONDUCTOR, AS PERMITTED IN SEC. **250.64(B)**
- **250.92(A)(3)**

**BONDING OF SERVICES
NEC 250.92(A)**

METHODS OF BONDING AT THE SERVICE 250.92(B)

Grounding Tip 126: Sections **250.92** and **250.94** covers the requirements for bonding metal raceways, cables, enclosures and service equipment on the supply side of the service per **250.142(A)(1)**.

Any one of the following methods shall be permitted to be utilized for the bonding of service equipment:

- Grounded service conductor
- Threaded connections
- Threadless couplings and connectors
- Other devices such as bonding type locknuts and bushings
- Bonding jumpers

See Figure 5-2 for a detailed illustration when applying these requirements.

METHOD OF BONDING AT THE SERVICE 250.92(B)(1) THRU (4)

A bonding conductor shall be used to connect and bond the raceway to the enclosure where smaller concentric or eccentric knockouts are removed, leaving larger ones. A metal bushing with a lug on the threaded raceway is used to terminate the bonding jumper to a lug connected to the enclosure or grounded busbar terminal. Any one of the following methods shall be permitted to be utilized for the bonding of service equipment:

- Grounded service conductor
- Threaded connections
- Threadless couplings and connectors
- Other devices

Bonding

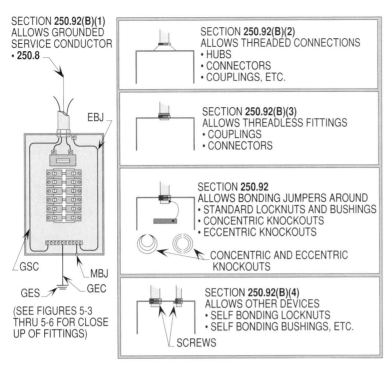

METHOD OF BONDING AT THE SERVICE NEC 250.92(B)(1) THRU (B)(4)

Figure 5-2. This illustration shows the proper methods for bonding raceways at the service equipment.

Grounding Tip 127: Threaded hubs shall not be required to be bonded with a bonding jumper.

Grounding Tip 128: Threaded fittings shall not be required to be bonded from raceway to raceway with bonding jumpers.

Grounding Tip 129: Remaining knockouts and standard locknuts and bushings shall be bonded properly with bonding jumpers attached to the raceway.

Grounding Tip 130: Self bonding locknuts and bushings shall not be required to be bonded with bonding jumpers unless there are punched out rings remaining around the raceway system.

GROUNDED SERVICE CONDUCTOR
250.92(B)(1)

The grounded service conductor shall be permitted for bonding equipment by one of the following methods per **250.8**:

- Exothermic welding
- Listed pressure connectors
- Listed clamps
- Other listed means

See Figure 5-3 for a detailed illustration when applying these requirements.

Grounding Tip 131: The grounded conductor (neutral or phase) can be used to bond metal cables, raceways and enclosures together where such conductor is used on the supply side of the service disconnecting means per **250.142(A)(1)**.

THREADED CONNECTIONS
250.92(B)(2)

Threaded connections, such as threaded couplings or threaded bosses shall be permitted to be installed where the service equipment is threaded into the meter base or the service equipment enclosure. Such threaded connections shall be made up wrench-tight. **(See Figure 5-4)**

THREADLESS COUPLINGS AND CONNECTORS
250.92(B)(3)

Threadless couplings and connectors installed for rigid metal conduit, intermediate metal conduit and electrical metallic tubing shall be made up tight. Standard locknuts and bushings shall not be installed for bonding the raceway to the enclosure. Only locknuts which are approved for bonding such as extra thick locknuts and bushings equipped with a set screw shall be permitted to be used for this purpose. **(See Figure 5-5)**

Grounding Tip 132: Threadless couplings and connectors are available in compression or set-screw type.

Figure 5-3. This illustration shows the fittings that shall be permitted to be used to bond and connect the grounded (neutral) conductor to the metal enclosure housing the components of the service.

GROUNDED SERVICE CONDUCTOR
NEC 250.92(B)(1)

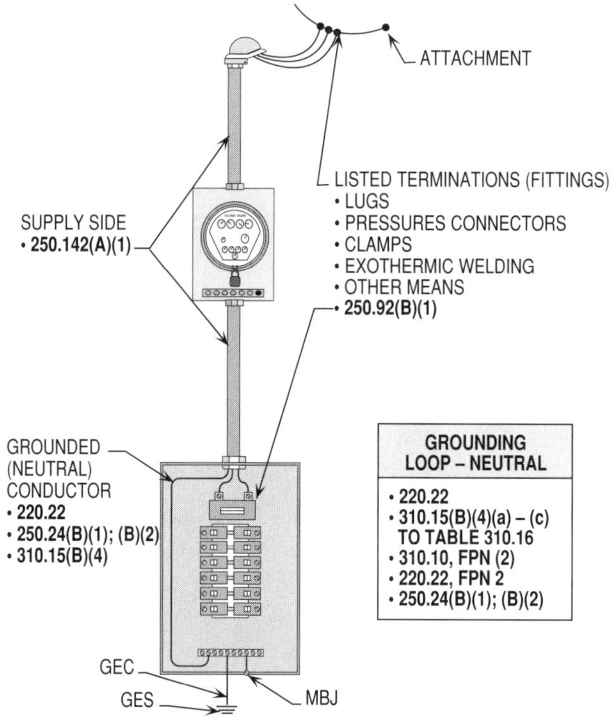

Figure 5-4. This illustration shows a threaded hub used as the bonding means for grounding the rigid metal conduit to the metal of the meter base enclosure.

Grounding Tip 133: Threaded hubs and couplings shall be considered self bonding fittings.

THREADED CONNECTIONS
NEC 250.92(B)(2)

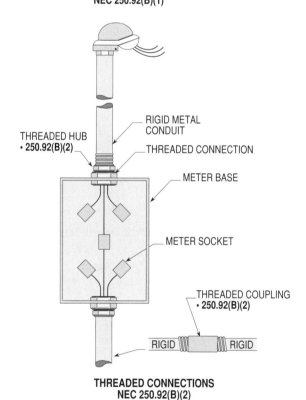

OTHER DEVICES 250.92(B)(4)

A wedge fitting (bushing) or locknut fitting is an application of other approved devices that shall be permitted to be used. A grounding wedge is equipped with a set-screw that tightens the metal conduit and ensures proper bonding of the service equipment enclosure.

Grounding Note: Two locknuts (one is self-bonding) with a bushing are an acceptable bonding means. However, two standard locknuts shall not be used as a bonding means per **250.92(B)(3)**. Note that bonding jumpers shall be sized by the provisions listed in **250.102(C)** and **Table 250.66**. **(See Figure 5-6)**

Figure 5-5. This illustration shows threadless connectors used as a bonding means for grounding the electrical metallic tubing to the metal panelboard.

Grounding Tip 134: Threadless fittings such as compression and set-screw couplings and connectors are considered self-bonding fittings.

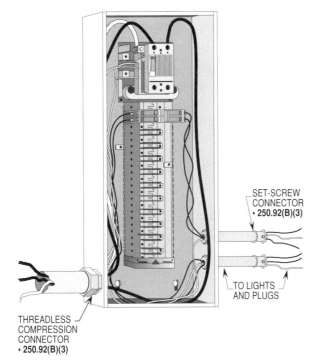

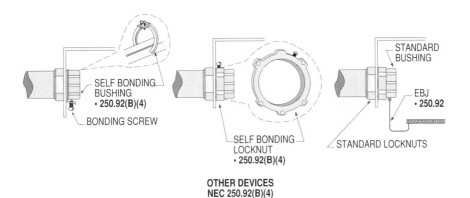

Figure 5-6. This illustration shows self-bonding devices such as locknuts and bushings used to bond and ground rigid metal conduit to the metal enclosure enclosing electrical components of equipment.

BONDING FOR OTHER SYSTEMS
250.94

One of the following accessible means external to enclosures shall be required to be provided at the service equipment and at the disconnecting means for any individual buildings or structures for connecting intersystem bonding and grounding conductors:

- Exposed nonflexible metallic service raceways
- Exposed grounding electrode conductor
- Approved means for the external connection of a copper or other corrosion-resistant bonding or grounding conductor connected to the service raceway or equipment

See Figure 5-7 for a detailed illustration when applying these requirements.

The disconnecting means at a separate building or structure per **250.32(D)** and the disconnecting means for a mobile home per **550.11(A)** shall be considered the service equipment when providing an accessible means for intersystem bonding. See Figures 7-20(a) through (d) on pages 7-15 and 7-16 for a detailed illustration for proper grounding of mobile homes for safety.

Grounding Note: An example of approved means for the external connection of bonding or grounding conductor to the service raceway or equipment is a 6 AWG copper conductor having one end bonded to the service raceway or equipment and with 6 in. (150 mm) or more of the other end accessible to the outside wall.

Figure 5-7. This illustration shows the procedures for bonding intersystems together such as communications lines, CATV lines, etc.

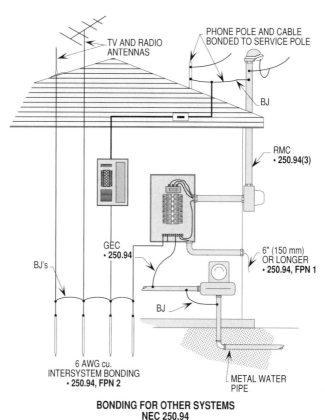

COMMUNICATIONS SYSTEMS - GROUNDING CONDUCTOR 250.94, FPN 2; 800.40(A)(1) THRU (A)(6)

Grounding Tip 135: Where a 14 AWG copper conductor is used to ground CATV or telephone systems to earth electrodes, they shall be bonded into the grounding electrode system of the service equipment with a 6 AWG copper bonding jumper.

Grounding Tip 136: For additional information pertaining to driven rods, see 9.2 in UL 467.

Communications systems are usually bonded and grounded to the existing grounding electrode system with at least a 14 AWG copper conductor.

The grounding conductor for communications systems shall have certain characteristics and be installed as follows:

- The grounding conductor shall be insulated and listed as suitable for the purpose.
- The grounding conductor shall be copper or other corrosion-resistive conductive material.
- The grounding conductor shall be stranded or solid.
- The grounding conductor shall not be smaller than 14 AWG or equivalent
- The grounding conductor shall be run to the grounding electrode in as straight a line as practical.

- Where necessary, the grounding conductor shall be guarded from physical damage.
- The grounding conductor shall be bonded at both ends where installed in metal raceway.

See Figure 5-8 for a detailed illustration when applying these requirements.

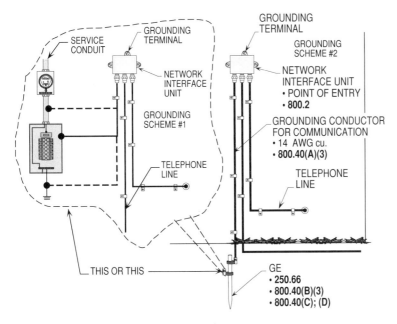

Figure 5-8. This illustration shows the methods in which communication systems shall be permitted to be grounded to an earth electrode for safety.

COMMUNICATIONS SYSTEMS - ELECTRODE
250.94, FPN 2; 800.40(B)(1(1)) THRU (7)

The grounding conductor shall be connected as follows to the nearest accessible location:

- The grounding conductor shall be permitted to be connected to the building or structure's grounding electrode system per **250.50**.
- The grounding conductor shall be permitted to be connected to the grounded interior metal water piping system per **250.52(A)(1)**.
- The grounding conductor shall be permitted to be connected to the power service accessible means external to enclosures per **250.94**.
- The grounding conductor shall be permitted to be connected to the metallic power service raceway.
- The grounding conductor shall be permitted to be connected to the service equipment enclosure.
- The grounding conductor shall be permitted to be connected to the grounding electrode or the grounding electrode conductor metal enclosure.
- The grounding conductor shall be permitted to be connected to the grounding conductor or the grounding electrode of a building or structure disconnecting means that is grounded to an electrode per **250.32**.

Grounding Tip 137: If the communication system is connected to the grounding electrode system of the facility, an additional grounding electrode shall not be needed.

Grounding Tip 138: Where the communications systems are grounded to the grounding electrode system of the electrical supply system, it is not necessary to earth ground it to its own grounding electrode.

If no grounding means is available for the building or structure per **800.40(B)(1)(1)** thru **(7)**, the grounding means is permitted to be an effectively grounded metal structure or a ground rod or pipe not less than 5 ft. in length and 1/2 in. (12.7 mm) in diameter. Communications wires and cables on buildings shall have a separation of at least 6 ft. (1.8 m) from lightning conductors per **800.13** and have a separation of at least 6 ft. (1.8 m) from electrodes of other systems.

See Figure 5-9 for a detailed illustration when applying these requirements.

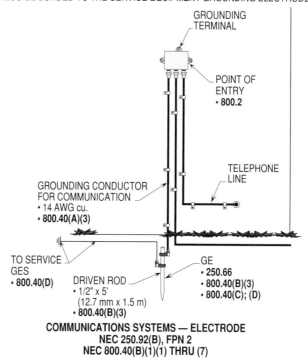

Figure 5-9. This illustration shows a 1/2 in. x 5 ft. (12.7 mm x 1.5 m) driven rod which shall be permitted to be used instead of a 1/2 in. x 8 ft. (13 mm x 2.5 m) driven rod that is required for grounding service equipment.

COMMUNICATIONS SYSTEMS - ELECTRODE CONNECTION 800.40(C)

Grounding Tip 139: For UL requirements pertaining to fittings, see Ch. 4 in UL 467.

Grounded fittings used to connect the grounding conductors and bonding jumpers to grounding electrodes shall be permitted to be any of the following:

- Listed connectors
- Listed clamps
- Listed fittings
- Listed lugs

Direct burial grounding clamps shall be utilized on grounding electrodes in the earth such as pipes, rods, rebar, concrete-encased electrodes, etc.

COMMUNICATIONS SYSTEMS - BONDING OF ELECTRODES 250.94, FPN 2; 800.40(D)

Communications systems grounded to a driven rod shall be bonded to the grounding electrode system with at least a 6 AWG copper conductor. **(See Figure 5-10)**

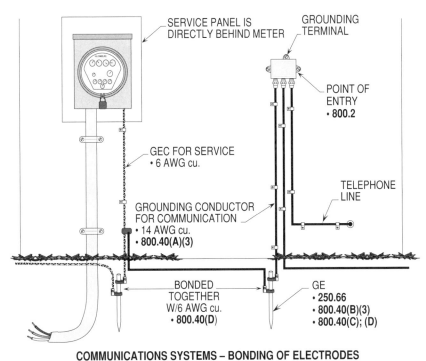

Figure 5-10. This illustration shows that at least a 6 AWG copper conductor shall be required to be used to bond the driven rod for communication systems to the driven rod of the service equipment.

Grounding Tip 140: The driven rods in Figure 5-10 have been bonded together to form an equipotential plane to reduce voltage differences between the grounded equipment supplied by each system.

RADIO AND TELEVISION EQUIPMENT - GROUNDING CONDUCTORS (RECEIVING STATIONS)
250.94, FPN 2; 810.21(A) thru (I)

The grounding conductors for receiving stations shall have certain characteristics and be installed as follows:

- The grounding conductors shall be of copper, aluminum, copper-clad steel, bronze or similar corrosion-resistant material

  - Where aluminum or copper-clad aluminum grounding conductors are in direct contact with masonry or the earth or subject to corrosive conditions, such conductors shall not be used

  - Where aluminum or copper-clad aluminum grounding conductors are used outside, such conductors shall be installed at least 18 in. from the earth

- The grounding conductors shall not be required to be insulated

- The grounding conductors shall be securely fastened-in-place and shall be permitted to be installed directly to the surface wired over without the use of insulating supports

- Where necessary, the grounding conductor shall be guarded from physical damage

- The grounding conductor shall be bonded at both ends where installed in metal raceway

- The grounding conductor shall be run to the grounding electrode in as straight a line as practical

- The grounding conductor shall be permitted to be installed either inside or outside the building

Grounding Tip 141: The NEC requires an outside satellite dish as shown in **Figure 5-11** to have its grounding electrode system bonded into the grounding electrode system of the service equipment for the facility that it serves.

- The grounding conductor shall be not smaller than 10 AWG copper, 8 AWG aluminum or 17 AWG copper-clad steel or bronze
- A single grounding conductor shall be permitted to be installed for both protective and operating purposes

See **Figure 5-11** for a detailed illustration when applying these requirements.

Figure 5-11. This illustration shows the proper procedure for grounding an outside satellite dish for safety.

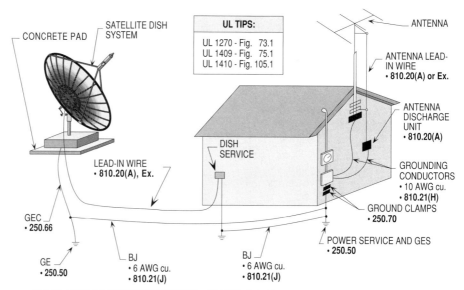

Grounding Tip 142: The bonding of the grounding electrode at the satellite dish to the grounding electrode system at the building forms an equipotential plane between the grounding electrodes.

RADIO AND TELEVISION EQUIPMENT - ELECTRODE (RECEIVING STATIONS)
250.94, FPN 2; 810.21(F)

Grounding Tip 143: Poles for radio and television antennas shall be required to be bonded to the grounding electrode system for the facility served to protect related equipment from lightning strikes.

The grounding conductor shall be connected and grounded as follows to the nearest accessible location:

- The grounding conductor shall be permitted to be connected to the building or structure's grounding electrode system per **250.50**
- The grounding conductor shall be permitted to be connected to the grounded interior metal water piping system per **250.52(A)(1)**.
- The grounding conductor shall be permitted to be connected to the power service accessible means external to enclosures per **250.94**.
- The grounding conductor shall be permitted to be connected to the metallic power service raceway.
- The grounding conductor shall be permitted to be connected to the service equipment enclosure.
- The grounding conductor shall be permitted to be connected to the grounding electrode or the grounding electrode conductor metal enclosure.

If no grounding means is available for the building or structure per **810.21(F)(1)a** thru **f**, the grounding means shall be permitted to be an effectively grounded metal structure or any of the individual electrodes per **250.50, 250.52** and **250.53**. **(See Figure 5-12)**

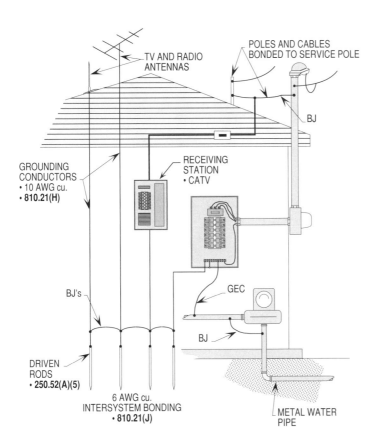

Figure 5-12. This illustration shows the procedure for bonding and grounding receiving stations for safety.

Grounding Tip 144: Figure 5-12 is also used to show the bonding of electrodes together to form an equipotential plane.

RADIO AND TELEVISION EQUIPMENT - BONDING OF ELECTRODES (RECEIVING STATIONS) 810.21(J)

Receiving stations grounded to a driven rod shall be bonded to the grounding electrode system with at least a 6 AWG copper conductor. **(See Figure 5-12)**

COMMUNITY ANTENNA TELEVISION AND RADIO DISTRIBUTION SYSTEMS - GROUNDING CONDUCTOR 250.94, FPN 2; 820.40(A)(1) THRU (6)

The grounding conductor for community antenna television and radio distribution systems shall have certain characteristics and be installed as follows:

- The grounding conductor shall be insulated and listed as suitable for the purpose.
- The grounding conductor shall be copper or other corrosion-resistive conductive material.
- The grounding conductor shall be stranded or solid.
- The grounding conductor shall not be smaller than 14 AWG or equivalent.
- The 14 AWG grounding conductor shall have a current-carrying capacity approximately equal to that of the outer conductor of the coaxial cable.
- The grounding conductor shall be run to the grounding electrode in as straight a line as practical.

Grounding Tip 145: Designers, installers and inspectors shall ensure that the size of the grounding conductor (can be 14 AWG) will carry about the same amount of current as the outer conductor of the coaxial cable serving the system.

- Where necessary, the grounding conductor shall be guarded from physical damage.
- The grounding conductor shall be bonded at both ends where installed in metal raceway.

See Figure 5-13 for a detailed illustration when applying these requirements.

Figure 5-13. This illustration shows the proper procedures for sizing and connecting the grounding conductors for community antenna television and radio distribution systems.

Grounding Tip 146: The earth ground electrodes shall be bonded together to create an equipotential plane which will reduce voltage differences between electrodes.

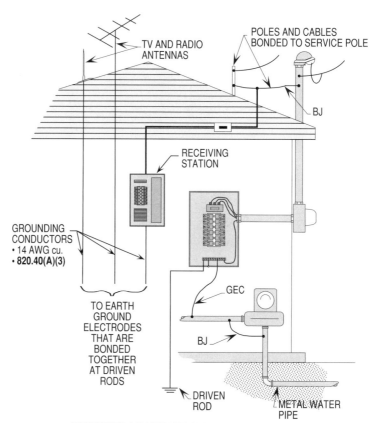

COMMUNITY ANTENNA TELEVISION AND RADIO DISTRIBUTION SYSTEMS - ELECTRODE
820.40(B)(1)(1) THRU (7)

The grounding conductor shall be connected and grounded as follows to the nearest accessible location:

Grounding Tip 147: Utilities normally will not permit television and radio antennas and communications systems to be connected to the metal enclosure of a meter base. Check with utility personnel before making such a connection.

- The grounding conductor shall be permitted to be connected to the building or structure grounding electrode system per **250.50**.
- The grounding conductor shall bes permitted to be connected to the grounded interior metal water piping system per **250.52(A)(1)**.
- The grounding conductor shall bes permitted to be connected to the power service accessible means external to enclosures per **250.94**.
- The grounding conductor shall be permitted to be connected to the metallic power service raceway.
- The grounding conductor shall be permitted to be connected to the service equipment enclosure.

- The grounding conductor shall be permitted to be connected to the grounding electrode or the grounding electrode conductor metal enclosure.
- The grounding conductor shall be permitted to be connected to the grounding conductor or the grounding electrode of a building or structure disconnecting means that is grounded to an electrode per **250.32**.

If no grounding means is available for the building or structure per **820.40(B)(1)(a)** thru **(g)**, the grounding means is permitted to be an effectively grounded metal structure or any of the individual electrodes per **250.50, 250.52** and **250.53**. **(See Figure 5-14)**

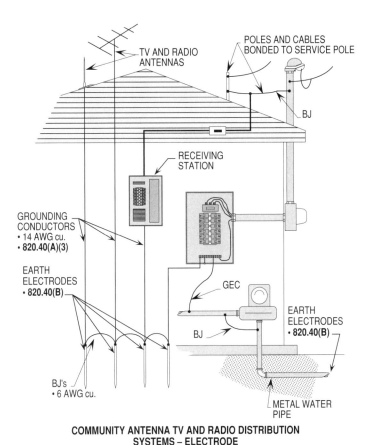

Figure 5-14. This illustration shows the proper procedure for sizing and connecting the equipment grounding conductors to earth ground for television and radio antennas and communications systems.

COMMUNITY ANTENNA TELEVISION AND RADIO DISTRIBUTION SYSTEMS - ELECTRODE CONNECTION 800.40(C)

Grounding fittings used to connect the grounding conductors and bonding jumpers to grounding electrodes shall be permitted to be any of the following:

- Listed connectors
- Listed clamps
- Listed fittings
- Listed lugs

COMMUNITY ANTENNA TELEVISION AND RADIO DISTRIBUTION SYSTEMS - BONDING OF ELECTRODES
250.94, FPN 2; 820.40(D)

Communications systems grounded to a driven rod shall be bonded to the grounding electrode system with at least a 6 AWG copper conductor. **(See Figure 5-15)**

Figure 5-15. This illustration shows the proper procedure for connecting grounding conductors to individual electrodes and then bonding such electrodes together.

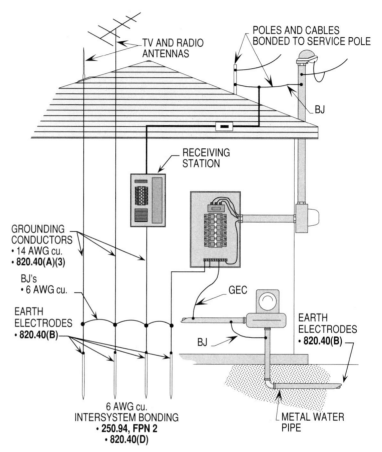

BONDING OTHER ENCLOSURES - GENERAL
250.96(A)

Metal raceways, cable trays, cable armor, etc. shall be bonded to metal enclosures, boxes, cabinets, etc. Metal raceways shall be installed as a complete run from a metal box to metal box per **300.10**. Such wiring methods shall be terminated with listed connectors which comply with **300.12, 300.13** and **300.15(A)**. Note that when metal raceways, armor cables, cable sheaths, enclosures, frames and other metal noncurrent-carrying metal parts which serve as the equipment grounding conductors, whether using a supplementary grounding conductor or not, proper bonding shall be required to ensure the safe handling of such fault currents that may be imposed and to ensure electrical continuity. Nonconductive paint or similar coatings shall be removed at the threads or other contact points to ensure clean surfaces. Fittings or bonding jumpers can be used for such a purpose. **(See Figure 5-16)**

Bonding

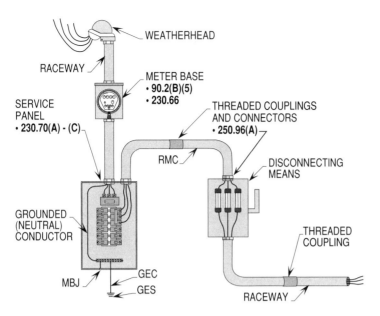

Figure 5-16. This illustration shows the metal weatherhead and raceway to the meter base, raceway between the panelboard, raceway to the disconnecting means and the raceway from the disconnect supplying the load is properly connected together to form a continuous unbroken grounding path.

ISOLATED GROUNDING CIRCUITS
250.96(B)

A nonmetallic spacer or fitting shall be permitted to be installed between the metal raceway supplying sensitive electronic equipment at the point of connection. The equipment grounding conductor shall be isolated and the metal raceway grounded. Electromagnetic interference that creates unwanted noise may be reduced in sensitive electronic equipment when this type of installation is applied. **(See Figure 5-17)**

Grounding Tip 148: For cord-and-plug connection of sensitive electronic equipment, see **Figure 7-16(d)** on page 7-13 in this book.

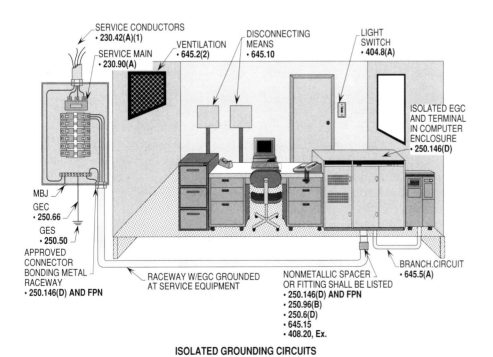

Figure 5-17. Reduction of electromagnetic interference that creates unwanted noise in sensitive electronic equipment may be reduced by installing a nonmetallic spacer or fitting in the metal raceway at the point of connection to such equipment.

BONDING FOR OVER 250 VOLTS
250.97

Grounding Tip 149: Section **250.97** covers rules for bonding and grounding the metal of enclosures, raceways, cables and other items where there is an OCPD ahead of the circuits they are associated with.

When on the load side, one of the following wiring methods shall be installed for bonding circuits over 250 volts-to-ground, not including services:

(1) Install the following wiring methods between sections of metal conduit, intermediate metal conduit, electrical metallic tubing or other type of metal raceways.
- Threaded couplings
- Threadless couplings

(2) Install the following wiring methods between metallic raceways, metallic boxes, cabinet enclosures, etc.
- Jumpers
- Two locknuts and bushing
- Threaded (bosses) hubs
- Other approved devices

Metal raceways and conduits enclosing conductors at more than 250 volts-to-ground (other than service conductors) shall have electrical continuity assured by one of the methods outlined in **250.92(B)(2)** thru **(B)(4)**. Note that the grounded (neutral) conductor shall not be used (shall be permitted for the supply side of services) except as permitted for separate buildings and separately derived systems per **250.24(A)(5), FPN, 250.92(B)(1)** and **250.142(A)(1), (A)(2)** and **(A)(3)**. **(See Figure 5-18)**

Figure 5-18. This illustration shows the methods that shall be permitted to be used to bond and ground metal conduits to metal enclosures. (See Figure 5-19 on page 5-17 for close up of fittings)

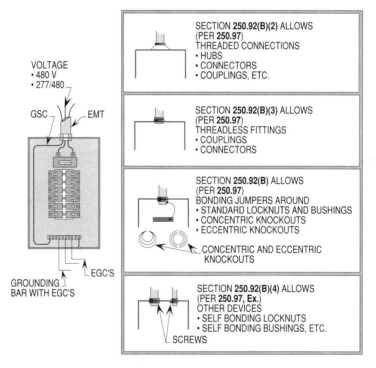

BONDING OVER 250 VOLTS
NEC 250.97

ECCENTRIC OR CONCENTRIC KNOCKOUTS
250.97, Ex.

If oversized eccentric or concentric knockouts are not encountered, the following items shall be permitted to be utilized:

- Threadless fittings
- Two locknuts
- One locknut
- Listed fittings

See Figure 5-19 for a detailed illustration when applying these requirements.

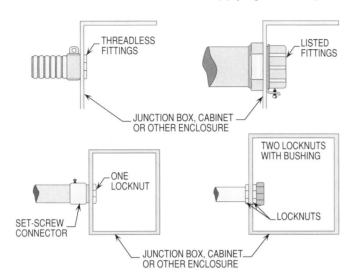

Grounding Tip 150: When the voltage-to-ground exceeds 250 volts as it does for 480/277 volt, three-phase, four-wire circuits or 480 volt, three-phase, three-wire corner grounded circuits, bonding jumpers shall be used to bond and ground metal conduits to metal enclosures. Note that this rule only applies if there are knockout rings left around conduits and wall of enclosure.

Figure 5-19. This illustration shows fittings that shall be permitted to be used to bond and ground the raceways and enclosures together to form a continuous grounding path.

Grounding Tip 151: This method shall be permitted to be used only if no knockout rings are left in the enclosure wall, around the conduit.

Grounding Tip 152: With knockout rings left around conduits, bonding jumpers shall be required for 480/277 volt, three-phase, four-wire or 480 volt, three-phase, three-wire circuits.

Grounding Tip 153: When knockouts are left around conduits, bonding jumpers shall not be required for 120 or 240 volt circuits if knockout rings are not damaged and loose.

THREADLESS FITTINGS
250.97, Ex. (a)

Threadless couplings and connectors shall be permitted to be used with metal sheath cables. However, they shall be listed as required per **300.15(A)** and **300.15(F)**.

TWO LOCKNUTS
250.97, Ex. (b)

Two locknuts shall be permitted to be used with rigid metal conduit or intermediate metal conduit, one locknut on the outside and one on the inside of the enclosure when entering boxes or cabinets. Appropriate bushings inside the enclosure, etc. shall be used to protect the insulation of the conductors. Grounding connections and circuit conductors that are not made-up tight are the cause of many arcing faults and electrical fires.

Grounding Tip 154: Care must be exercised when bonding metal conduits and cables around knockouts left in the walls of the enclosure enclosing conductors rated at 277/480 volts. Circuits of 277 volts have a restrike voltage of 250 volts which is hard to interrupt.

ONE LOCKNUT
250.97, Ex. (c)

One locknut on the inside of boxes or cabinets provides the proper bonding where the fitting shoulders seat firmly on the box or cabinet. These include electric metallic tubing connectors, cable connectors and flexible metal conduit connectors, etc.

LISTED FITTINGS
250.97, Ex. (d)

A wedge fitting (bushing) or locknut fitting is an application of other approved devices that shall be permitted to be used. A grounding wedge is equipped with a set-screw that tightens the metal conduit and ensures proper bonding of the equipment enclosure.

BONDING LOOSELY JOINTED METAL RACEWAYS
250.98

Equipment bonding jumpers or other means shall be installed to provide electrical continuity where expansion fittings and telescoping sections of metal raceways are used. **(See Figure 5-20)**

Figure 5-20. This illustration shows an equipment bonding jumper used to bond and ground the metal raceway around the flexible metal conduit between the raceway run.

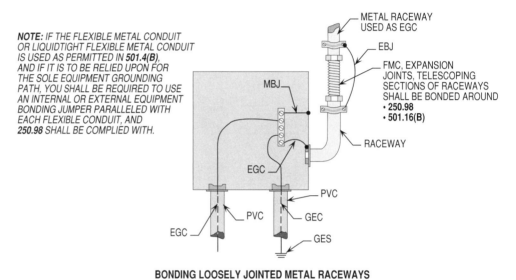

BONDING IN HAZARDOUS (CLASSIFIED) LOCATIONS
250.100

Grounding Tip 155: When bonding and grounding in Class I, Divisions 1 and 2 locations, refer to **501.16(A)** and **(B)**.

In hazardous locations, regardless of the voltage, the electrical continuity of raceways and boxes shall be assured by one of the methods in **250.92(B)(1)** thru **(B)(4)**. However, this does not cover the continuity in hazardous locations in its entirety. The requirement for the use of a specific bonding fitting used in a hazardous location shall be looked up and the correct one selected.

For bonding between sections of metal conduit or electrical metallic tubing, one of the following methods shall be permitted to be used:

- Threaded couplings for conduit
- Threadless couplings made up tight for electrical metallic tubing or conduit

For bonding between metallic raceway and metallic boxes, cabinets and the like one of the following shall be permitted to be used:

- Threaded connections
- Bonding jumpers
- Other devices approved for the use

See Figures 5-21(a) and (b) for a detailed illustration of bonding in hazardous (classified) locations.

Bonding

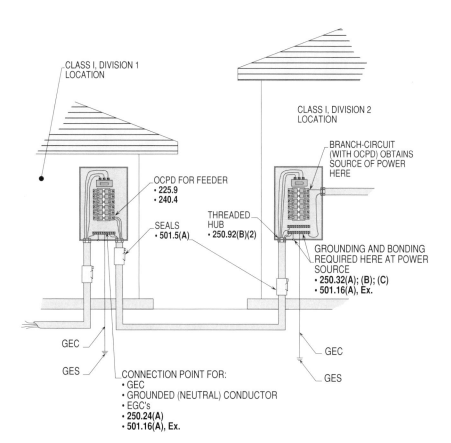

Figure 5-21(a). This illustration shows the bonding and grounding at the power source in a Class I, Division 2 location.

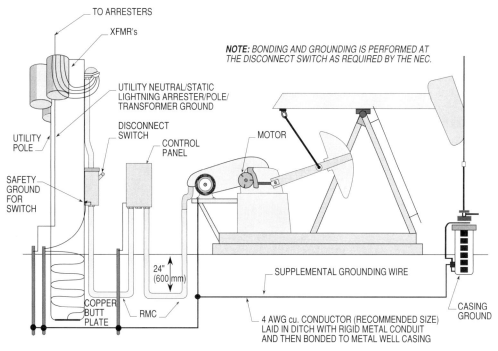

Figure 5-21(b). This illustration shows the bonding and grounding of an oil rig (rocking horse) with a motor for pumping oil.

5-19

Quick Calc
If a 1 AWG, 90°C, 150 amp copper equipment bonding jumper is used, what size aluminum equipment bonding jumper is assumed to be equivalent?   • Sizing alu. EBJ **250.102(C); Table 310.16** 150 A requires 2/0 AWG alu. per 90°C column   • The size EBJ alu. equal to 150 amps is 2/0 AWG.

EQUIPMENT BONDING JUMPERS - MATERIAL AND ATTACHMENT
250.102(A); (B)

Copper or other noncorrosive material shall be used for equipment bonding jumpers. This eliminates aluminum in many places. Remember that if other than copper is used, it shall be sized to an equivalent to what is required for copper bonding jumpers.

When a screw is used as the main bonding jumper, it shall be identified by being colored green per **250.28(B)**. It must also be visible after installation. Section **250.8** specifies the manner in which main and equipment bonding jumper shall be attached for circuits and equipment. Section **250.70** specifies the manner of attachment for grounding electrodes. Section **250.102(A)** and **(B)** shall be reviewed very carefully. Note that soldering shall not be permitted as the sole connection.

SIZE - EQUIPMENT BONDING JUMPER ON SUPPLY SIDE OF SERVICE
250.102(C)

There has always been considerable confusion as to when to use **Table 250.66** or **Table 250.122**. By reviewing the headings of each Table there should be no confusion. **Table 250.66** is used for sizing grounding electrode conductors for AC systems and are sized based on the fact that no OCPD is ahead of the conductors. **Table 250.122** is used for sizing equipment grounding conductors for grounding raceways and equipment and are sized based on an OCPD ahead of the circuit conductors.

Grounding Tip 156: For requirements pertaining to testing and safe use of bushings, see Sec.'s 6.1 through 6.6 in UL 467.

Problem 5-22(A). What size copper equipment bonding jumper is required to bond the service raceway (having 3 - 250 KCMIL THWN conductors) to the grounded busbar in a panelboard?

Step 1: Sizing EBJ on supply side
250.102(C); Table 250.66
250 KCMIL THWN cu. requires 2 AWG cu.

Solution: **The size equipment bonding jumper required on the supply side is 2 AWG copper.**

See **Figure 5-22(a)** for an exercise problem when sizing an individual (separate) equipment bonding jumper on the supply side.

Grounding Tip 157: For requirements pertaining to devices used for the connection of conduits to metal enclosures, see Sec. 4.1.1 in UL 467.

Problem 5-22(b). What size individual equipment bonding jumper is required to bond each service raceway (RMC) containing 3 - 700 KCMIL THWN copper conductors paralleled three times per phase?

Step 1: Sizing EBJ on supply side
250.102(C); Table 250.66
700 KCMIL requires 2/0 AWG cu.

Solution: **The size equipment bonding jumper for each conduit on the supply side is 2/0 AWG copper.**

See **Figure 5-22(b)** for an exercise problem when sizing individual (separate) equipment bonding jumper on the supply side.

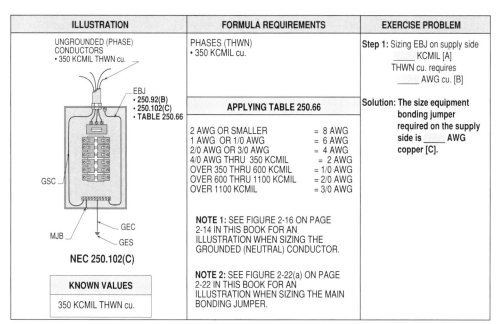

Figure 5-22(a). This illustration shows an exercise problem for sizing an individual equipment bonding jumper to bond and ground the metal conduit to the grounded busbar in the panelboard.

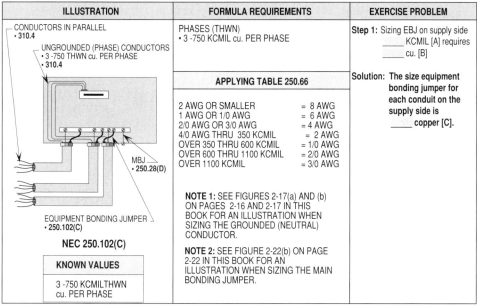

Figure 5-22(b). This illustration shows an exercise problem for sizing an individual equipment bonding jumper to bond and ground each separate conduit to the grounded busbar in the panelboard.

Problem 5-22(c). What size equipment bonding jumper (single conductor) is required to bond all service raceways (RMC) containing 3 - 700 KCMIL THWN copper conductors paralleled three times per phase?

- **Step 1:** Sizing EBJ on supply side
  250.102(C)
  700 KCMIL x 3 = 2100 KCMIL

- **Step 2:** Finding KCMIL to size EBJ
  2100 KCMIL x .125 = 262.5 KCMIL

- **Step 3:** Sizing EBJ
  Table 8 to Ch. 9
  262.5 KCMIL requires 300 KCMIL cu.

- **Solution:** **The size equipment bonding jumper for all service raceways on the supply side is 300 KCMIL copper.**

See Figure 5-22(c) for an exercise problem when sizing the equipment bonding jumper (single conductor) on the supply side

Grounding Tip 158: When selecting 300 KCMIL copper conductor, convert the 262.5 KCMIL to CM by multiplying by 1000 (262.5 KCMIL x 1000 = 262,500 CM) and select the 300 KCMIL copper conductor at 300,000 CM per **Table 8 in Ch. 9**.

Figure 5-22(c). This illustration shows an exercise problem for sizing an individual equipment bonding jumper to be used for bonding and grounding all the conduits to the grounded busbar in the panelboard.

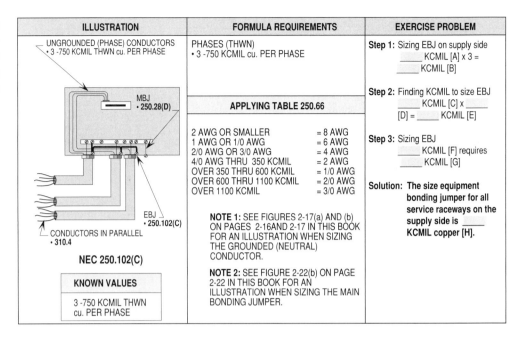

Grounding Note: The locknut-bushing and double-locknut types of connection shall not be depended upon for proper bonding purposes, but bonding jumpers with listed fittings or other approved methods of bonding shall be used. This method of bonding applies to all intervening raceways, fittings, boxes, enclosures, etc. between the point of grounding of the service equipment in Class I locations which extends back to the grounding point at the building disconnecting means.

SIZE - EQUIPMENT BONDING JUMPER ON LOAD SIDE OF SERVICE 250.102(D)

Grounding Tip 159: Table 250.122 shall be used to size the equipment bonding jumper when there is an OCPD ahead of the circuit conductors in the metal conduit or cable.

Grounding Tip 160: It is recommended to use an individual equipment bonding jumper to bond and ground each raceway instead of connecting (looping) one equipment bonding jumper to each raceway. The looping method requires, in most cases, a larger equipment bonding jumper which is too large to handle when terminating properly. This type of installation method is the reason electricians will down size the equipment bonding jumper and thus be in violation of the NEC. (See **Figures 5-22(b)** and **(c)**)

Table 250.122 lists the size of the equipment bonding jumpers on the load side of the service OCPD's. The bonding jumper shall be a single conductor sized per **Table 250.122** for the largest OCPD that supplies the electrical circuits. If the bonding jumper supplies two or more raceways or cables, the same rules apply and proper sized conductors shall be selected. Note that it is not necessary to size the equipment bonding jumpers larger than the ungrounded (phase) conductors, but in no case shall it be smaller than 14 AWG per **250.102(D)**. Also, review the discussion in **250.102(C)** on when to use Table 250.66 or Table 250.122.

Problem 5-23(a). What size equipment bonding jumper is required to bond the raceway of a feeder-circuit having a 225 amp OCPD ahead of its conductors?

Step 1: Sizing EBJ on load side
250.102(D); Table 250.122
225 A OCPD requires 4 AWG cu.

Solution: The size equipment bonding jumper required on the load side is 4 AWG copper.

See **Figure 5-23(a)** for an exercise problem when sizing the equipment bonding jumper on the load side of an electrical system.

Bonding

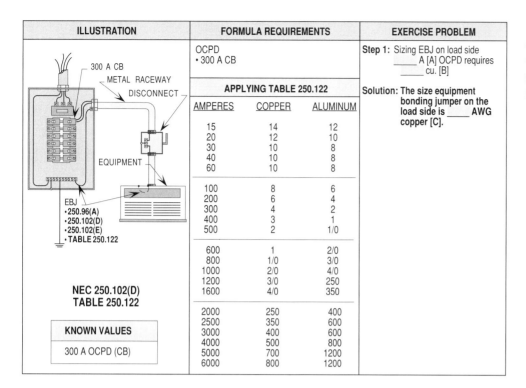

Figure 5-23(a). This illustration shows an exercise problem for sizing the equipment bonding jumper to bond and ground the metal of the equipment supplied by the branch-circuit.

Problem 5-23(b). What size copper equipment bonding jumper is required to bond each raceway of a feeder-circuit having a 1200 amp OCPD ahead of its conductors?

Step 1: Sizing EBJ on load side
250.102(D); Table 250.122
1200 A OCPD requires 3/0 AWG cu.

Solution: **The size equipment bonding jumper required on the load side is 3/0 AWG copper to bond each conduit.**

See Figure 5-23(b) for an exercise problem when sizing the equipment bonding jumper on the load side for each conduit.

Grounding Tip 161: Table 250.122 lists the size of the equipment bonding jumpers on the load side of the service overcurrent protection devices. The bonding jumper is to be a single conductor sized according to **Table 250.122** for the largest overcurrent protection device that is provided to supply the circuits. If the bonding jumper supplies two or more raceways or cables, the same rule applies as if a single equipment bonding jumper is used.

Problem 5-23(c). What size copper equipment bonding jumper is required to bond all raceways of a feeder-circuit having a 1200 amp OCPD ahead of its conductors?

Step 1: Sizing EBJ on load side
250.102(D); Table 250.122
1200 A OCPD requires 3/0 AWG cu.

Solution: **The size equipment bonding jumper required on the load side is 3/0 AWG copper to bond all conduits.**

See Figure 5-23(c) for an exercise problem when sizing the equipment bonding jumper on the load side for all conduits.

Figure 5-23(b). This illustration shows an exercise problem for sizing individual equipment bonding jumper to bond and ground each metal conduit to the grounding busbar in the subpanel.

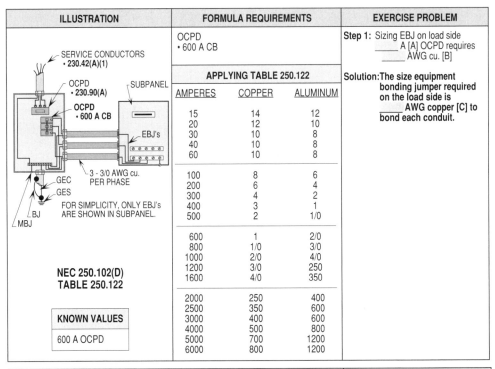

Figure 5-23(c). This illustration shows an exercise problem for sizing an individual equipment bonding jumper to bond and ground all the conduits to the grounding busbar in the subpanel.

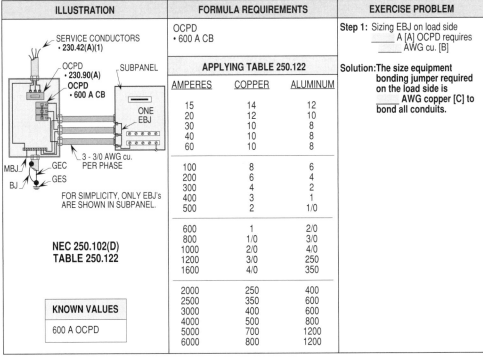

EQUIPMENT BONDING JUMPER - INSTALLATION 250.102(E)

The equipment bonding jumper shall be permitted to be installed either inside or outside of the raceway or enclosure. When it is installed outside of the raceway or enclosure, it shall not be over 6 ft. (1.8 m) in length and shall be routed with raceway or enclosure. If the equipment bonding jumper is inside the raceway, refer to **310.12(B)**, **250.119** and **250.148** for the conductor identification requirements. Note that if such equipment bonding jumper is routed on the outside of the raceway, listed grounding fittings shall be used to terminate the raceway and equipment bonding jumper.

BONDING OF PIPING SYSTEMS AND EXPOSED STRUCTURAL STEEL
250.104

The rules of this section requires that certain piping systems and structural steel to be bonded and grounded properly to ensure the safety of the electrical system.

METAL WATER PIPING - GENERAL
250.104(A)(1)

Regardless of whether the water piping is supplied by metallic or nonmetallic pipe to the building or structure, proper bonding of the interior metal water piping or copper tubing to the service equipment enclosure shall be required. Metal water piping that is installed in or attached to a building or structure shall be bonded to the service equipment enclosure, the grounded (neutral) conductor at the service, the grounding electrode conductor where of sufficient size or to the one or more grounding electrodes used. The bonding jumper to the metal water piping shall be sized per Table 250.66, except as permitted in **250.104(A)(2)** and **(A)(3)**. **(See Figure 5-24)**

Grounding Tip 162: Electrical appliances such as dishwashers, washing machines, etc., can develop a partial fault through the windings of its motor which will energize the metal piping with a voltage and current flow. This type of situation is dangerous and such metal piping shall be bonded to the grounding electrode system. Proper bonding of the metal piping will help reduce the threat of electrical shock and fire hazards.

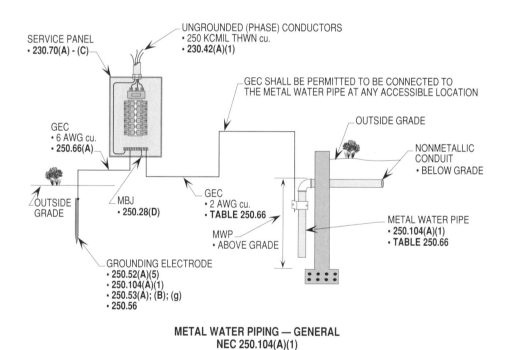

Figure 5-24. This illustration shows the procedure for bonding and grounding the metal water pipe for safety.

METAL WATER PIPING - BUILDINGS OF MULTIPLE OCCUPANCY
250.104(A)(2)

In multiple occupancies, if the interior piping is metal and the piping systems of all occupancies are not tied together, the metal piping shall be isolated from all other occupancies due to the use of nonmetallic pipe as a supply line. Each individual metal water piping system shall be required to be bonded separately to the panelboard or switchboard enclosure in each apartment. Bonding jumper points of attachment shall be required to be accessible and sized per **Table 250.122**. This ensures that the metal piping system in each occupancy is at ground potential and the threat of electrical shock and fire hazards are eliminated. **(See Figure 5-25)**

Grounding Tip 163: Sections **250.50** and **250.52(A)(1)** cover the requirements when the metal water piping is considered a grounding electrode. Sections **250.104(A)(1)** and **(A)(2)** cover the requirements for metal water piping and copper tubing when they are not considered a grounding electrode.

Figure 5-25. This illustration shows the procedure for sizing the bonding jumper to bond and ground the copper tubing or metal water piping to the grounded busbar in the subpanel in apartment units 1 and 2.

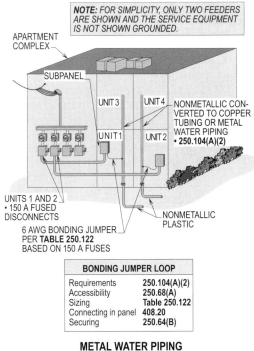

METAL WATER PIPING - MULTIPLE BUILDINGS OR STRUCTURES SUPPLIED FROM A COMMON SERVICE 250.104(A)(3)

The interior metal water piping system shall be bonded to the building or structure disconnecting means enclosure where located at the building or structure, or to the equipment grounding conductor run with the supply conductors, or to the one or more grounding electrodes used. The bonding jumper shall be sized per **Table 250.66**, bsed on the size of the feeder or branch-circuit conductors that supply the building. the bonding jumper shall not be required to be larger than the largest ungrounded (phase) feeder or branch0circuit conductors supplying the building. **(See Figure 5-26)**

Figure 5-26. This illustration shows the procedure for sizing the bonding jumper to connect the equipment grounding conductor (run with the feeder-circuit conductors) and the grounded busbar to the concrete-encased electrode of the separate building supplied from a common service.

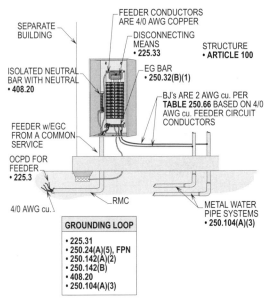

METAL WATER PIPING - SEPARATELY DERIVED SYSTEMS 250.104(A)(4)

When there is a separately derived system utilizing a grounding electrode near an available point of metal water pipe which is in the area supplied by such derived system, the metal water pipe system shall be bonded to the derived systems grounded (neutral) conductor. Each bonding jumper shall be sized per **Table 250.66**. See **250.30(A)(4)(2)** for such rules explained in greater detail. **(See Figure 5-27)**

METAL WATER PIPING — SEPARATELY DERIVED SYSTEMS
NEC 250.104(A)(4)

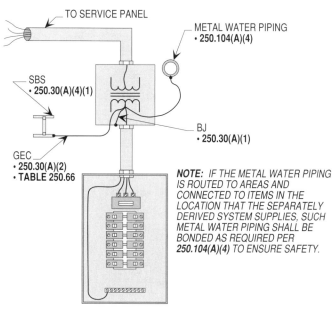

Figure 5-27. This illustration shows that metal water pipe located in the area of the separately derived system shall be bonded and grounded to the grounded (neutral) conductor located in the transformer's grounding point.

OTHER METAL PIPING 250.104(B)

Each aboveground portion of a gas piping system upstream from the equipment shutoff valve shall be electrically continuous and bonded to any grounding electrode, as defined by **250.50** and **250.52**. Note that gas piping shall not be used as a grounding electrode per **250.52(A)(1)** and NFPA 54, Sec. 3.14(B). **(See Figure 5-28)**

Grounding Tip 164: For a better understanding of the rules concerning bonding and using the metal gas pipe, see Sec. 's 3.14 through 3.16 in NFPA 54.

METAL GAS PIPING
NEC 250.104(B)

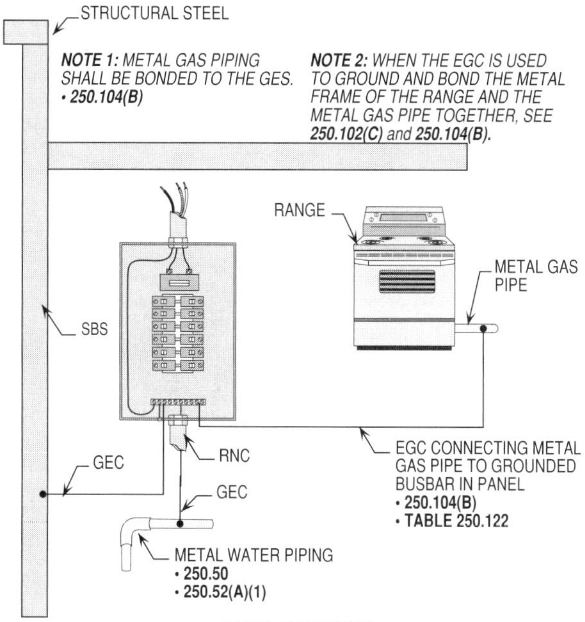

Figure 5-28. This illustration shows the procedure for bonding and grounding the metal gas pipe above ground to the grounding electrode system.

ELECTRICAL CIRCUITS
NFPA 54, 3.15

Electrical circuits shall not utilize gas piping or components as conductors. However, the Ex. to 3.15 allows low-voltage (50 volts or less) control circuits, ignition circuits and electronic flame detection device circuits to make use of piping or components as a part of an electric circuit.

OTHER METAL PIPING
250.104(B)

All interior metal piping systems which may be subjected to being energized shall be bonded to the service equipment ground at the service enclosure to the common grounding conductor, if it is of sufficient size, or to one or more of the other grounding electrodes. This bonding jumper shall be size according to **Table 250.122**. **(See Figure 5-29)**

Grounding Note: With circuit conductors that could energize metal piping, it shall be permitted to use the equipment grounding conductor run with circuit conductors to bond and ground such piping system.

STRUCTURAL STEEL
250.104(C)

Exposed building steel frames that are not intentionally grounded shall be bonded to either the service enclosure, the neutral bus, the grounding electrode conductor or to a grounding electrode. The bonding conductor shall be sized according to **Table 250.66** and installed according to **250.64(A), (B)** and **(E)**. **(See Figure 5-30)**

Grounding Note: For additional safety, it is recommended to bond all metal air ducts and metal piping systems on the premises to reduce voltage differences.

Grounding Tip 165: For a better understanding of the requirements pertaining to lightning down conductors used to divert lightning strikes, see Sec.'s 3.9.14 thru 3.9.11 in NFPA 780.

LIGHTNING PROTECTION SYSTEMS
250.106

Noncurrent-carrying metal parts of electric equipment located within 6 ft. (1.8 m) of lightning down conductors shall be required to bonded together. **(See Figure 5-31)**

USE OF LIGHTNING RODS
250.60

Lightning down conductors and made electrodes for grounding of lightning rods shall be used in place of made electrodes for grounding wiring systems and equipment. However, they may be bonded together, in fact, the NEC recommends the bonding of these electrodes to limit the difference of potential that might appear between them. See **250.50 thru 250.53**.

To assist in the application of the above requirements, see **250.50, 800.40(D), 810.21(J)** and **820.40(D)** for required bonding procedures. Difference of potential between two different grounding systems on the same building, if tied together, will reduce the potential difference between them, for instance, lightning rods and grounding electrodes per **250.52**. **(See Figure 5-31)**

Bonding

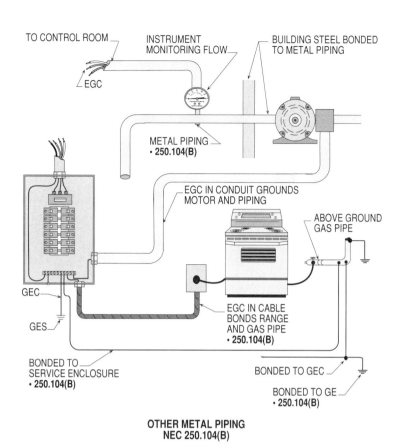

Figure 5-29. This illustration shows the procedure for bonding and grounding other metal piping above grade with an equipment grounding conductor.

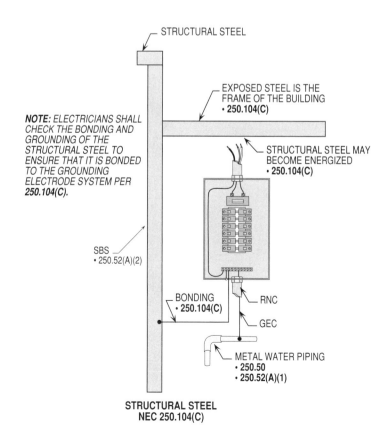

Figure 5-30. This illustration shows that exposed structural steel that forms the structural frame of a building shall be bonded and grounded into the grounding electrode system.

Figure 5-31. This illustration shows the procedure for bonding and grounding the grounding electrodes together to create an equalpotential plane.

Grounding Tip 166: For a better understanding of installing and maintaining ground rods that are used for lightning protection, see Sec.'s 3.13.1 thru 3.13.1.2 in NFPA 780.

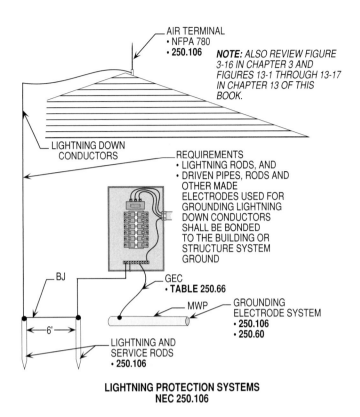

Name _____   Date _____

Chapter 5
Bonding

	Section	Answer

1. Bonding is the permanent joining of metallic parts to form an electrically conductive path that will assure electrical continuity and the capacity to conduct safely any current likely to be imposed. _____ T   F

2. Standard locknuts and bushings shall be permitted to be installed for bonding the raceway to the enclosure. _____ T   F

3. A wedge fitting (bushing) or locknut fitting shall be permitted to be installed for bonding the raceway to the enclosure. _____ T   F

4. Nonconductive paint or similar coatings shall be removed at the threads or other contact points to ensure clean surfaces for bonding of raceways and enclosures. _____ T   F

5. Threadless couplings shall be permitted to be installed for bonding circuits over 250 volts-to-ground. _____ T   F

6. Two locknuts shall be permitted to be installed for circuits over 250 volts-to-ground if oversized eccentric or concentric knockouts are encountered. _____ T   F

7. Equipment bonding jumpers shall not be required to be installed where an expansion fitting and telescoping section of a metal raceway is used. _____ T   F

8. It shall be permitted to install the equipment bonding jumper either inside or outside of the raceway or enclosure. _____ T   F

9. If a building or structure is supplied by nonmetallic piping, the interior metal water piping shall not be required to be bonded to the service equipment enclosure. _____ T   F

10. Each aboveground portion of a gas piping system upstream from the equipment shutoff valve shall be electrically continuous and bonded to any grounding electrode of the electrical system. _____ T   F

11. Communications systems are usually bonded and grounded to the existing grounding electrode system with at least a _____ AWG copper conductor. _____ _____

12. The grounding conductor for communications systems shall be run to the grounding electrode in as _____ a line as practical. _____ _____

13. Communications systems grounded to a driven rod shall be grounded to the grounding electrode system with at least a _____ AWG copper conductor. _____ _____

14. Where aluminum or copper-clad aluminum grounding conductors are used outside for radio and television equipment, such conductors shall be installed at least _____ in. from the earth. _____ _____

Section   Answer

15. The grounding conductor for radio and television equipment shall not be smaller than _____ AWG copper.

16. The grounding conductor for community antenna television and radio distribution systems shall not be smaller than _____ AWG or equivalent.

17. The grounding conductor for community antenna television and radio distribution systems shall be _____ at both ends where installed in metal raceway.

18. A _____ spacer or fitting shall be permitted to be installed between the metal raceway supplying sensitive electronic equipment at the point of connection.

19. Where installed on the raceway or enclosure, the equipment bonding jumper shall not be over _____ ft. in length and shall be routed with raceway or enclosure.

20. Noncurrent-carrying metal parts of electric equipment located within _____ ft. of lightning down conductors shall be required to be bonded together.

21. What size copper equipment bonding jumper shall be required to bond the service raceway (having 3 - 300 KCMIL THWN conductors) to the grounded busbar in a panelboard?

22. What size copper equipment bonding jumper is required to bond the service raceway (having 3 - 500 KCMIL THWN conductors) to the grounded busbar in a panelboard?

23. What size equipment bonding jumper (single conductor) is required to bond all service raceways (RMC) containing 4 - 500 KCMIL THWN copper conductors paralleled three times per phase?

24. What size equipment bonding jumper (single conductor) is required to bond all service raceways (RMC) containing 3 - 600 KCMIL THWN copper conductors paralleled three times per phase?

25. What size equipment bonding jumper is required to bond the raceway of a feeder-circuit having a 200 amp OCPD ahead of its conductors?

26. What size equipment bonding jumper is required to bond the raceway of a feeder-circuit having a 250 amp OCPD ahead of its conductors?

27. What size equipment bonding jumper is required to bond each raceway of a feeder-circuit having an 800 amp OCPD ahead of its circuit conductors?

28. What size equipment bonding jumper is required to bond each raceway of a feeder-circuit having a 1000 amp OCPD ahead of its circuit conductors?

29. What size equipment bonding jumper is required to bond all raceways of a feeder-circuit having an 800 amp OCPD ahead of its circuit conductors?

30. What size equipment bonding jumper is required to bond all raceways of a feeder-circuit having a 1000 amp OCPD ahead of its circuit conductors?

6

EQUIPMENT GROUNDING and EQUIPMENT GROUNDING CONDUCTORS

An equipment grounding conductor is a conductor that carries current only in cases where there is a ground-fault condition. The equipment grounding conductor is the conductor used to ground the noncurrent-carrying metal parts of equipment. It keeps the equipment elevated as close as possible above ground, at zero potential, and provides a low impedance path for the ground-fault current. Equipment grounding conductors protect elements of circuits and equipment and also ensure the safety of personnel from electrical shock. Proper grounding also protects the equipment and facility from fire hazards.

EQUIPMENT FASTENED-IN-PLACE OR CONNECTED BY PERMANENT WIRING METHODS (FIXED)
250.110

Under any of the following conditions, exposed noncurrent-carrying metal parts of fixed equipment that are likely to become energized shall be grounded:

- Vertically or horizontally
- Wet or damp locations
- Electrical contact with metal
- Class I, Divisions 1 and 2
- Class II, Divisions 1 and 2
- Class III, Divisions 1 and 2
- Intrinsically safe systems
- Class I, Zones 0, 1 and 2
- Commercial garages, repair and storage
- Aircraft hangars
- Gasoline dispensing and service stations
- Bulk storage plants
- Spray application, dipping and coating processes
- Health care facilities
- Wiring methods with equipment grounding conductors
- Over 150 volts-to-ground

VERTICALLY OR HORIZONTALLY
250.110(1)

Exposed metal parts of fixed equipment shall be grounded where within 8 ft. (2.5 m) vertically or 5 ft. (1.5 m) horizontally of ground or grounded metal objects and where subject to contact by persons. **(See Figure 6-1)**

Figure 6-1. This illustration shows the vertical and horizontal clearances that shall be applied for ungrounded metal parts of fixed equipment such as luminaires (lighting fixtures).

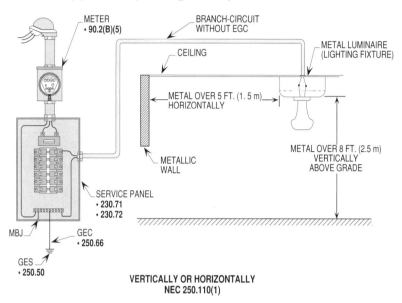

WET OR DAMP LOCATIONS
250.110(2)

Exposed metal parts of fixed equipment shall be grounded where located in wet or damp locations and not isolated. **(See Figure 6-2)**

Figure 6-2. This illustration shows weatherproof enclosures that are installed outside and exposed to the elements of the weather.

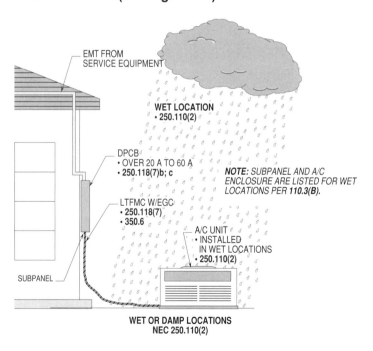

ELECTRICAL CONTACT WITH METAL
250.110(3)

Exposed metal parts of fixed equipment shall be grounded where in electrical contact with metal. **(See Figure 6-3)**

Equipment Grounding and Equipment Grounding Conductors

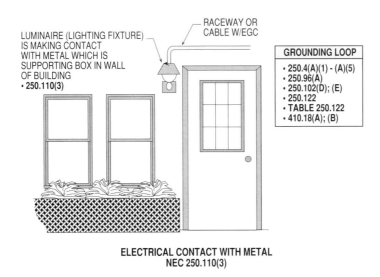

Figure 6-3. This illustration shows a luminaire (lighting fixture) with metal parts that are grounded by an equipment grounding conductor routed with the circuit conductors in the raceway or cable system.

CLASS I, DIVISIONS 1 AND 2
250.110(4); 501.16(B); 501.4(B); 250.102

Exposed metal parts of fixed equipment shall be grounded where installed in Class I, Divisions 1 and 2 locations. Flexible metal conduit or liquidtight flexible metal conduit shall be permitted to be used per **501.4(B)** as the sole equipment grounding path. Flexible metal conduit and liquidtight flexible metal conduit shall be required to be installed with internal or external bonding jumpers in parallel with each conduit. See **250.102(D)** and **(E)** for equipment bonding jumpers where they are installed as internal or external bonding jumpers.

The bonding jumper for liquidtight flexible metal conduit in Class I, Division 2 locations shall not be required to be installed where the following conditions are complied with:

- Where listed liquidtight flexible metal conduit is installed in 6 ft. (1.8 m) or less lengths with listed fittings for grounding.

- The overcurrent protection in the circuit is limited to 10 amps or less.

- The load is not a power utilization load.

See Figure 6-4 for a detailed illustration when applying these requirements.

Grounding Tip 167: When armored metal is used as an equipment grounding conductor means in flexible cable, see the requirements in Sec.'s 8.1.1 thru 8.3 in UL 467.

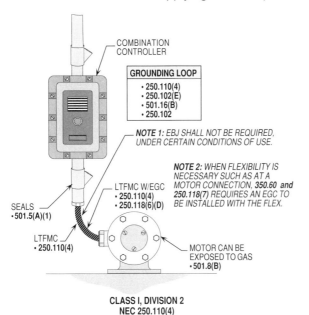

Figure 6-4. This illustration shows an equipment bonding jumper used with metal-clad in the liquidtight flexible metal conduit to bond and ground the motor to disconnect and combination controller.

Grounding Tip 168: When flexible metal conduit is used to connect equipment requiring flexibility, the equipment bonding jumper which is shown in Figure 6-5 in this book shall be required. Otherwise, no equipment bonding jumper is required per 348.60 and 250.118(6) of the NEC.

CLASS II, DIVISIONS 1 AND 2
250.110(4); 502.16(B); 502.4; 250.102

Exposed metal parts of fixed equipment shall be grounded where installed in Class II, Divisions 1 and 2 locations. Flexible metal conduit or liquidtight flexible metal conduit shall be permitted to be used per **502.4**. Flexible metal conduit or liquidtight flexible metal conduit shall be required to be installed with internal or external bonding jumpers in parallel with each conduit. See **250.102(D)** and **(E)** for equipment bonding jumpers where they are installed as internal or external bonding jumpers.

The bonding jumper for liquidtight flexible metal conduit in Class I, Division 2 locations shall not be required to be installed where the following conditions are complied with:

- Where listed liquidtight flexible metal conduit is installed in 6 ft. (1.8 m) or less lengths with listed fittings for grounding.
- The overcurrent protection in the circuit is limited to 10 amps or less.
- The load is not a power utilization load.

See Figure 6-5 for a detailed illustration when applying these requirements.

Figure 6-5. This illustration shows the procedure for installing flex with an equipment bonding jumper run on the outside of the flex to bond and ground the motor to the disconnecting means.

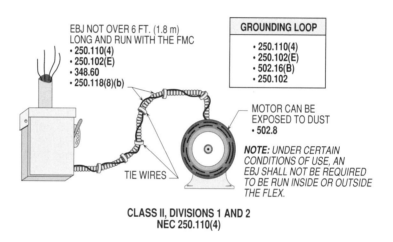

Grounding Tip 169: Where liquidtight flexible metal conduit is used to connect equipment (such as motors) requiring flexibility, the equipment bonding jumper that is shown in Figure 6-6 in this book shall be required per 350.60 and 250.118(7) in the NEC.

CLASS III, DIVISIONS 1 AND 2
250.110(4); 503.16(B); 503.3; 250.102

Exposed metal parts of fixed equipment shall be grounded where installed in Class III, Divisions 1 and 2 locations. Flexible metal conduit or liquidtight flexible metal conduit shall be permitted to be used per **503.3**. Flexible metal conduit or liquidtight flexible metal conduit shall be required to be installed with internal or external bonding jumpers in parallel with each conduit. See **250.102(D)** and **(E)** for equipment bonding jumpers where they are installed as internal or external bonding jumpers.

The bonding jumper for liquidtight flexible metal conduit in Class I, Division 2 locations shall not be required to be installed where the following conditions are complied with:

- Where listed liquidtight flexible metal conduit is installed in 6 ft. (1.8 m) or less lengths with listed fittings for grounding.
- The overcurrent protection in the circuit is limited to 10 amps or less.
- The load is not a power utilization load.

See Figure 6-6 for a detailed illustration when applying these requirements.

Equipment Grounding and Equipment Grounding Conductors

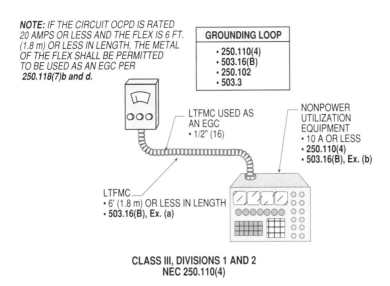

Figure 6-6. This illustration shows the metal clad of the flex used as an equipment grounding conductor to bond and ground the nonpower utilization equipment.

INTRINSICALLY SAFE SYSTEMS
250.110(4); 504.50(A)

Exposed metal parts of fixed equipment shall be grounded where installed for intrinsically safe systems such as intrinsically safe apparatus, associated apparatus, cable shields, enclosures and raceways related to such equipment.

CLASS I, ZONES 0, 1 AND 2
250.110(4); 505.25; 505.15(C); 250.102

Exposed metal parts of fixed equipment shall be grounded where it is installed in Class I, Zones 0, 1 and 2 locations. Flexible metal conduit or liquidtight flexible metal conduit shall be permitted to be used per **505.15(C)** as the sole equipment grounding path. Flexible metal conduit and liquidtight flexible metal conduit shall be required to be installed with internal or external bonding jumpers in parallel with each conduit. See **250.102(D)** and **(E)** for equipment bonding jumpers where they are installed as internal or external bonding jumpers. **(See Figure 6-7)**

Grounding Tip 170: The requirements for grounding shunt diode barriers can be found in Sec. 12.7.5 in UL 913.

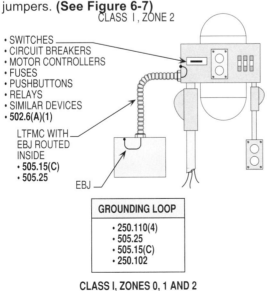

Figure 6-7. This illustration shows that all metal parts shall be bonded and grounded with metal conduits or a combination metal conduits and equipment grounding conductors.

Grounding Tip 171: Before installing flexible metal conduit without an equipment bonding jumper routed either inside or outside the flex, review **250.102** and **250.118(6)** in the NEC.

The bonding jumper for liquidtight flexible metal conduit in Class I, Division 2 locations shall not be required to be installed where the following conditions are complied with:

- Where listed liquidtight flexible metal conduit is installed in 6 ft. (1.8 m) or less length with listed fittings for grounding.
- The overcurrent protection in the circuit is limited to 10 amps or less.
- The load is not a power utilization load.

COMMERCIAL GARAGES, REPAIR AND STORAGE
250.110(4); 511.16; 501.16(B); 501.4(B); 250.102

Exposed metal parts of fixed equipment shall be grounded where installed in commercial garages, repair and storage facilities. All noncurrent-carrying metal parts of fixed or portable electrical equipment shall be grounded for safety, regardless of the voltage. Flexible metal conduit or liquidtight flexible metal conduit shall be permitted to be used per **501.4(B)** as the sole equipment grounding path. Flexible metal conduit and liquidtight flexible metal conduit shall be required to be installed with internal or external bonding jumpers in parallel with each conduit. See **250.102(D)** and **(E)** for equipment bonding jumpers where they are installed as internal or external bonding jumpers. **(See Figure 6-8)**

Figure 6-8. This illustration shows a diagnostic machine wired with an equipment bonding jumper routed inside the flex with the circuit conductors.

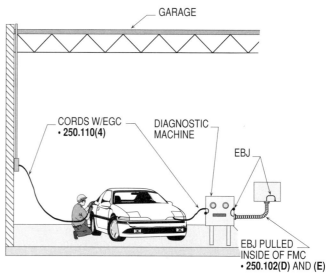

COMMERCIAL GARAGES, REPAIR AND STORAGE
NEC 250.110(4)

Grounding Tip 172: Before installing liquidtight flexible metal conduit without an equipment bonding jumper routed either inside or outside the flex, review **250.102** and **250.118(7)** in the NEC.

The bonding jumper for liquidtight flexible metal conduit in Class I, Division 2 locations shall not be required to be installed where the following conditions are complied with:

- Where listed liquidtight flexible metal conduit is installed in 6 ft. (1.8 m) or less lengths with listed fittings for grounding.
- The overcurrent protection in the circuit is limited to 10 amps or less.
- The load is not a power utilization load.

AIRCRAFT HANGERS
250.110(4); 513.16; 501.16(B); 501.4(B); 250.102

Exposed metal parts of fixed equipment shall be grounded where installed in aircraft hangers. All noncurrent-carrying metal parts of fixed or portable electrical equipment shall be grounded for safety, regardless of the voltage. Flexible metal conduit or liquidtight flexible metal conduit shall be permitted to be used per **501.4(B)** as the sole equipment grounding path. Flexible metal conduit and liquidtight flexible metal conduit shall be required to be installed with internal or external bonding jumpers in parallel with each conduit. See **250.102(D)** and **(E)** for equipment bonding jumpers where they are installed as internal or external bonding jumpers. **(See Figure 6-9)**

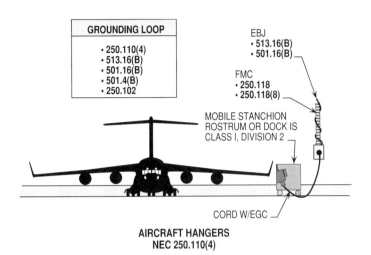

Figure 6-9. This illustration shows flexible metal conduit or liquidtight flexible metal conduit with an outside equipment bonding jumper routed around the flex supplying a receptacle used to cord-and-plug connect a mobile stanchion rostrum.

The bonding jumper for liquidtight flexible metal conduit in Class I, Division 2 locations shall not be required to be installed where the following conditions are complied with:

- Where listed liquidtight flexible metal conduit is installed in 6 ft. (1.8 m) or less lengths with listed fittings for grounding.
- The overcurrent protection in the circuit is limited to 10 amps or less.
- The load is not a power utilization load.

Grounding Tip 173: Before installing flexible metal conduit or liquidtight flexible metal conduit without an equipment grounding conductor routed inside or outside the flex, review **250.102** and **250.118(6)** and **(7)** in the NEC.

GASOLINE DISPENSING AND SERVICE STATIONS
250.110(4); 514.16; 501.16(B); 501.4(B); 250.102

Exposed metal parts of fixed equipment shall be grounded where installed in gasoline dispensing and service stations. All noncurrent-carrying metal parts of fixed or portable electrical equipment shall be grounded for safety, regardless of the voltage. Flexible metal conduit or liquidtight flexible metal conduit shall be permitted to be used per **501.4(B)** as the sole equipment grounding path. Flexible metal conduit and liquidtight flexible metal conduit shall be required to be installed with internal or external bonding jumpers in parallel with each conduit. See **250.102(D)** and **(E)** for equipment bonding jumpers where they are installed as internal or external bonding jumpers. **(See Figure 6-10)**

Grounding Tip 174: For wiring methods installed in gasoline dispensing and service stations, see **501.4(A)** and **(B)** and **514.8** with exceptions in the NEC.

The bonding jumper for liquidtight flexible metal conduit in Class I, Division 2 locations shall not be required to be installed where the following conditions are complied with:

- Where listed liquidtight flexible metal conduit is installed in 6 ft. (1.8 m) or less lengths with listed fittings for grounding.
- The overcurrent protection in the circuit is limited to 10 amps or less.
- The load is not a power utilization load.

Figure 6-10. This illustration shows that an equipment grounding conductor shall be routed inside the rigid nonmetallic conduit with the circuit conductors.

Grounding Tip 175: Where rigid nonmetallic conduit is installed in the ground between the rigid metal conduit and elbows, an equipment grounding conductor shall be required to be routed with the circuit conductors. A factory lug shall be required to be provided on the dispenser motor or equipment to terminate the equipment grounding conductor.

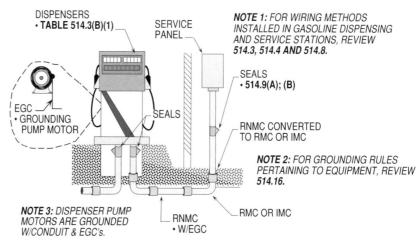

GASOLINE DISPENSING AND SERVICE STATIONS
NEC 250.110(4)

BULK-STORAGE PLANTS
250.110(4); 515.16; 501.16(B); 501.4(B); 250.102

Exposed metal parts of fixed equipment shall be grounded where installed in bulk storage plants. All noncurrent-carrying metal parts of fixed or portable electrical equipment shall be grounded for safety, regardless of the voltage. Flexible metal conduit or liquidtight flexible metal conduit shall be permitted to be used per **501.4(B)** as the sole equipment grounding path. Flexible metal conduit and liquidtight flexible metal conduit shall be required to be installed with internal or external bonding jumpers in parallel with each conduit. See **250.102(D)** and **(E)** for equipment bonding jumpers where they are installed as internal or external bonding jumpers. **(See Figure 6-11)**

The bonding jumper for liquidtight flexible metal conduit in Class I, Division 2 locations shall not be required to be installed where the following conditions are complied with:

- Where listed liquidtight flexible metal conduit is installed in 6 ft. (1.8 m) or less lengths with listed fittings for grounding.
- The overcurrent protection in the circuit is limited to 10 amps or less.
- The load is not a power utilization load.

Figure 6-11. This illustration shows rigid metal conduit with an equipment grounding conductor routed with the circuit conductors and used to establish the grounding means for the electrical equipment associated with the bulk-storage tank.

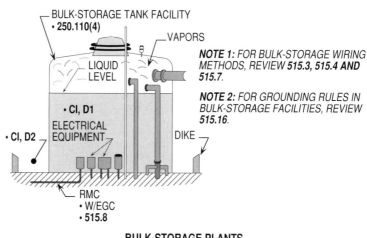

BULK-STORAGE PLANTS
NEC 250.110(4)

SPRAY APPLICATION, DIPPING AND COATING PROCESSES
250.110(4); 516.16; 501.16(B); 501.4(B); 250.102

Exposed metal parts of fixed equipment shall be grounded where installed in spray application, dipping and coating processes. All noncurrent-carrying metal parts of fixed or portable electrical equipment shall be grounded for safety, regardless of the voltage. Flexible metal conduit or liquidtight flexible metal conduit shall be permitted to be used per **501.4(B)** as the sole equipment grounding path. Flexible metal conduit and liquidtight flexible metal conduit shall be required to be installed with internal or external bonding jumpers in parallel with each conduit. See **250.102(D)** and **(E)** for equipment bonding jumpers where they are installed as internal or external bonding jumpers. **(See Figure 6-12)**

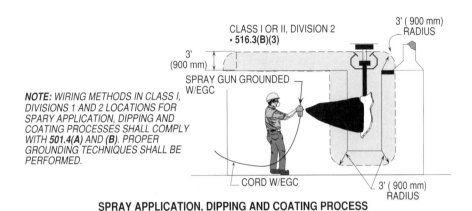

Figure 6-12. This illustration shows that proper wiring methods and grounding techniques shall be used for spray application, dipping and coating process.

The bonding jumper for liquidtight flexible metal conduit in Class I, Division 2 locations shall not be required to be installed where the following conditions are complied with:

- Where listed liquidtight flexible metal conduit is installed in 6 ft. (1.8 m) or less lengths with listed fittings for grounding.
- The overcurrent protection in the circuit is limited to 10 amps or less.
- The load is not a power utilization load.

HEALTH CARE FACILITIES
250.110(4)

There are specific grounding rules that shall be applied in health care facilities. Particular areas and types and use of equipment determine the severity of these grounding requirements. The following areas in health care facilities are key areas where grounding shall be considered by designers, installers and inspectors:

- Grounding of receptacles and fixed electric equipment in patient care areas per **517.13(A)**
- Grounding of wiring methods for branch-circuits serving patient care areas per **517.13(B)**
- Bonding panelboards with a 10 AWG copper conductor per **517.14**
- Grounding receptacles with insulated grounding terminals per **517.16**
- Grounding equipment in general care areas per **517.18(B)**
- Grounding equipment in critical care areas per **517.19(C), (D), (F)** and **(G)**

Grounding Tip 176: In areas used for patient care, the grounding terminals of all receptacles and noncurrent conductive surfaces of fixed equipment likely to become energized shall be grounded with an insulated copper equipment grounding conductor per **517.13(A)**.

- Grounding equipment in wet locations per **517.20**
- Grounding equipment in anesthetizing locations per **517.61(A)** and **(B)**
- Grounding X-ray equipment per **517.71** and **517.72(C)**
- Grounding equipment in isolated power systems per **517.160(A)** and **(B)**

See Figures 6-13(a) through (d) for a detailed illustration of grounding requirements in health care facilities for the locations mentioned.

Figure 6-13(A). This illustration shows specific grounding requirements for critical care areas in a health care facility.

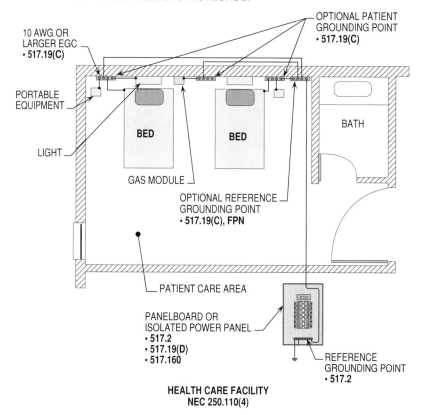

Figure 6-13(B). This illustration shows the grounding requirements when using metal raceways and cable assemblies to supply loads located in patient care areas.

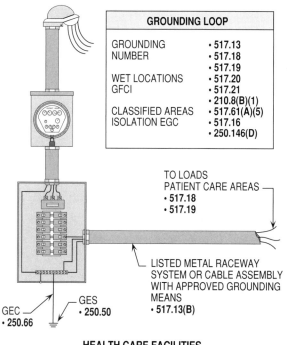

Equipment Grounding and Equipment Grounding Conductors

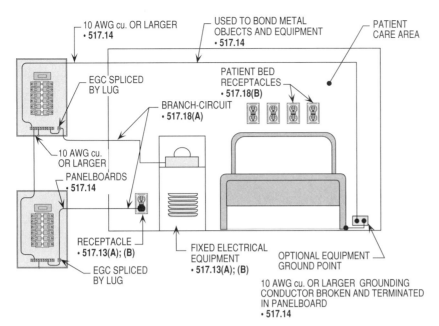

Figure 6-13(c). This illustration shows the panelboard bonding requirements when using a 10 AWG or larger copper conductor.

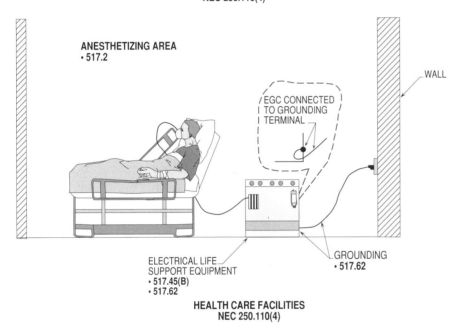

Figure 6-13(d). This illustration shows the grounding requirements for electrical equipment used in anesthetizing areas in a health care facility.

WIRING METHODS WITH EQUIPMENT GROUNDING CONDUCTORS
250.110(5)

Exposed metal parts of fixed equipment shall be grounded where supplied by wiring methods with an equipment grounding conductor such as metal-clad, metal-sheathed and metal raceways. Short sections of metal raceways shall not be required to be grounded where such raceway provides protection of cable assemblies from physical damage per **250.86, Ex. 2**. **(See Figure 6-14)**

OVER 150 VOLTS-TO-GROUND
250.110(6)

Exposed metal parts of fixed equipment shall be grounded where equipment operates at over 150 volts-to-ground on any terminal or lug.

Grounding Tip 177: These cables have an additional equipment grounding conductor inside their jacket and metal-clad or metal-sheath to redundantly ground (in some cases) the metal frames of electrical equipment.

Figure 6-14. This illustration shows that metal raceways shall be bonded and grounded for safety.

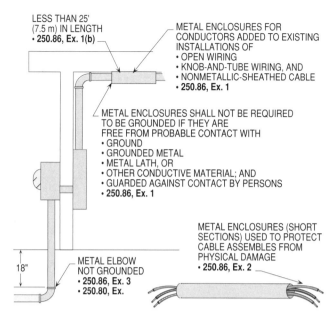

WIRING METHODS WITH EQUIPMENT GROUNDING CONDUCTORS
NEC 250.110(5)

Grounding Tip 178: Equipment shall be considered within reach of a person who, while touching the equipment could at the same time be in contact with ground or a "grounded object". The "grounded object" could be a water faucet in a bathroom or kitchen. Note that if the electrical equipment is close enough to a pipe, faucet, etc., so that a person could touch both the electrical equipment and the pipe, faucet, etc., at the same time, the equipment shall be grounded regardless of the type of wiring.

FIXED EQUIPMENT NOT REQUIRED TO BE GROUNDED
250.110, Ex.'s 1 THRU 3

Exposed metal parts of fixed equipment shall not be required to be grounded when complying with the following conditions:

- Metal frames of electrically heated appliances
- Distribution apparatus
- Listed equipment

METAL FRAMES OF ELECTRICALLY HEATED APPLIANCES
250.110, Ex. 1

Metal frames of electrically heated appliances shall not be required to be grounded where, exempted by special permission, the frames are permanently and effectively insulated from ground.

Grounding Tip 179: A transformer bank that is located on an elevated platform which is mounted to a utility pole is an example of distribution apparatus per **250.110, Ex. 2**.

DISTRIBUTION APPARATUS
250.110, Ex. 2

Distribution apparatus such as transformer and capacitor cases shall not be required to be grounded where mounted on wooded poles at a height of at least of 8 ft. (2.5 m) above ground or grade level. **(See Figure 6-15)**

LISTED EQUIPMENT
250.110, Ex. 3

Listed equipment which is distinctively marked shall not be required to be grounded where protected by a system of double insulation, or its equivalent. **(See Figure 6-16)**

Equipment Grounding and Equipment Grounding Conductors

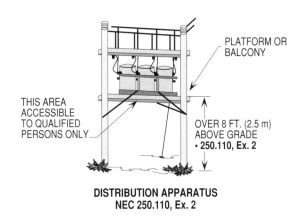

Figure 6-15. This illustration shows a transformer bank installation with cases that are not required to be grounded.

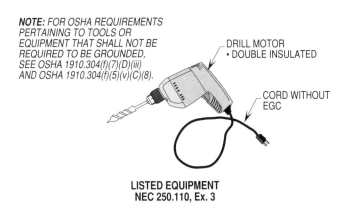

Figure 6-16. This illustration shows a drill motor that is not required to be grounded.

FASTENED-IN-PLACE OR CONNECTED BY PERMANENT WIRING METHODS (FIXED) - SPECIFIC 250.112

Regardless of voltage, all of the following exposed noncurrent-carrying metal parts of fixed equipment (specific) shall be grounded:

- Switchboard frames and structures
- Pipe Organs
- Motor frames
- Enclosures for motor controllers
- Elevators and cranes
- Garages, theaters and motion picture studios
- Electric signs
- Motion picture projection equipment
- Power-limited remote-control, signaling and fire alarm circuits
- Luminaires (lighting fixtures)
- Skid-mounted equipment
- Motor-operated water pumps
- Metal well casings

Grounding Tip 180: There are types of equipment that are required to be grounded for the safety and protection of personnel and property per **250.112(A)** thru **(M)**.

SWITCHBOARD FRAMES AND STRUCTURES 250.112(A)

Exposed noncurrent-carrying metal parts of switchboard frames and structures supporting switching equipment shall be grounded. Frames for two-wire DC switchboards shall not be required to be grounded where effectively insulated from ground. **(See Figure 6-17)**

Grounding Tip 181: For frequency grounding tips, see Chapter 29 in NFPA 70B.

Figure 6-17. This illustration shows the grounded (neutral) conductor and equipment grounding conductor connections for the service switchboard and subfeed switchboard.

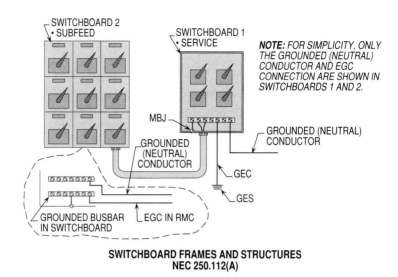

PIPE ORGANS
250.112(B)

Generator and motor frames in an electrically operated pipe organ shall be grounded. Generator and motor frames in an electrically operated pipe organ shall not be required to be grounded where effectively insulated from ground and the motor driving it. **(See Figure 6-18)**

Figure 6-18. This illustration shows the bonding and grounding of an electric pipe organ.

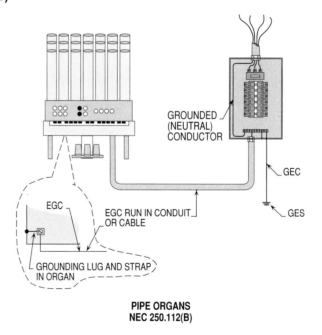

Grounding Tip 182: When connecting a wiring method to a motor and flexibility is needed, an equipment grounding conductor shall be routed with the circuit conductors per **250.102(D)** and **(E)**.

MOTOR FRAMES
250.112(C); 430.142

Under any of the following conditions, the frames of stationary motors shall be grounded:

- Where supplied by metal-enclosed wiring
- Where installed in a wet location and not isolated or guarded
- If installed in a hazardous (classified) location, see **250.110(4)**
- If the motor operates at over 150 volts-to-ground with any terminal

See Figure 6-19 for a detailed illustration when applying these requirements.

Equipment Grounding and Equipment Grounding Conductors

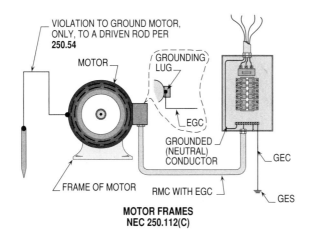

Figure 6-19. This illustration shows the bonding and grounding for the metal frame of a motor when installing rigid metal conduit with an equipment grounding conductor as the grounding means.

ENCLOSURES FOR MOTOR CONTROLLERS
250.112(D)

Enclosures for motor controllers shall be grounded. Enclosures shall not be required to be grounded where attached to ungrounded portable equipment. **(See Figure 6-20)**

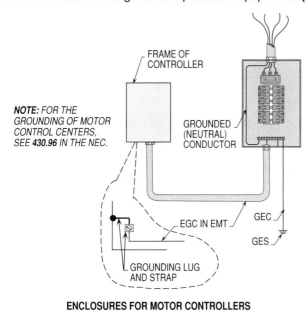

Figure 6-20. This illustration shows the metal frame of a controller being bonded and grounded with a combination of an equipment grounding conductor and electrical metallic tubing.

ELEVATORS AND CRANES
250.112(E)

Exposed noncurrent-carrying metal parts for elevators and cranes shall be grounded. **(See Figure 6-21)**

Grounding Tip 183: For grounding requirements pertaining to overhead electric cranes, see CMAA 70, "Specifications for Electrical Overhead Traveling Cranes".

GARAGES, THEATERS AND MOTION PICTURE STUDIOS
250.112(F)

Exposed noncurrent-carrying metal parts for electric equipment in commercial garages, theaters and motion picture studios shall be grounded. Pendant lampholders installed in commercial garages, theaters and motion picture studios shall not be required to be grounded where the circuit is not over 150 volts-to-ground. **(See Figure 6-22)**

Figure 6-21. This illustration shows the bonding and grounding techniques used to ground the overhead crane motor for safety.

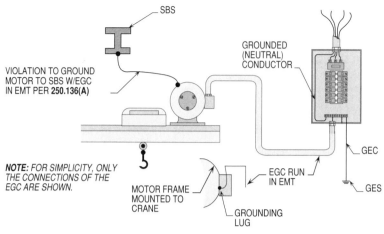

Figure 6-22. This illustration shows the bonding and grounding techniques used to safely ground theater related equipment.

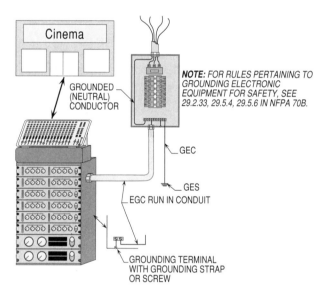

Grounding Tip 184: For UL requirements pertaining to bonding and grounding electrical signs, see Sec.'s 18.1 thru 18.21 in UL 48.

ELECTRIC SIGNS
250.112(G); 600.7; 600.32

Electric signs and metal equipment of outline lighting systems shall be grounded. Small metal parts of less than 2 in. (50 mm) in any dimension that are not likely to become energized and that are spaced at least 3/4 in. (19 mm) from neon tubing shall not be required to be bonded. Listed flexible metal conduit or listed liquidtight flexible metal conduit shall be permitted to be used as a bonding jumper in lengths less than 100 ft. (30 m). A bonding jumper shall be required to be routed separately and remotely from nonmetallic conduit and be spaced at least 1 1/2 in. (38 mm) from such conduit when the circuit operates at 100 Hz or 1 3/4 in. (45 mm) if the circuit operates over 100 Hz. Bonding jumpers shall be sized at least 14 AWG copper or larger. Metal parts of a building shall not be permitted to be used as a secondary return conductor or equipment grounding conductor.

Nonmetallic conduit or flexible nonmetallic conduit shall be permitted to be used to enclose the high-voltage secondary wiring on neon systems. However, nonmetallic conduit or flexible nonmetallic conduit shall be spaced at least 1 1/2 in. (38 mm) from grounded or bonded metal parts when the transformer output is operating at 100 Hz or less. If the output operates over 100 Hz, the spacing shall be at least 1 3/4 in. (45 mm) from such parts. Metal parts of a structure or building shall not be used as a secondary return conductor or equipment grounding conductor. **(See Figures 6-23(a), (b) and (c))**

Equipment Grounding and Equipment Grounding Conductors

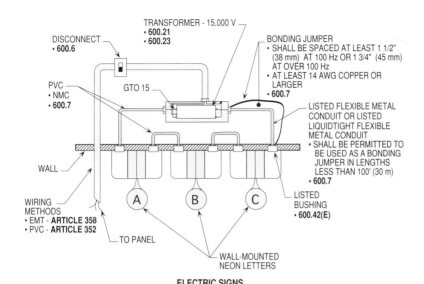

Figure 6-23(a). Electric signs and metal equipment of outline lighting systems shall be grounded. Metal parts of a building shall not be permitted to be used as a secondary return conductor or equipment grounding conductor.

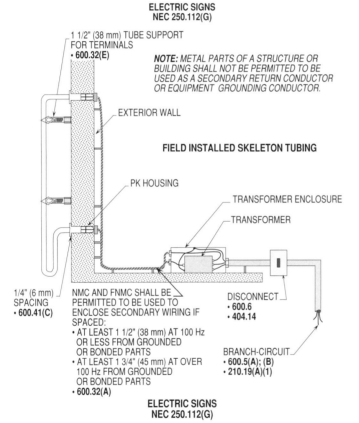

Figure 6-23(b). Nonmetallic conduit or flexible nonmetallic conduit shall be permitted to be used for high-voltage wiring except when used with electric discharge lighting or neon tubing in wet or damp locations if not adequately spaced.

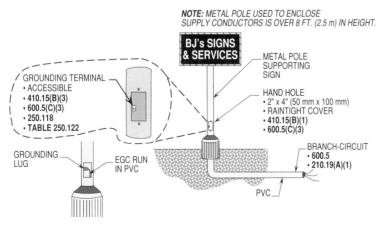

Figure 6-23(c). A handhole not less than 2 in. x 4 in. (50 mm x 100 mm) shall be required with a raintight cover for a metal pole. A grounding terminal shall be accessible from the handhole.

MOTION PICTURE PROJECTION EQUIPMENT
250.112(H)

Exposed noncurrent-carrying metal parts of motion picture projection equipment shall be grounded. **(See Figure 6-24)**

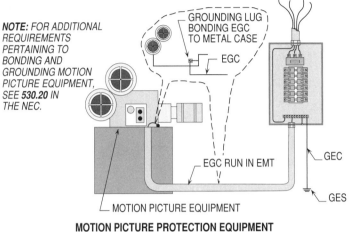

Figure 6-24. This illustration shows the bonding and grounding techniques required to safely ground a motion picture projector.

POWER-LIMITED REMOTE-CONTROL SIGNALING AND FIRE ALARM CIRCUITS
250.112(I)

A Class 1, Class 2 or Class 3 remote-control or signaling transformer that is rated not over 1000 VA has a grounded secondary conductor bonded to the metal case of the transformer and no grounding electrode conductor shall be required. However, the metal transformer case shall be properly grounded by a grounded metal raceway or equipment grounding conductor which supplies its primary or by means of an equipment grounding conductor that connects the case back to the grounding electrode of the primary system.

At least a 14 AWG copper conductor shall be used to bond and ground the transformer secondary to the transformer frame per **250.30(A)(1), Ex. 2**. This leaves the supply raceway, equipment grounding conductor or both to the transformer to provide the path to ground back to the common service grounding point. **(See Figure 6-25)**

Grounding Tip 185: The exception to **250.30(A)(2)** permits a smaller equipment grounding conductor per **Table 250.122** to be used for grounding a small control transformer. Otherwise, an 8 AWG copper or larger grounding electrode conductor per **Table 250.66** would have to be used which could not be terminated to the small control transformer.

Figure 6-25. This illustration shows a 14 AWG copper equipment grounding conductor used to bond and ground the transformer supplying a Class 1 control circuit.

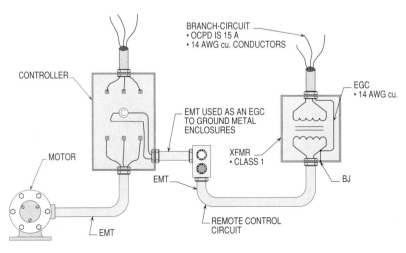

LUMINAIRES (LIGHTING FIXTURES)
250.112(J); 410.18(A); (B); 410.20; 410.21

In residential, commercial and industrial locations, exposed metal parts of luminaires (lighting fixtures) shall be required to be grounded with an equipment grounding conductor if the branch-circuit is provided with an equipment grounding conductor. The equipment grounding conductor shall be selected from any of the wiring methods listed in **250.118** and sized per **Table 250.122** based upon the size OCPD. **(See Figure 6-26)**

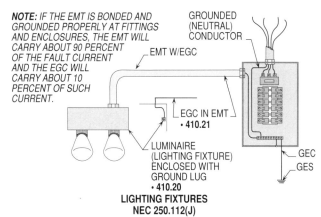

Figure 6-26. This illustration shows a luminaire (lighting fixture) bonded and grounded by an equipment grounding conductor routed through the electrical metallic tubing with circuit conductors. Note that the electrical metallic tubing is also a means to ground the luminaire (lighting fixture) per **250.118(4)**.

Grounding Tip 186: In Figure 6-26, the equipment grounding conductor will only carry about 30 percent of the fault current if the electrical metallic tubing carries 70 percent of the fault current due to loose connections.

Grounding Note: Section 410-92 of the 1971 NEC and earlier editions required lighting fixtures with metal parts to be used with metallic wiring systems such as metal conduit or metal clad cables. The metal clad AC cable (BX) and metal clad cables (MC) were used to ground the exposed metal of lighting fixtures. The metal of metal conduits such as rigid metal conduit and electrical metallic tubing was mostly used as a grounding means. Copper or aluminum conductors were also pulled in conduits and utilized as an additional grounding means. Under certain conditions of use, flexible metal conduit and liquidtight flexible metal conduit was used for such grounding means.

The branch-circuit wiring in older sites of residential, commercial and industrial locations does not have an equipment grounding conductor in the nonmetallic-sheathed cable (romex) or in knob-and-tube wiring systems to ground the exposed metal parts of lighting fixtures. Lighting fixtures in older facilities were not required to be grounded with an equipment grounding conductor.

Note that lighting fixtures were not referred to as luminaires until the 2002 NEC was published.

Grounding Note: Lighting fixtures were not required by the NEC to be grounded with an equipment grounding conductor or grounding means until the 1975 edition of the Code.

Luminaires (lighting fixtures) with exposed metal parts shall not be used to replace luminaires (lighting fixtures) on existing wiring systems that are not equipped with an equipment grounding means.

Grounding Note: Section 410-18(b) in the 1975 NEC as well as previous editions require a lighting fixture with an insulated material to be used to replace an existing metal fixture in an existing wiring system. If luminaires (lighting fixtures) with exposed metal parts are used for replacements, a branch-circuit with an equipment grounding conductor shall be provided per **410.18(B)** or installed per **410.18(B), Ex.**

SKID-MOUNTED EQUIPMENT
250.112(K)

Exposed noncurrent-carrying metal parts of permanently mounted electrical equipment and skids shall be grounded with an equipment bonding jumper size per **250.122**. **(See Figure 6-27)**

Grounding Tip 187: When wiring and working on skids, the electrical equipment needs to be checked for proper bonding to the skid.

Figure 6-27. This illustration shows a skid bonded and grounded for safety using rigid metal conduit with an equipment grounding conductor routed inside with circuit conductors.

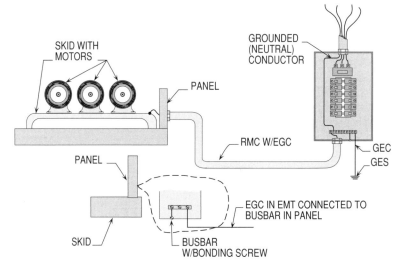

SKID-MOUNTED EQUIPMENT
NEC 250.112(K)

Grounding Note: A lot of electrical equipment is being mounted on skids and delivered to construction sites. There is no requirement in the NEC at present to ground or bond the skids. This can create a potential safety hazard when a person is standing on the skid and touching the building steel or standing on the ground and touching the skid during a ground-fault. Skids shall be bonded and grounded for safety.

MOTOR-OPERATED WATER PUMPS
250.112(L)

Motor-operated water pumps including the submersible type shall be grounded. **(See Figure 6-28)**

Grounding Tip 188: Care must be exercised when bonding the water pump motor to the metal well casing. If the water pump motor and metal well casing are not bonded and grounded properly, a dangerous shock hazard can exist.

Figure 6-28. This illustration shows the procedure for bonding and grounding the metal of the pump motor to the metal well casing.

Grounding Tip 189: Safety from electrical shock can be reduced greatly if the metal casing and frame of the motor are bonded and grounded together to form an equipotential plane between elements.

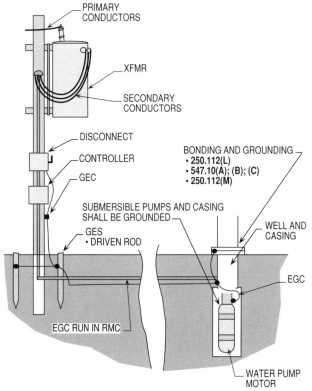

MOTOR-OPERATED WATER PUMPS - METAL WELL CASINGS
NEC 250.112(L); (M)

METAL WELL CASINGS
250.112(M)

The metal well casing enclosing a submersible pump shall be bonded to the equipment grounding conductor of the power circuit supplying the pump. **(See Figure 6-28)**

EQUIPMENT CONNECTED BY CORD-AND-PLUG
250.114

Under any of the following conditions, exposed noncurrent-carrying metal parts of cord-and-plug connected equipment likely to become energized shall be grounded:

- Class I, Divisions 1 and 2
- Class II, Divisions 1 and 2
- Class III, Divisions 1 and 2
- Commercial garages, repair and storage
- Aircraft hangers
- Bulk storage plants
- Health care facilities
- Over 150 volts-to-ground
- In residential occupancies
- In other than residential occupancies
- Tools used in wet or conductive locations
- Portable handlamps

Grounding Tip 190: Under certain conditions of use, cord-and-plug connections shall be permitted to be used to connect the power supply to electrical equipment and associated apparatus.

CLASS I, DIVISIONS 1 AND 2
250.114(1); 501.11; 501.12

A flexible cord shall be permitted to be used for connection between a portable lamp or other portable utilization equipment and that fixed portion of its supply circuit. The following conditions of use shall be considered where such connections are used:

- Be of a type that is listed for extra hard-usage.

- Contain, in addition to the conductors of the circuit, an equipment grounding conductor conforming to **250.119** and **400.23**, that is, of green color and used for no other purpose than for grounding.

- Be connected to terminals or to supply conductors in an approved manner.

- Be supported by clamps or other suitable means in such a manner that there will be no tension on the terminal connections.

- Have suitable seals provided where the flexible cord enters boxes, fittings or enclosures of the explosionproof type.

See Figure 6-29 for a detailed illustration when applying these requirements.

In Class I, Division 1 locations, flexible cord shall be permitted to be used for that portion of the circuit where the fixed wiring method per **501.4(A)(1)(a) thru (d)** will not provide the necessary degree of movement for fixed and mobile electrical utilization equipment. However, proper maintenance and supervision shall be provided and such equipment shall be located in an industrial establishment. **(See Figure 6-30)**

Grounding Tip 191: Flexible cord shall be permitted to be used to connect electrical equipment to their supply circuit in both Class I, Divisions 1 and 2 locations per **501.4(A)(2)** and **501.11**.

Grounding Tip 192: When flexible cord is used in hazardous locations, review the following sections very carefully:

- **250.100**
- **501.4(A)(2)**
- **501.4(B)**
- **501.11**
- **501.11, Ex.**
- **501.3(B)(6)**
- **501.12**
- **501.16**

Figure 6-29. This illustration shows the bonding and grounding of a portable luminaire (lighting fixture) used in a Class I, Division 2 location.

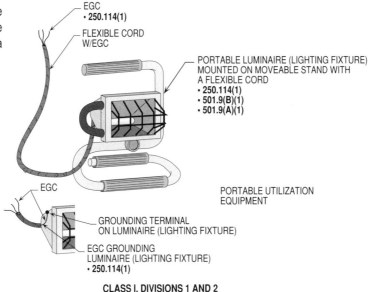

Figure 6-30. This illustration shows the bonding and grounding of a motor installed in a Class I, Division 1 location where the motor requires a certain amount of flexibility.

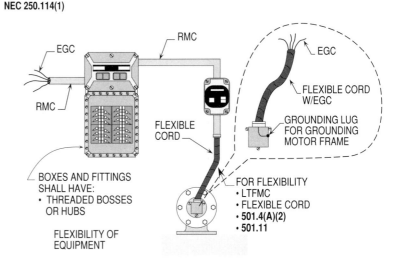

Electric submersible pumps with means for removal without entering the wet pit shall be considered portable utilization equipment. The extension of the flexible cord between the wet pit and power source shall be permitted to be enclosed in a suitable raceway. **(See Figure 6-31)**

Figure 6-31. Flexible cord shall be permitted to be used as a wiring method to connect the branch-circuit conductors to the submersible pump motor in the wet pit. Note that the flexible cord shall contain an equipment grounding conductor to bond and ground the metal frame of the pump for safety.

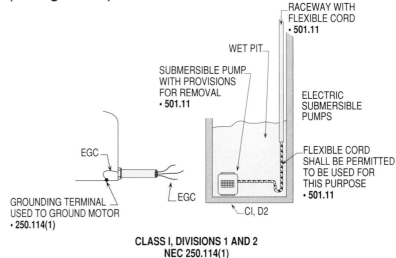

Electric mixers traveling in and out of open-type mixing tanks or vats shall be considered as portable utilization equipment so that wiring procedures won't have to be too restrictive. **(See Figure 6-32)**

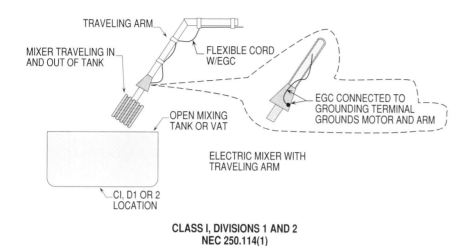

Figure 6-32. This illustration shows a flexible cord with an equipment grounding conductor in a flexible cord that is used to bond and ground the electric mixer for safety.

To facilitate replacements, process control instruments shall be permitted to be connected through flexible cord, attachment plug and receptacle per **501.3(B)(6)**. However, the following shall be provided:

- A switch complying with the rules of **501.3(B)(1)** is provided so that the attachment plug is not depended on to interrupt current.

- The current does not exceed 3 amps at 120 volts.

- The power supply cord does not exceed 3 ft. (900 mm) and is listed for extra-hard usage or for hard usage if protected by location and is supplied through an attachment plug and receptacle of the locking and grounding type.

- Only necessary receptacles are provided.

- The receptacle carries a label warning against unplugging under load.

Where provisions must be made for a flexible connection, such as motors, flexible cord shall be permitted to be used but additional grounding shall be provided around these flexible connections to ensure an effective ground path for clearing short-circuits and ground-faults per **501.4(B)(2)**.

Arcing at exposed contacts must be prevented in Class I, Divisions 1 and 2 locations. To accomplish this, receptacles are designed so that plug contacts are safely within an explosionproof enclosure when they are electrically engaged, confining arcing, if any, to the receptacle interior per **501.12**.

With this configuration, the plug cannot be inserted unless the switch is in the OFF position and cannot be withdrawn with the receptacle in the ON position. The reason for this, is the receptacle contacts are interlocked with a switch located in an explosionproof enclosure. Therefore, arcing does not occur outside the enclosure due to mated parts being deenergized during plug insertion and removal. **(See Figure 6-33)**

Receptacles without switches rely on a mechanical means which provides a delayed action to confine arcing to the receptacle interior during plug insertion and withdrawal. The design used in these receptacles prevents removal of the plug until any flame, spark or hot metal from an arc has cooled sufficiently to prevent ignition of the surrounding explosive atmosphere. **(See Figure 6-34)**

Grounding Tip 193: See **501.11** and **501.12** in the NEC for requirements pertaining to the use and installation of flexible cords and receptacles.

Grounding Tip 194: To prevent arcing, the grounding prong is usually the last of the prongs of the circuit to be deenergized from the receptacle outlet.

Figure 6-33. This illustration shows a flexible cord with an equipment grounding conductor being used to bond and ground the electric equipment that it supplies.

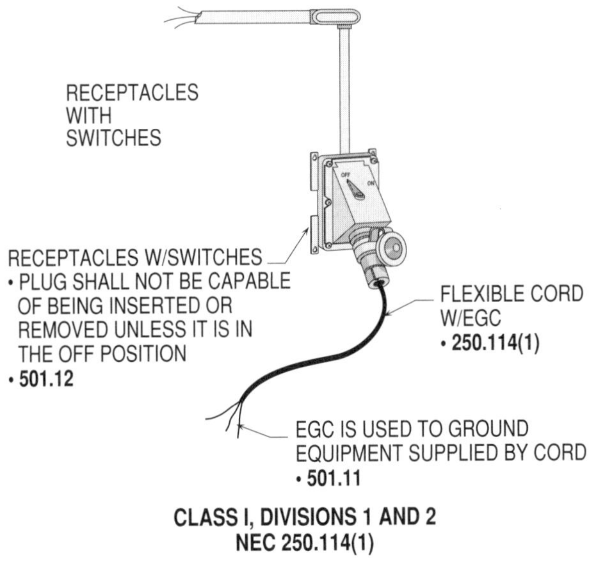

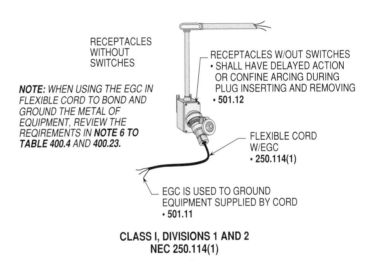

Figure 6-34. This illustration shows a flexible cord with an equipment grounding conductor and attachment cap being used to bond and ground the metal of the equipment that is supplied by the flexible cord.

Grounding Tip 195: When using an equipment grounding conductor in a flexible cord, review **Note 6 to Table 400.4** and **400.23(A)** and **(B)**.

CLASS II, DIVISIONS 1 AND 2
250.114(1); 502.12; 502.13(A); (B)

Under the following conditions of use, flexible cords shall be permitted to be used in Class II, Divisions 1 and 2 locations:

- Be of a type listed for extra hard-usage.

- Contain, in addition to the conductors of the circuit, an equipment grounding conductor conforming to **250.119** and **400.23**, that is, of green color and used for no other purpose than for grounding.

- Be connected to terminals or to supply conductors in an approved manner.

- Be supported by clamps or other suitable means in such a manner that there will be no tension on the terminal connections.

- Have suitable seals provided where the flexible cord enters boxes, fittings or enclosures of the dust-ignitionproof type.

Arcing at exposed contacts must be prevented in Class II, Division 1 locations. To accomplish this, receptacles are designed so that plug contacts are safely within an explosionproof enclosure when they are electrically engaged, confining arcing, if any, to the receptacle interior per **502.13(A)**.

Receptacles and attachment plugs installed in Class II, Division 2 locations shall be designed so that connection to the supply circuit cannot be made or broken while live parts are exposed per **502.13(B)**. **(See Figure 6-35)**

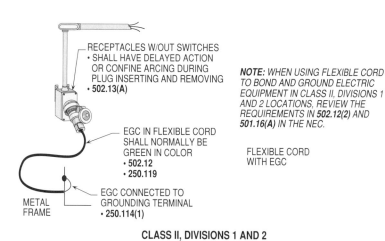

Figure 6-35. This illustration shows flexible cord with an equipment grounding conductor to bond and ground electrical equipment installed in Class II, Divisions 1 and 2 locations.

CLASS III, DIVISIONS 1 AND 2
250.114(1); 503.10; 503.11

Under the following conditions of use, flexible cords shall be permitted to be used in Class III, Divisions 1 and 2 locations:

- Be of a type listed for extra hard-usage.

- Contain, in addition to the conductors of the circuit, an equipment grounding conductor conforming to **250.119** and **400.23** that is of green color and used for no other purpose than for grounding.

- Be connected to terminals or to supply conductors in an approved manner.

- Be supported by clamps or other suitable means in such a manner that there will be no tension on the terminal connections.

- Have suitable means to prevent the entrance of fibers or flyings where the cord enters boxes or fittings.

Receptacles and attachment plugs installed in Class III, Divisions 1 and 2 locations shall be of the grounding type and designed so as to minimize the accumulation or the entry of fibers and flyings, and prevent the escape of sparks or molten particles per **503.11**.

Where only moderate accumulations of fibers and flyings will collect in the vicinity of a receptacle in Class III, Divisions 1 or 2 locations, the authority having jurisdiction may permit general-purpose grounding-type receptacles to be installed where such receptacle is readily accessible for routine cleaning so as to minimize the entry of fibers or flyings per **503.11, Ex. (See Figure 6-36)**

Grounding Tip 196: See **503.10** and **503.16(A)** and **(B)** for requirements pertaining to the bonding and grounding of electrical equipment installed in Class III, Division 1 and 2 locations.

Grounding Tip 197: To connect an equipment grounding conductor to a Class III, Divisions 1 and 2 piece of equipment, there shall be a grounding terminal lug or means for connection of the equipment grounding conductor.

Figure 6-36. This illustration shows flexible cord with an equipment grounding conductor to bond and ground the metal of electrical equipment used in Class III, Divisions 1 and 2 locations.

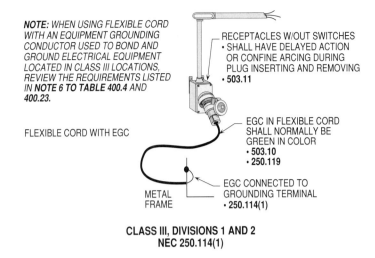

COMMERCIAL GARAGES, REPAIR AND STORAGE
250.114(1); 511.4(B); 511.10(B)(3)

Attachment plug receptacles installed in commercial garages shall be installed and located in a fixed position above any location that is classified as a Class I location or shall be approved for such location. **(See Figure 6-37)**

Figure 6-37. This illustration shows flexible cord with an equipment grounding conductor to bond and ground a diagnostic machine located in a commercial garage.

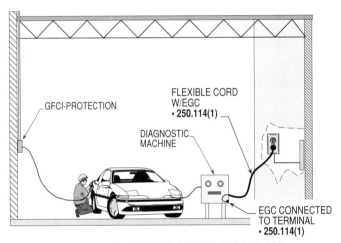

Grounding Tip 198: Wiring methods installed in Class I locations that are located in aircraft hangers shall be required to comply with **513.16(A)** and **(B)** in the NEC.

AIRCRAFT HANGERS
250.114(1); 513.16(A)

A flexible cord can only be used for connection between a portable lamp or other portable utilization equipment and that fixed portion of its supply circuit. The following conditions of use shall be considered where such connections are used:

- Be of a type listed for extra hard-usage.

- Contain, in addition to the conductors of the circuit, an equipment grounding conductor conforming to **250.119** and **400.23**, that is of green color and used for no other purpose than for grounding.

- Be connected to terminals or to supply conductors in an approved manner.

Equipment Grounding and Equipment Grounding Conductors

- Be supported by clamps or other suitable means in such a manner that there will be no tension on the terminal connections.

- Have suitable seals provided where the flexible cord enters boxes, fittings or enclosures of the explosionproof type.

In Class I, Division 1 locations, flexible cord shall be permitted to be used for that portion of the circuit where the fixed wiring method per **501.4(A)(1)(a)** thru **(d)** cannot provide the necessary degree of movement for fixed and mobile electrical utilization equipment.

Arcing at exposed contacts must be prevented in Class I, Divisions 1 and 2 locations. To accomplish this, receptacles are designed so that plug contacts are safely within an explosionproof enclosure when they are electrically engaged, confining arcing, if any, to the receptacle interior per **501.12**.

Flexible cord shall be suitable for the type of service and approved for extra-hard usage including a separate equipment grounding conductor where used for portable utilization equipment and lamps per **513.10(E)(1)** and **(E)(2)** and **513.7(B)**.

Flexible cord shall be suitable for the type of service and approved or identified for extra-hard usage including a separate equipment grounding conductor where used for mobile servicing equipment with electric components. Attachment plugs and receptacles shall be identified for the location and also provide for the connection of the equipment grounding conductor per **501.12**. **(See Figure 6-38)**

Grounding Tip 199: For inspection and preventative maintenance requirements pertaining to bonding and grounding, see Sec.'s 5-6.1 and 5-6.2 in NFPA 410.

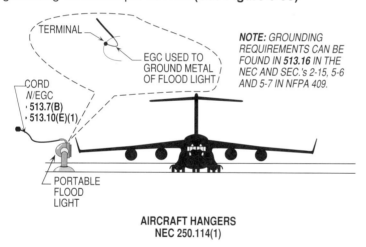

Figure 6-38. This illustration shows a portable flood light bonded and grounded by an equipment grounding conductor per **513.7(B)** and **513.10(E)(1)**.

BULK STORAGE PLANTS
250.114(1); 515.4

A flexible cord shall only be permitted to be used for connection between a portable lamp or other portable utilization equipment and that fixed portion of its supply circuit. The following conditions of use shall be considered where such connections are used:

- Be of a type approved for extra hard-usage.

- Contain, in addition to the conductors of the circuit, an equipment grounding conductor conforming to **250.119** and **400.23**, that is, of green color and used for no other purpose than for grounding.

- Be connected to terminals or to supply conductors in an approved manner.

- Be supported by clamps or other suitable means in such a manner that there will be no tension on the terminal connections.

Grounding Tip 200: For grounding requirements pertaining to bulk storage plants, see **515.16** in the NEC and OSHA 1910.304(F).

- Have suitable seals provided where the flexible cord enters boxes, fittings or enclosures of the explosionproof type.

In Class I, Division 1 locations, flexible cord shall be permitted to be used for that portion of the circuit where the fixed wiring method per **501.4(A)(1)** cannot provide the necessary degree of movement for fixed and mobile electrical utilization equipment.

Grounding Tip 201: For bonding and grounding requirements using the equipment grounding conductor in flexible cords, see **Note 6** to **Table 400.4** and **400.23** in the NEC.

Arcing at exposed contacts shall be prevented in Class I, Divisions 1 and 2 locations. To accomplish this, receptacles are designed so that plug contacts are safely within an explosionproof enclosure when they are electrically engaged, confining arcing, if any, to the receptacle interior per **501.12**.

HEALTH CARE FACILITIES
250.114(1); 517.12

Metal frames of an electrical apparatus used in patient care areas, which is cord-and-plug connected, shall be grounded by an equipment grounding conductor run inside the flexible cord supplying such apparatus. **(See Figure 6-39)**

Figure 6-39. This illustration shows the equipment, which is used to care for the patient, bonded and grounded by the equipment grounding conductor in the flexible cord connecting such equipment to receptacle outlet.

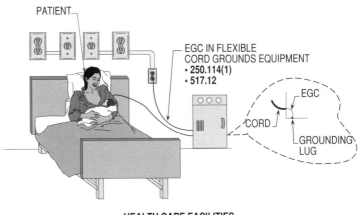

OVER 150 VOLTS-TO-GROUND
250.114(2)

Exposed noncurrent-carrying metal parts of cord-and-plug connected equipment shall be grounded where operating at over 150 volts-to-ground.

IN RESIDENTIAL OCCUPANCIES
250.114(3)

Grounding Tip 202: When cord-and-plug connecting a refrigerator to an existing circuit without an equipment grounding conductor, either a new circuit with an equipment grounding conductor shall be installed or an equipment grounding conductor shall be routed as permitted per **250.130(C)**.

Under any of the following conditions of use, exposed noncurrent-carrying metal parts of cord-and-plug connected equipment shall be grounded in residential occupancies:

- Refrigerators, freezers and air-conditioners
- Clothes-washing, clothes-drying, dish-washing machines
- Information technology equipment
- Sump pumps and electrical aquarium equipment
- Hand-held motor-operated tools, stationary and fixed motor-operated tools and light industrial motor-operated tools

- Motor-operated appliances such as hedge clippers, lawn mowers, snow blowers and wet scrubbers
- Portable handlamps

See **Figure 6-40** for a detailed illustration pertaining to these requirements.

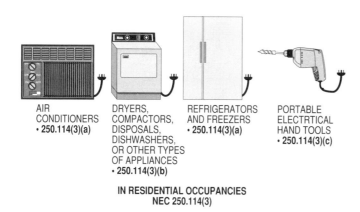

Figure 6-40. This illustration shows electric appliances and hand tools that shall be required to be bonded and grounded for safety of the user.

IN OTHER THAN RESIDENTIAL OCCUPANCIES
250.114(4)

Under any of the following conditions of use, exposed noncurrent-carrying metal parts of cord-and-plug connected equipment shall be grounded in other than residential occupancies:

- Refrigerators, freezers and air-conditioners
- Clothes-washing, clothes-drying, dish-washing machines
- Information technology equipment
- Sump pumps and electrical aquarium equipment
- Hand-held motor-operated tools, stationary and fixed motor-operated tools and light industrial motor-operated tools
- Motor-operated appliances such as hedge clippers, lawn mowers, snow blowers and wet scrubbers
- Cord-and-plug connected appliances used in damp or wet locations
- Cord-and-plug connected appliances used by persons standing on the ground or metal floors or working inside of metal tanks or boilers
- Tools likely to be used in wet or conductive locations
- Portable handlamps

See **Figure 6-41** for a detailed illustration pertaining to these requirements.

Grounding Tip 203: Receptacle outlets supplied by a wiring method without an equipment grounding conductor shall be permitted to have a GFCI receptacle installed per **406.3(D)(2)**.

NOT REQUIRED TO BE GROUNDED
250.114(1) THRU (4), Ex.'s

Exposed noncurrent-carrying metal parts of cord-and-plug connected equipment shall not be required to be grounded when complying with the following conditions:
- Listed equipment
- Motors - where guarded
- Metal frames of electrically heated appliances
- Tools and portable handlamps

Grounding Tip 204: If electric hand tools are not grounded with an equipment grounding conductor, they shall be double insulated per **250.110, Ex. 3; 250.114, Ex.** and OSHA 1910.305 (f)(5)(v)(C)(8).

Figure 6-41. This illustration shows the bonding and grounding requirements for such equipment in other than residential occupancies.

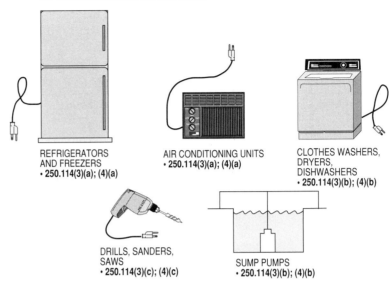

LISTED EQUIPMENT
250.114, Ex.

Listed equipment which is distinctively marked shall not be required to be grounded where protected by a system of double insulation, or its equivalent.

Grounding Tip 205: When working on electrical equipment and the noncurrent-carrying metal parts of such equipment are not bonded and grounded, safety steps must be applied to alleviate shock hazards.

MOTORS - WHERE GUARDED
250.114(2), Ex. 1

The exposed noncurrent-carrying metal parts of cord-and-plug connected motors operating at over 150 volts-to-ground, where guarded, shall not be required to be grounded.

Grounding Tip 206: When applying this permissive rule, care must be exercised not to confuse 250.110, Ex. 1 which covers requirements for equipment fastened-in-place or connected by permanent wiring methods.

METAL FRAMES OF ELECTRICALLY HEATED APPLIANCES
250.114(2), Ex. 2

Metal frames of electrically heated appliances operating at over 150 volts-to-ground shall not be required to be grounded where, exempted by special permission, the frames are permanently and effectively insulated from ground.

TOOLS AND PORTABLE HANDLAMPS
250.114(4), Ex.

Tools and portable handlamps shall not be required to be grounded where used in wet or conductive locations where supplied through an isolating transformer which has an ungrounded secondary of not over 50 volts. **(See Figure 6-42)**

Equipment Grounding and Equipment Grounding Conductors

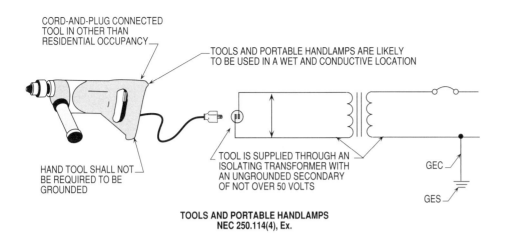

Figure 6-42. This illustration shows a hand tool that is not required to be bonded and grounded due to an isolated supply circuit of 50 volts or less.

NONELECTRIC EQUIPMENT
250.116

Under the following conditions of use, the metal parts of nonelectric equipment shall be grounded:

- Frames and tracks of electrically operated cranes and hoists
- Frames of nonelectrically driven elevator cars to which electrical conductors are attached
- Hand-operated metal shifting ropes or cables of electric elevators

See Figure 6-43 for a detailed illustration pertaining to these requirements.

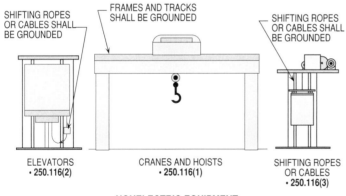

Figure 6-43. This illustration shows the elements of elevators, cranes and hoists which shall be bonded and grounded for safety.

TYPES OF EQUIPMENT GROUNDING CONDUCTORS
250.118(1) THRU (14)

The equipment grounding conductor run with or enclosing the circuit conductors shall be one or more or a combination of the following:

- A copper, aluminum or copper-clad aluminum conductor
- Rigid metal conduit (RMC)
- Intermediate metal conduit (IMC)
- Electrical metallic tubing (EMT)
- Flexible metal conduit (FMC)
- Listed liquidtight flexible metal conduit (LTFC)
- Flexible metallic tubing (FMT)

- Type AC cable (BX)
- Mineral-insulated sheathed cable (MI)
- Type MC cable (MC)
- Cable trays
- Cablebus framework
- Other electrically continuous metal raceways

See Figure 6-44 for a detailed illustration of wiring methods that shall be permitted to be installed as an equipment grounding means.

Figure 6-44. This illustration shows the various types of wiring methods that shall be permitted to be used as an equipment grounding means.

TYPES OF EQUIPMENT GROUNDING CONDUCTORS
NEC 250.118(1) THRU (14)

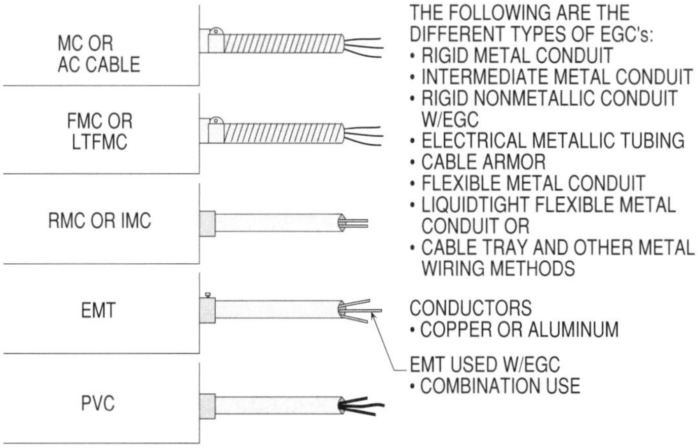

Grounding Tip 207: If two luminaires (lighting fixtures) are connected together with flexible metal conduit, the total length shall not exceed 6 ft. (1.8 m) unless an equipment grounding conductor is routed through the flex.

Sections **250.118(6)(a)** thru **(d)** and **250.118(7)(a)** thru **(e)** permits flexible metal conduit and liquidtight flexible metal conduit to be used as follows:

- The maximum length that shall be permitted to be used in any combination of flexible metal con- duit, flexible metallic tubing and liquidtight flexible metal conduit in any grounding run is 6 ft. (1.8 m) or less.
- The circuit conductors in a 6 ft. (1.8 m) run shall be protected by overcurrent devices not to exceed 20 amps, but that shall also be permitted to be less than 20 amps. The flexible metal conduit or tubing shall be terminated only in fittings listed for grounding. **(See Figure 6-45)**

Figure 6-45. This illustration shows the maximum length permitted for listed flexible metal conduit and listed liquidtight flexible metal conduit to be used as an equipment grounding means.

FLEXIBLE METAL CONDUIT
AND LIQUIDTIGHT FLEXIBLE METAL CONDUIT
NEC 250.118(6)(a) THRU (d)
NEC 250.118(7)(a) THRU (e)

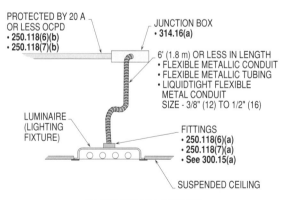

Liquidtight flexible metal conduit shall be permitted to be used in sizes up to 1 1/4 in. (35) that have a total length not to exceed 6 ft. (1.8 m) in the ground return path. Fittings for terminating the flexible conduit shall be listed for grounding. Liquidtight flexible metal conduit 3/8 in. (12) and 1/2 in. (16) trade size shall be protected by overcurrent protection devices rated at 20 amps or less, and sizes of 3/4 in. (21) through 1 1/4 in. (35) in trade size shall be protected by overcurrent protection devices rated at 60 amps or less. **(See Figure 6-46)**

Equipment Grounding and Equipment Grounding Conductors

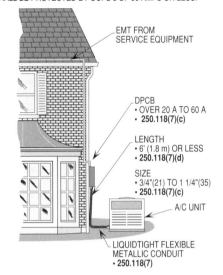

Figure 6-46. This illustration shows the maximum length that listed liquidtight flexible metal conduit shall be permitted to be used as a grounding means without an equipment grounding conductor routed with the circuit conductors inside the flex.

IDENTIFICATION OF EQUIPMENT GROUNDING CONDUCTORS
250.119

Equipment grounding conductors shall be permitted to be installed as bare, covered or insulated, unless otherwise permitted elsewhere in the NEC. Equipment grounding conductors smaller than 6 AWG copper or aluminum shall have a continuous outer finish that is either green or green with one or more yellow stripes, unless otherwise permitted elsewhere in the NEC.

Grounding Tip 208: Equipment grounding conductors smaller than 6 AWG shall be required to be stripped bare or the insulation colored green or green with one or more yellow stripes. In other words, equipment grounding conductors smaller than 6 AWG shall not have their insulation color coded green where it appears exposed in the electrical system.

CONDUCTORS LARGER THAN 6 AWG
250.119(A)

Equipment grounding conductors that are insulated or covered shall be permitted to be installed larger than 6 AWG copper or aluminum if color coded at each end and at every point where the conductor is accessible. The following methods shall be permitted to be used to identify an equipment grounding conductor larger than 6 AWG:

- Stripping the insulation or covering from the entire exposed length
- Coloring the exposed insulation or covering green
- Marking the exposed insulation or covering with green colored tape or green adhesive labels

See Figure 6-47 for a detailed illustration when applying these requirements.

Grounding Tip 209: Copper conductors that are 4 AWG and larger with black insulation shall be permitted to be color coded (marked) with green tape, etc.

Figure 6-47. This illustration shows the methods of identifying and using the equipment grounding conductor when its insulation is other than green in color.

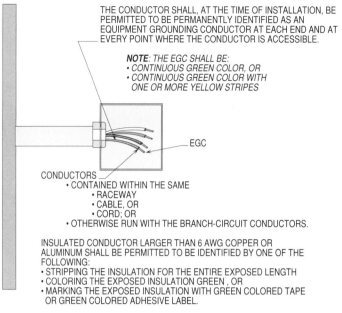

Grounding Tip 210: When installing multiconductor cables and there are not an adequate number of equipment grounding conductors to ground the equipment, other conductors shall be permitted to be identified and used as equipment grounding conductors, if certain conditions of use are met.

MULTICONDUCTOR CABLE 250.119(B)

Where maintenance and supervision is performed only by qualified persons, one or more insulated conductors in a multiconductor cable may at the time of installation be permanently marked to indicate it is an equipment grounding conductor. Note that this shall be done at both ends and at any point where there is access to the equipment grounding conductor. The following methods shall be permitted to be used to identify an equipment grounding conductor in multiconductor cables:

- By removing the insulation along the entire length of the exposed conductor
- Coloring the exposed insulation green
- By a painted or otherwise permanent color such as green tape where the equipment grounding conductor is exposed

Grounding Note: Section **250.28** covers bonding jumpers and the requirements thereof, and **400.7** and **400.23** covers the use of flexible cords for fixed equipment.

See Figure 6-48 for a detailed illustration when applying these requirements.

Figure 6-48. This illustration shows the requirements when other conductors are marked and used as equipment grounding conductors in a multiconductor cable.

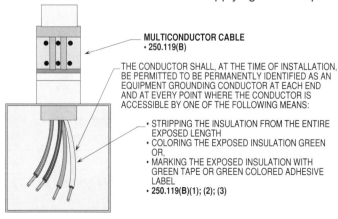

FLEXIBLE CORD
250.119(C)

An uninsulated equipment grounding conductor shall be permitted to be installed in flexible cord but, only, if it's individually covered and the covering shall have a continuous outer finish that is either green or green with one or more yellow stripes. **(See Figure 6-49)**

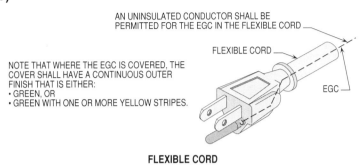

Figure 6-49. This illustration shows the marking requirements for an equipment grounding conductor when used in a flexible cord.

EQUIPMENT GROUNDING CONDUCTOR INSTALLATION
250.120(A) thru (C)

When an equipment grounding conductor is used in a raceway, cable tray, cable armor or cable sheath, and where it is routed in a raceway or cable, the installation shall conform to the applicable provisions of the NEC based upon this type wiring system.

All terminations and joints shall be listed and approved for such use. All joints and fittings shall be made up tight by using tools suitable for the purpose.

Sections **250.130(A)** and **(B)** provides the requirements for using the equipment grounding conductor as a separate conductor.

If the equipment grounding conductors are routed in the hollow spaces of a building that is protected, sizes smaller than 6 AWG shall not be required to be in raceways. **(See Figure 6-50)**

Problem: What size copper equipment grounding conductor is required for a branch-circuit having a 40 A OCPD ahead of its circuit conductors?

Step 1: Sizing EGC
Table 250.122
40 A OCPD requires 10 AWG cu.

Solution: A 10 AWG copper equipment grounding conductor is required per Table 250.122.

Note: See Figure 6-51 for an illustration of this problem.

TABLE 250.122. MINIMUM SIZE EQUIPMENT GROUNDING CONDUCTORS FOR GROUNDING RACEWAY AND EQUIPMENT		
RATING OR SETTING OF AUTOMATIC OVERCURRENT DEVICE IN CIRCUIT AHEAD OF EQUIPMENT, CONDUIT, ETC., NOT EXCEEDING (AMPERES)	SIZE (AWG OR KCMIL)	
	COPPER	ALUMINUM OR COPPER-CLAD ALUMINUM*
15	14	12
20	12	10
30	10	8
40	10	8
60	10	8
100	8	6
200	6	4
300	4	2
400	3	1
500	2	1/0
600	1	2/0
800	1/0	3/0
1000	2/0	4/0
1200	3/0	250
1600	4/0	350
2000	250	400
2500	350	600
3000	400	600
4000	500	800
5000	700	1200
6000	800	1200

Figure 6-50. This illustration shows the Table that shall be used to size equipment grounding conductors based on the size OCPD ahead of the circuit conductors supplying the electrical apparatus.

Grounding Tip 211: Table 250.122 is used when there is an OCPD ahead of the circuit conductors supplying the equipment.

SIZING OF EQUIPMENT GROUNDING CONDUCTORS 250.122; TABLE 250.122

When an equipment grounding conductor is used in a raceway, cable tray, cable armor or cable sheath, and where it is routed in a raceway or cable, the installation and size of the conductor shall conform to the applicable provisions in the NEC. The size of the equipment grounding conductor shall be based on the size of the overcurrent protection device (OCPD) protecting the circuit conductors. **(See Figure 6-51)**

Figure 6-51. This illustration shows the equipment grounding conductor is sized based on the OCPD ahead of the circuit conductors supplying the electrical equipment.

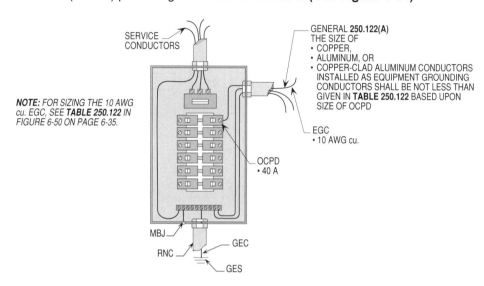

SIZE OF EQUIPMENT GROUNDING CONDUCTORS
NEC 250.122

Grounding Tip 212: If all electrical apparatus is connected to a grounding grid of 1 ohm, there would be zero potential and no voltage differences between equipment enclosures. However, if the service ground is 5 ohms and equipment ground is 10 ohms, there is a voltage difference of 5 ohms (10R − 5R = 5R)

The equipment grounding conductor is the conductor used to ground the noncurrent-carrying metal parts of equipment. It keeps the equipment elevated above ground, at zero potential, and provides a path for the ground-fault current. Equipment grounding conductors protect elements of circuits and equipment and also ensure the safety of personnel from electrical shock. The size of the equipment grounding conductor shall be based on the size of the overcurrent protection device protecting the circuit conductors.

> **Problem 6-52.** What size copper equipment grounding conductor is required to supply a subpanel if the OCPD protecting the feeder-circuit is rated at 400 amps?
>
> **Step 1:** Sizing EGC
> **Table 250.122**
> 400 A OCPD requires 3 AWG cu.
>
> **Solution: The size copper equipment grounding conductor required is 3 AWG copper.**

See Figure 6-52 for an exercise problem when sizing the equipment grounding conductor used to ground the metal frame of a specific type of electrical equipment.

GENERAL
250.122(A)

Grounding Tip 213: Where necessary to comply with **250.4(A)(1) thru (A)(5)**, the equipment grounding conductor may have to be sized larger than **Table 250.122**. For installation restrictions, see **250.120**.

Equipment grounding conductors 6 AWG and smaller shall have their insulation for its entire length colored green or green with one or more yellow stripes. Note that the equipment grounding conductor shall be permitted to be bare under certain conditions of use. The insulation or covering shall be stripped from its entire length or the length exposed.

Equipment Grounding and Equipment Grounding Conductors

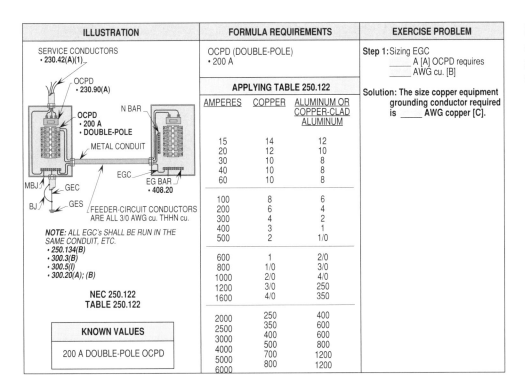

Figure 6-52. This illustration shows an exercise problem for sizing an equipment grounding conductor based on the OCPD ahead of the circuit conductors.

Equipment grounding conductors that are insulated or covered shall be permitted to be installed larger than 6 AWG copper or aluminum if color coded at each end and at every point where the conductor is accessible. The following methods shall be permitted to be used to identify an equipment grounding conductor larger than 6 AWG:

- Stripping the insulation or covering from the entire length exposed.
- Coloring the exposed insulation or covering green.
- Marking the exposed insulation or covering with green colored tape or green adhesive labels.

See Figure 6-53 for a detailed illustration when applying these requirements.

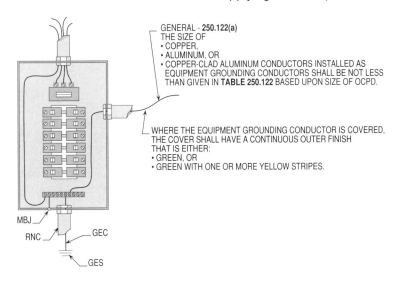

Figure 6-53. This illustration shows the marking techniques that shall be permitted to be used when identifying the equipment grounding conductor routed through a raceway or cable.

INCREASED IN SIZE
250.122(B)

Where ungrounded (phase) conductors are increased in size, equipment grounding conductors (if installed) shall be increased and sized proportionately according to the circular mil area of the ungrounded (phase) conductors.

6-37

Grounding Tip 214: Poor voltage drop (VD) to a piece of equipment can cause the components to overheat and deteriorate its working elements. For example, a motor is not supposed to have the voltage at its terminals drop below 10 percent. For this reason, when the ungrounded (phase) conductors are increased due to poor voltage drop problems, the equipment grounding conductor shall also be increased in size.

Problem 6-54. What size equipment grounding conductor is required for a feeder-circuit which has poor voltage drop when applying the following known values:

- 175 A OCPD
- 2/0 AWG THWN copper conductors
- 140 A continuous load

Step 1: Calculating EGC
250.122(B); Table 250.122
175 A OCPD requires 6 AWG cu.

Step 2: Selecting conductors based on OCPD
Table 310.16
175 A load requires 2/0 AWG cu.
Increased to 3/0 AWG due to VD

Step 3: Finding CM of conductors
250.122(B); Table 8, Ch. 9
2/0 AWG cu. (175 A OCPD) = 133,100 CM
3/0 AWG cu. (due to VD) = 167,800 CM
6 AWG (EGC) = 26,240 CM

Step 4: Finding multiplier based on VD
250.122(B)
Multiplier = 167,800 CM ÷ 133,100 CM
Multiplier = 1.2607062

Step 5: Calculating EGC based on VD
EGC = 26,240 CM x 1.2607062
EGC = 33,081 CM

Step 6: Selecting EGC based on VD
Table 8, Ch. 9
33,081 CM requires 4 AWG cu. (41,740 CM)

Solution: **The size equipment grounding conductor based upon voltage drop is 4 AWG copper and not 6 AWG copper, which is based on size of OCPD.**

See Figure 6-54 for an exercise problem when determining the size equipment grounding conductor required for a feeder-circuit with poor voltage drop.

MULTIPLE CIRCUITS
250.122(C)

Single equipment grounding conductors that are run with multiple circuits in the same raceway shall be permitted to be sized, based on the largest OCPD protecting the circuit conductors in the raceway or cable.

Grounding Tip 215: When selecting circular mil (CM) or KCMIL ratings of conductors, refer to Table 8, Ch. 9 in the NEC.

Problem 6-55. What size equipment grounding conductor is required for multiple circuits in the same raceway with the following known values:

- 3 - 10 AWG copper conductors (30 A OCPD)
- 2 - 12 AWG copper conductors (20 A OCPD)
- 3 - 14 AWG copper conductors (15 A OCPD)

Step 1: Sizing EGC
250.122(C)
30 A OCPD requires 10 AWG cu.

Solution: **A 10 AWG copper equipment grounding conductor is required for multiple circuits in the same raceway.**

See Figure 6-55 for an exercise problem when sizing the equipment grounding conductor required for multiple circuits in the same raceway.

Equipment Grounding and Equipment Grounding Conductors

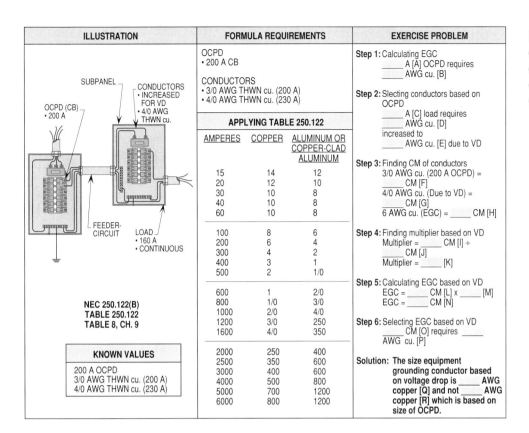

Figure 6-54. This illustration shows an exercise problem for increasing the size of the equipment grounding conductor due to poor voltage drop in a feeder-circuit.

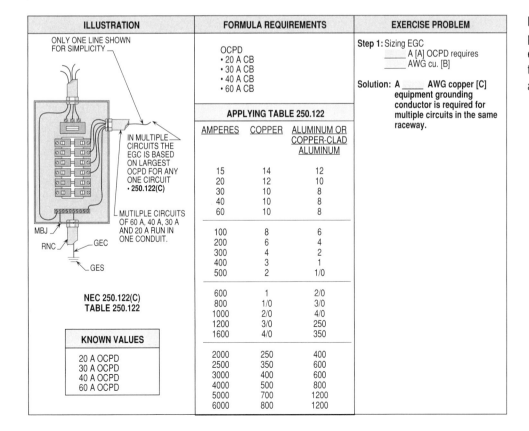

Figure 6-55. This illustration shows the procedure for sizing and selecting the equipment grounding conductor when there are multiple circuits routed through a raceway system.

6-39

MOTOR CIRCUITS
250.122(D)

If the OCPD is a motor short-circuit protector or an instantaneous trip breaker, see **430.52** and **Table 430.52** for sizing such device. The size of the equipment grounding conductor in such cases shall be permitted to be based upon the rating of the protective device. In any design, the equipment grounding conductor never has to be larger than the largest ungrounded (phase) conductor supplying the motor. **(See Figure 6-56)**

Figure 6-56. This illustration shows that under certain design conditions that the size of the equipment grounding conductor shall be permitted to be determined by the rating of the Motor Overload Protective Device (MOLPD) in the controller.

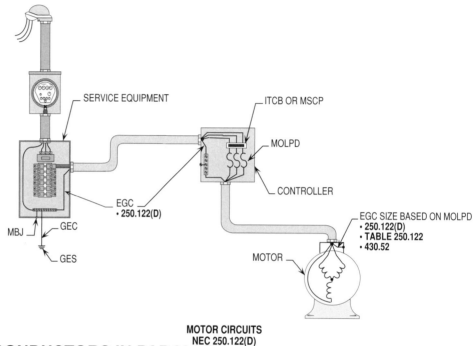

CONDUCTORS IN PARALLEL
250.122(F)

Grounding Tip 216: It is never permitted to down size the equipment grounding conductor in a circuit, even if such conductor is used as a circuit bonding and grounding conductor in series, in parallel or as an individual conductor.

Circuit conductors that are connected in parallel and routed through separate conduits shall have a separate equipment grounding conductor run in each conduit. The equipment grounding conductor shall be sized from **Table 250.122**, based on the rating of the OCPD protecting the circuit.

Problem 6-57. What size copper equipment grounding conductors are required to ground the metal parts of a piece of equipment supplied by three 3/0 AWG THWN copper conductors per phase connected to a 600 amp OCPD in the panelboard?

Step 1: Sizing EGC's
250.122(F); Table 250.122
600 A OCPD requires 1 AWG cu.

Solution: **The size equipment grounding conductors are required to be 1 AWG copper for conductors which are installed in parallel.**

See Figure 6-57 for an exercise problem when sizing the equipment grounding conductors required for conductors which are installed in parallel.

Note that the circular mil rating of the equipment grounding conductor must not be divided by the number of conductors in parallel and this CM value run in each conduit. A full size equipment grounding conductor shall be pulled in each conduit along with the ungrounded (phase) conductors and grounded (neutral) conductor if used.

Equipment Grounding and Equipment Grounding Conductors

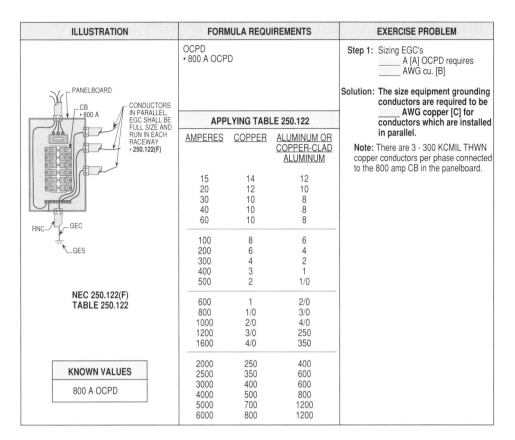

Figure 6-57. This illustration shows an exercise problem for sizing the equipment grounding conductors when conductors are installed in parallel.

EQUIPMENT GROUNDING CONDUCTOR CONTINUITY
250.124(A); (B)

Grounding type plugs shall be so constructed that the grounding prong of the plug will make contact before the circuit conductors make and break contact after the circuit conductors break. A grounding prong longer than the circuit prongs would satisfy these requirements. Interlocked equipment so constructed as to accomplish the same result is also acceptable per **250.124(A)**.

This requirement would cover the unusual case of an automatic disconnect being placed in the equipment grounding conductor. (The equipment grounding conductor should not be confused with the grounded circuit conductor, the neutral) If such a disconnect is placed in the equipment grounding conductor, the disconnect shall open all circuit conductors at the same time that it opens the equipment grounding conductor per **250.124(B)**. **(See Figure 6-58)**

IDENTIFICATION OF WIRING DEVICE TERMINALS
250.126

The terminal for the connection of the equipment grounding conductor shall be identified by one of the following means:

- A green, not readily removable terminal screw with a hexagonal head
- A green, hexagonal, not readily removable terminal nut
- A green pressure wire connector. If the terminal for the equipment grounding conductor is not visible, the conductor entrance hole shall be marked with the word green or ground, the letters G or GR or otherwise identified by a distinctive green color. If the terminal for the equipment grounding conductor is readily removable, the area adjacent to the terminal shall be similarly marked.

Grounding Tip 217: When identifying the terminals for the grounded (neutral) conductor, see **200.9**. When identifying the terminals for the equipment grounding conductor, see **250.126**. To ensure that circuit conductors are connected to terminals to provide proper polarity, see **200.10**.

Figure 6-58. This illustration shows the requirements for making and breaking continuity of the equipment grounding conductor in an electric circuit along with its associated apparatus.

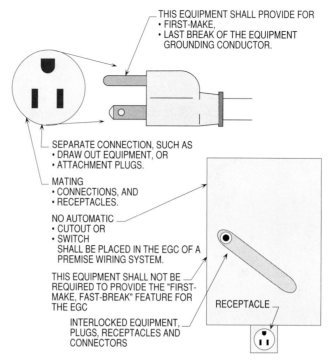

EQUIPMENT GROUNDING CONDUCTOR CONTINUITY
NEC 250.124(A); (B)

Grounding Tip 218: Care must be exercised when connecting circuit conductors to the terminals of receptacles. For example, the equipment grounding conductor may be accidentally connected to the neutral terminal or the grounded (neutral) conductor connected to the ungrounded (phase) conductor terminal. When this happens, the proper polarity is not maintained and this is not only dangerous, but can cause equipment to malfunction.

Grounding Note: Terminals for equipment grounding conductors shall be identified. In general, green hexagonal (six-sided) screws or nuts shall be acceptable as are entrance holes marked "G," "GR," or marked with the standard grounding symbol as shown below.

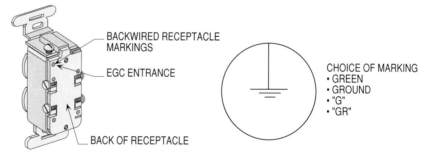

See **Figure 6-59** for a detailed illustration when applying these requirements.

Figure 6-59. This illustration shows the methods in which equipment grounding terminals shall be identified so as to ensure correct termination of the equipment grounding conductor to a receptacle.

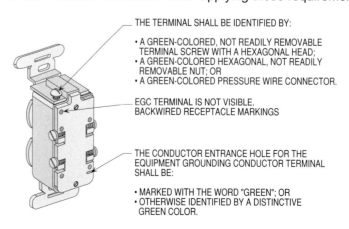

IDENTIFICATION OF WIRING DEVICE TERMINALS
NEC 250.126

Name _____   Date _____

Chapter 6
Equipment Grounding and Equipment Grounding Conductors

	Section	Answer

1. Exposed metal parts of fixed equipment shall be grounded where installed in Class I, Divisions 1 and 2 locations. _____   T   F

2. Exposed metal parts of fixed equipment shall not be required to be grounded where installed for intrinsically safe systems. _____   T   F

3. All noncurrent-carrying metal parts of fixed or portable electrical equipment for commercial garages shall be grounded, regardless of the voltage. _____   T   F

4. Flexible metal conduit shall be permitted to be used as the sole equipment grounding path where installed in service stations. _____   T   F

5. Listed equipment which is distinctively marked shall be required to be grounded, where protected by a system of double insulation, or its equivalent. _____   T   F

6. Frames for two-wire DC switchboards shall be required to be grounded, where effectively insulated from ground. _____   T   F

7. Exposed noncurrent-carrying metal parts for elevators and cranes shall be required to be grounded. _____   T   F

8. Exposed noncurrent-carrying metal parts of motion picture projection equipment shall not be required to be grounded. _____   T   F

9. Motor-operated pumps including the submersible type shall be required to be grounded. _____   T   F

10. Flexible cords installed in Class II, Division 1 and 2 locations shall be of a type approved for hard-usage. _____   T   F

11. A flexible cord can only be used for connection between a portable lamp or other portable utilization equipment where installed in bulk storage plants. _____   T   F

12. Exposed noncurrent-carrying metal parts of cord-and-plug connected equipment shall be required to be grounded where operating at over 50 volts-to-ground. _____   T   F

13. Tools and portable handlamps shall be required to be grounded where used in wet or conductive locations where supplied through an isolating transformer which has an ungrounded secondary of not over 50 volts. _____   T   F

14. Equipment grounding conductors shall be permitted to be installed as a bare conductor. _____   T   F

15. The size of the equipment grounding conductor shall be based on the size of the overcurrent protection device protecting the circuit conductors. _____   T   F

6-43

Section	Answer	
_____	T F	**16.** Equipment grounding conductors smaller than 6 AWG shall have their insulation for its entire length colored green or green with one or more yellow stripes.
_____	T F	**17.** Equipment grounding conductors larger than 6 AWG copper shall not be required to be colored coded at each end and every point where the conductor is accessible.
_____	T F	**18.** Single equipment grounding conductors that are run with multiple circuits in the same raceway shall be permitted to be sized based on the smallest OCPD protecting the circuit conductors in the raceway.
_____	T F	**19.** Grounding type plugs shall be so constructed that the grounding prong of the plug will make contact before the circuit conductors make and break contact after the circuit conductors break.
_____	T F	**20.** A green pressure wire connector shall be permitted to be installed for the connection of the equipment grounding conductor.
_____	_____	**21.** Exposed metal parts of fixed equipment shall be grounded where within _____ ft. vertically or 5 ft. horizontally of ground.
_____	_____	**22.** Exposed metal parts of fixed equipment shall be grounded where located in _____ or damp locations.
_____	_____	**23.** A bonding jumper for liquidtight flexible metal conduit shall not be required where installed in _____ ft. or less lengths with listed fittings for grounding in Class I, Divisions 1 and 2 locations.
_____	_____	**24.** A bonding jumper for liquidtight flexible metal conduit shall not be required where the overcurrent protection device in the circuit is limited to _____ amps or less in Class II, Divisions 1 and 2 locations.
_____	_____	**25.** The equipment grounding terminal buses or the normal and essential branch-circuit panelboards serving the same individual patient vicinity shall be bonded together with an insulated continuous conductor not smaller than _____ AWG copper.
_____	_____	**26.** Exposed metal parts of fixed equipment shall be grounded where equipment operates at over _____ volts-to-ground with any terminal or lug.
_____	_____	**27.** Distribution apparatus such as transformer and capacitor cases shall not be required to be grounded where, mounted on wooden poles, at a height of at least _____ ft. above ground or grade level.
_____	_____	**28.** The frames of stationary motors shall not be required to be grounded where the motor operates at over _____ volts-to-ground with any terminal.
_____	_____	**29.** Pendant lampholders installed in motion picture studios shall not be required to be grounded where the circuit is not over _____ volts-to-ground.

	Section	Answer

30. Small metal parts for outline lighting systems of less than _____ in. in any direction that are not likely to become energized and that are spaced at least 3/4 in. from neon tubing shall not be required to be grounded.

31. Listed flexible metal conduit shall be permitted to be used as a bonding jumper for outline lighting systems where installed in lengths less than __ ft.

32. At least a _____ AWG copper equipment grounding conductor shall be used to bond and ground the transformer supplying a Class 1 control circuit.

33. Liquidtight flexible metal conduit shall be permitted to be used and sized up to _____ in. if the total length does not exceed 6 ft. in the ground return path.

34. Liquidtight flexible metal conduit 3/8 in. to 1/2 in. trade size shall be protected by overcurrent protection devices rated at _____ amps or less.

35. Liquidtight flexible metal conduit 3/4 in. to 1 1/4 in. in trade size shall be protected by overcurrent protection devices rated at _____ amps or less.

36. Equipment grounding conductors smaller than _____ AWG copper or aluminum are required to have a continuous outer finish that is either green or green with one or more yellow stripes.

37. Equipment grounding conductors that are insulated or covered shall be permitted to be installed in sizes No. _____ or larger copper if color coded at each end and at every point where the conductor is accessible.

38. Equipment grounding conductors smaller than _____ AWG shall be protected from physical damage by a raceway or cable armor.

39. No _____ cutout or switch shall be placed in the equipment grounding conductor of a premises wiring system unless the opening of the cutout or switch disconnects all sources of energy.

40. A ___, not readily removable terminal screw with a hexagonal head shall be permitted to be used for connection of the equipment grounding conductor.

41. What size copper equipment grounding conductor is required to ground a piece of equipment when supplied by a 20 amp OCPD?

42. What size copper equipment grounding conductor is required to ground a piece of equipment when supplied by a 50 amp OCPD?

43. What size copper equipment grounding conductor is required to supply a subpanel if the OCPD protecting the feeder-circuit is rated at 100 amps?

44. What size copper equipment grounding conductor is required to supply a subpanel is the OCPD protecting the feeder-circuit is rated at 300 amps?

Section   Answer

_____ _____ **45.** What size copper equipment grounding conductor is required for a feeder-circuit which has poor voltage drop when applying the following known values:

- 100 A OCPD
- 3 AWG THWN copper conductors
- 80 A continuous load

_____ _____ **46.** What size copper equipment grounding conductor is required for a feeder-circuit which has poor voltage drop when applying the following known values:

- 150 A OCPD
- 1/0 AWG THWN copper conductors
- 120 A continuous load

_____ _____ **47.** What size equipment grounding conductor is required for multiple circuits in the same raceway with the following known values:

- 3 - 8 AWG copper conductors (50 A OCPD)
- 3 - 10 AWG copper conductors (30 A OCPD)
- 3 - 14 AWG copper conductors (15 A OCPD)

_____ _____ **48.** What size equipment grounding conductor is required for multiple circuits in the same raceway with the following known values:

- 3 - 6 AWG copper conductors (60 A OCPD)
- 3 - 10 AWG copper conductors (30 A OCPD)
- 3 - 12 AWG copper conductors (20 A OCPD)

_____ _____ **49.** What size copper equipment grounding conductors are required to ground the metal parts of a piece of equipment supplied by two 3/0 KCMIL THWN copper conductors per phase connected to a 400 amp OCPD in the panelboard?

_____ _____ **50.** What size copper equipment grounding conductors are required to ground the metal parts of a piece of equipment supplied by two 250 KCMIL THWN copper conductors per phase connected to a 500 amp OCPD in the panelboard?

7

METHODS OF EQUIPMENT GROUNDING

The grounding of equipment and conductor enclosures as well as the grounding of one conductor of an electrical power source has been practiced since the use of electricity. However, the use of ungrounded systems have been used since electricity was first used in industrial plants.

Ungrounded systems unlike grounded systems are subjected to relatively severe transient overvoltages that are two to three times normal to ground. If one system has one conductor grounded (grounded system), the value of these transient overvoltages, as they develop, are greatly reduced.

On a grounded and ungrounded system, the equipment supplied shall have its enclosure connected to ground with an equipment grounding conductor means to keep such enclosure as near ground potential as possible and help reduce electric shock and fire hazards to a minimum. This chapter covers grounding techniques, when applied, will help to provide a degree of such protection.

EQUIPMENT GROUNDING CONDUCTOR CONNECTIONS
250.130

Equipment grounding conductor connections for a separately derived system and for service equipment shall be made in accordance with **250.30(A)(1)**. The equipment grounding conductor connections shall be made at service equipment and separately derived systems as follows:

- For grounded systems
- For ungrounded systems

FOR GROUNDED SYSTEMS
250.130(A)

The equipment grounding conductor shall be bonded to the grounded (neutral) conductor and the grounding electrode conductor at the service equipment neutral busbar terminal. The grounded conductor (may be a neutral) shall be installed and connected to the grounding electrode system per **250.52(A)(1)** thru **(A)(6)**. Note that the grounded (neutral) conductor shall be connected to the busbar where the grounding electrode conductor is terminated per **250.24(A)**. **(See Figure 7-1)**

Figure 7-1. This illustration shows the utility transformer and the service panel both connected to earth ground for safety.

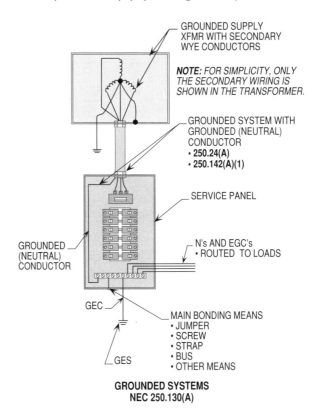

FOR UNGROUNDED SYSTEMS
250.130(B)

The equipment grounding conductor shall be bonded to the grounding electrode conductor at the service equipment neutral busbar terminal. A grounded (neutral) conductor is not present to be connected to ground. Equipment grounding conductors installed per **250.118** shall be permitted to be used to bond and connect all metal noncurrent-carrying parts of the wiring system to the grounding electrode conductor at the service equipment. **(See Figure 7-2)**

Grounding Tip 219: There are two exceptions in the NEC that permits an equipment grounding conductor to be routed outside the raceway or cable and not routed with the circuit conductors. These rules can be found in **250.102(E)** and **250.130(C)**.

NONGROUNDING RECEPTACLE REPLACEMENT OR BRANCH-CIRCUIT EXTENSIONS
250.130(C)(1) THRU (C)(5)

The basic rule for replacement of receptacles is that grounding-type receptacles shall be used to replace existing receptacles where there is an equipment grounding conductor routed with the branch-circuit. A nongrounding-type receptacle shall be used with a branch-circuit that has no equipment grounding conductor or grounding means. The equipment grounding conductor shall connect the green grounding terminal of the receptacle to the grounding electrode conductor or to the closest metal water pipe per **250.104(A), 250.50** and **250.50(A)(1)**. **(See Figure 7-3)**

Methods of Equipment Grounding

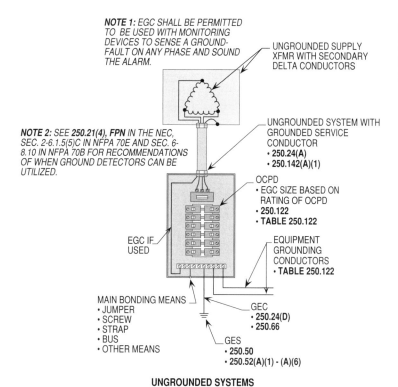

Figure 7-2. This illustration shows an ungrounded transformer being used to supply uninterrupted power to an industrial plant.

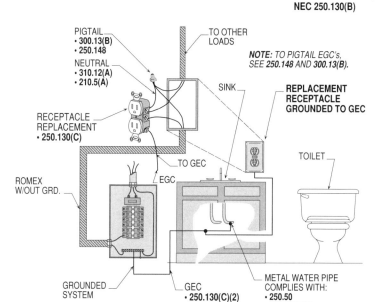

Figure 7-3. This illustration shows one method in which a grounding type receptacle shall be permitted to be grounded to the grounding electrode conductor. Note that the branch-circuit is not equipped with an equipment grounding conductor.

The NEC permits five methods by which a nongrounding type receptacle shall be permitted to be replaced with a circuit not having an equipment grounding conductor and they are as follows:

- Install a nongrounding type receptacle at each receptacle location.
- A GFCI receptacle protecting each outlet at receptacle location.
- A GFCI receptacle protecting a single outlet and additional outlets downstream.
- An equipment grounding conductor routed from the receptacle to a metal water pipe per **250.50**.
- A GFCI CB protecting all outlets.

See Figure 7-4 for a detailed illustration when applying these requirements.

Grounding Tip 220: It is permissible to install grounding type receptacles downstream from a fed-thru GFCI receptacle that replaces a nongrounding type receptacle upstream per **406.3(D)(3)(c)**. Wiring method of branch-circuit does not have an equipment grounding conductor means.

Figure 7-4. This illustration shows methods which shall be permitted to be used to replace receptacles when there is no equipment grounding conductor available in the branch-circuit wiring method.

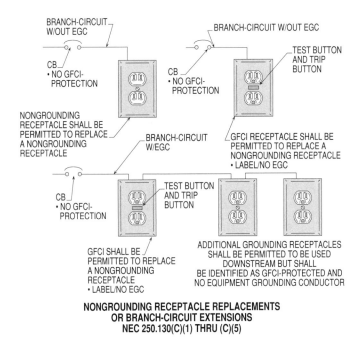

SHORT SECTIONS OF RACEWAYS
250.132

Where required to be grounded, isolated sections of metal raceways or cable armor shall be grounded in accordance with **250.134**.

EQUIPMENT FASTENED-IN-PLACE OR CONNECTED BY PERMANENT WIRING METHODS (FIXED) - GROUNDING
250.134

Grounding Tip 221: Generally, when a piece of electrical equipment is permanently wired, bolted or screwed down, it's considered as fastened-in-place.

Noncurrent-carrying metal parts of equipment, raceways, cables or other enclosures shall be grounded by one of the following methods where required to be grounded for the safety of circuits, equipment and personnel:

- Equipment grounding conductors
- Metal raceways
- Metal clad cables, etc.

See Figure 7-5 for a detailed illustration when applying these requirements.

EQUIPMENT GROUNDING CONDUCTOR TYPES
250.134(A)

The following wiring methods shall be permitted to be installed as the equipment grounding conductor to ground electrical enclosures and equipment cases:

- Rigid metal conduit
- Electrical metallic tubing
- Armored cable (BX)
- Approved metal conduit or cables

See Figure 7-6 for a detailed illustration when applying these requirements.

Methods of Equipment Grounding

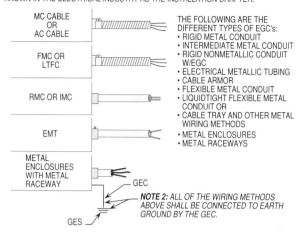

Figure 7-5. This illustration shows wiring methods that shall be permitted to be used to earth ground equipment which is fastened-in-place.

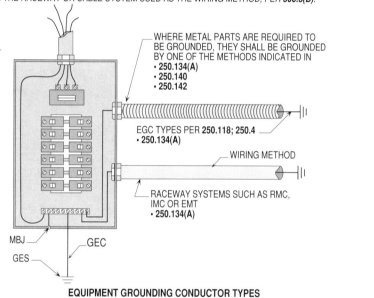

Figure 7-6. This illustration shows wiring methods that shall be permitted to be used to earth ground the noncurrent-carrying metal parts of electrical related equipment and enclosures.

WITH CIRCUIT CONDUCTORS
250.134(B)

An equipment grounding conductor contained within the same raceway, cable or cord shall be permitted to be carried along with circuit conductors and installed to ground electrical enclosures and equipment. Equipment grounding conductors shall be permitted to be bare, covered or insulated. Insulated equipment grounding conductors shall be identified green or green with one or more yellow stripes. For more detailed information on this subject (with circuit conductors), see **300.3(B), 300.5(I) and 300.20**. **(See Figure 7-7)**

Grounding Tip 222: Section **250.102(E)** permits an equipment grounding conductor to be routed outside the raceway that houses the circuit conductors. However, this grounding conductor is used in this manner as an equipment bonding jumper in lengths not to exceed 6 ft. (1.8 m).

EQUIPMENT CONSIDERED EFFECTIVELY GROUNDED
250.136

Equipment that is bolted or otherwise fastened to a metal rack or other metal support and the rack is grounded. The equipment is considered to be effectively grounded by the rack. Likewise, if an elevator machine is grounded, the elevator cab frame is considered to be effectively grounded through the hoisting cables which are grounded.

Figure 7-7. This illustration shows the methods by which the equipment grounding conductors shall be routed to ensure a low-impedance path for clearing ground-faults.

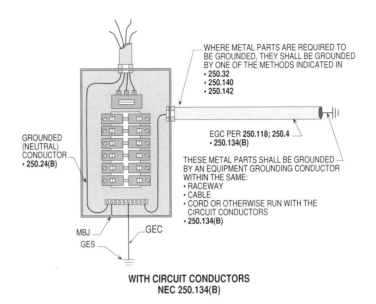

EQUIPMENT SECURED TO GROUNDED METAL SUPPORTS
250.136(A)

Even if electrical equipment is secured and making good electrical contact with metal racks or supports, it shall be grounded by one of the means indicated in **250.134(A)**. Structural steel shall not be used as a conductor for equipment grounding. Note that it is required that the equipment grounding conductor be routed with the circuit conductors per **250.134(B)**. **(See Figure 7-8)**

Figure 7-8. This illustration shows metal racks grounded by an equipment grounding conductor and the racks used to ground the metal of enclosures mounted to the racks.

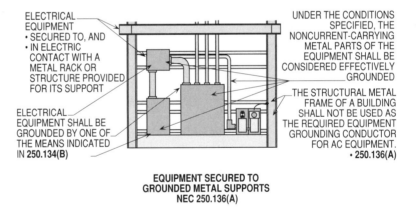

METAL CAR FRAMES
250.136(B)

The metal car frames of elevators and similar devices are considered to be adequately grounded if attached to a metal cable running over or attached to a metal drum that is well grounded according to the grounding requirements of **250.134**. **(See Figure 7-9)**

Figure 7-9. This illustration shows grounded metal cables used to ground the metal frame of the elevator car.

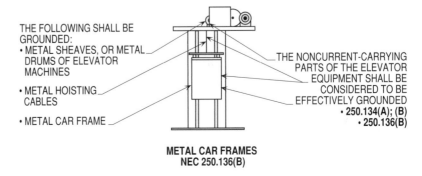

CORD-AND-PLUG CONNECTED EQUIPMENT
250.138

The metal cases of portable cord-and-plug connected stationary equipment shall be permitted to be grounded by any one of the following three wiring methods:

- By means of metal enclosure
- By means of an equipment grounding conductor
- By means of a separate flexible wire or strap

Metal enclosures housing conductors shall be permitted to be used as a grounding means if an approved grounding type attachment plug is utilized with one fixed member making contact for the purpose of bonding and grounding the metal enclosure. However, the attachment plug shall be of a type which is approved for grounding from a grounding type receptacle and such cord is required to carry an equipment grounding conductor. One end shall be attached to the grounding terminal of the attachment plug and the other end shall terminate to the frame of the portable equipment. For proper grounding, it is important that the metal enclosure of such conductors be attached to the attachment plug and to the equipment by approved connectors.

BY MEANS OF AN EQUIPMENT GROUNDING CONDUCTOR
250.138(A)

The equipment grounding conductor in a cord is used to connect the grounding prong of a plug to ground the equipment. This equipment grounding conductor shall be permitted to be bare or insulated with green color or green with one or more yellow stripes. Such grounding conductor shall be routed in the cable or flexible cord assembly, provided that it terminates in an approved grounding type attachment plug having a fixed grounding type contact member. **(See Figure 7-10)**

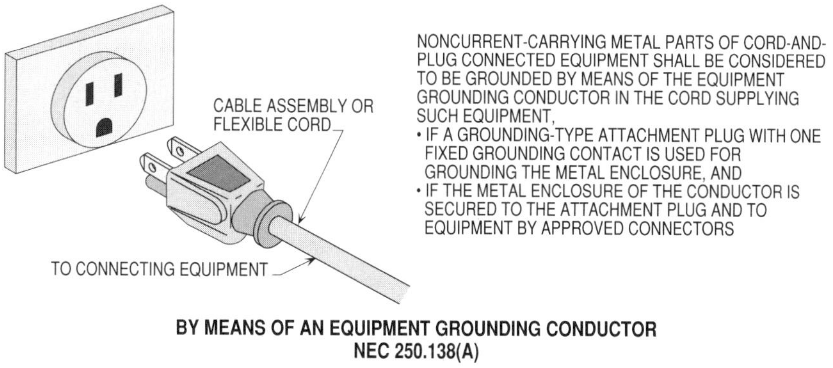

Grounding Tip 223: When using the equipment grounding conductor inside a flexible cord to ground equipment, such equipment grounding conductor shall comply with the requirements in **Table 400.4** and **Note 6** to the Table. Also, review **400.23** and **400.33**.

Figure 7-10. This illustration shows a flexible cord with an attachment cap and equipment grounding conductor used to ground the connected equipment to the receptacle and its supply.

The **Ex.** to **250.138(A)** permits a movable self-restoring type such as an approved attachment plug that is equipped with a hinged grounding prong with a spring. Note that this spring can be folded out of the way and still restore itself back to the normal position on the grounding type receptacle.

BY MEANS OF A SEPARATE FLEXIBLE WIRE OR STRAP
250.138(B)

A separate flexible wire or strap shall be permitted to be used to ground portable equipment. This strap should be protected from physical damage as well as practical to ensure proper bonding and grounding. **(See Figure 7-11)**

Figure 7-11. This illustration shows a separate flexible wire or strap used to ground equipment under certain conditions of use.

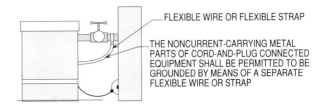

BY MEANS OF A SEPARATE FLEXIBLE WIRE OR STRAP
NEC 250.138(B)

Grounding Tip 224: The grounded (neutral) conductor can only be used as a neutral plus an equipment grounding conductor when the branch-circuit wiring method is existing.

FRAMES OF RANGES AND CLOTHES DRYERS
250.140

Existing branch-circuit installations for the frames or ranges, cooktops, ovens, clothes dryers and junction boxes or outlet boxes that are part of the circuit shall be permitted to be bonded and grounded with the grounded (neutral) conductor under the following conditions:

- The supply circuit is 120/240 volt, single-phase, three-wire or 208Y/120 volt derived from a three-phase, four-wire, wye-connected system.

- The grounded (neutral) conductor installed is no smaller than 10 AWG copper or 8 AWG aluminum.

- The grounded (neutral) conductor is insulated and part of NM cable (romex) or PVC conduit.

- The grounded (neutral) conductor is bare and part of a service-entrance cable.

A new branch-circuit installation for ranges, cooktops, ovens, clothes dryers, including junction boxes or outlet boxes that are part of the circuit shall be permitted to be bonded and grounded with an equipment grounding conductor. However, an isolated grounded (neutral) conductor shall also be installed. For further information concerning this type of installation, see **250.114, 250.134** and **250.138**. **(See Figure 7-12)**

Figure 7-12. This illustration shows the procedure for installing a range, cooktop, oven or dryer on existing and new installations.

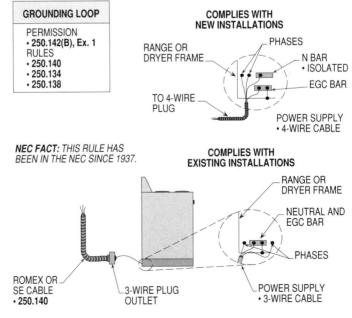

USE OF GROUNDED CIRCUIT CONDUCTOR FOR GROUNDING EQUIPMENT
250.142

Under certain conditions, all metal parts of enclosures used to install the service equipment shall be permitted to be grounded by the grounded (neutral) conductor on the supply side of the system. The service weatherhead, service raceway, service meter base and the service equipment enclosure are included when using this type of grounded system.

The grounded (neutral) conductor shall be installed and isolated from the other system circuit conductors and from metal enclosures or metal conduits when installed on the load side of the electrical system.

SUPPLY-SIDE EQUIPMENT
250.142(A)(1) THRU (A)(3)

The grounded (phase or neutral) conductor shall be permitted to be used as a current-carrying conductor and grounding means on the supply side of the service disconnecting means and secondary side of a separately derived system as follows:

- On the supply side of the service equipment per **250.142(A)(1)** and **250.24(B)(1)**.
- On the supply side of the main service disconnecting means for separate buildings and structures per **250.142(A)(2)** and **250.32(B)(1)** and **(B)(2)**.
- On the supply side of the disconnect or overcurrent protection device of a separately derived system per **250.142(A)(3)** and **250.30(A)(6)(a)**.

See Figure 7-13 for a detailed illustration when applying these requirements.

Grounding Tip 225: For the grounded (phase or neutral) conductor to be used as a current-carrying neutral and equipment grounding conductor to clear ground-faults, there shall not be an OCPD ahead of the supply conductors from the utility transformer.

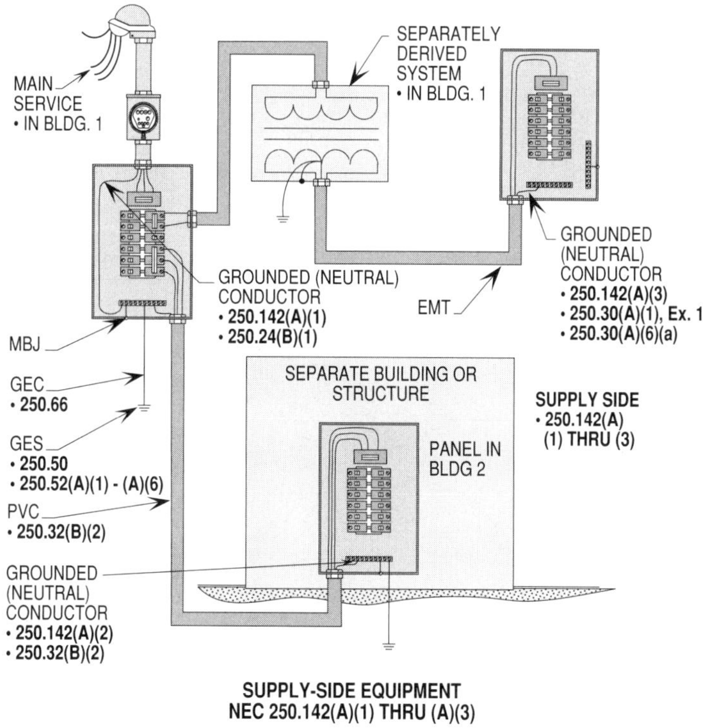

Figure 7-13. This illustration shows three installations where the grounded (neutral) conductor shall be permitted to be used as a current-carrying neutral plus equipment grounding conductor.

Grounding Tip 226: Where the grounded (neutral) conductor is used as a current-carrying neutral conductor plus an equipment grounding conductor, there must be an OCPD ahead of the supplying conductors on the load side.

LOAD-SIDE EQUIPMENT
250.142(B)

The grounded (neutral) conductor shall not be used as an equipment grounding conductor on the load side except as follows:

- Frame of ranges, wall-mounted ovens, counter-mounted cooking units and clothes dryers per **250.140**. (Only for existing branch-circuits)

- Where one or more buildings or structures are supplied from a common AC grounded service, each grounded service at each individual building or structure shall be separately grounded per **250.32**.

- If no service ground-fault protection is provided, all meter enclosures located near the service disconnecting means shall be permitted to be grounded by the grounded (neutral) conductor on the load side of the service disconnect. The grounded (neutral) conductor shall not be smaller than the size specified in **Table 250.122**.

- DC systems shall be permitted to be grounded at the first disconnecting means or overcurrent protection device per **250.164(B)(2)**.

See **Figure 7-14** for a detailed illustration when applying these requirements.

Figure 7-14. This illustration shows an installation where the grounded (neutral) conductor is used as a neutral conductor plus an equipment grounding conductor.

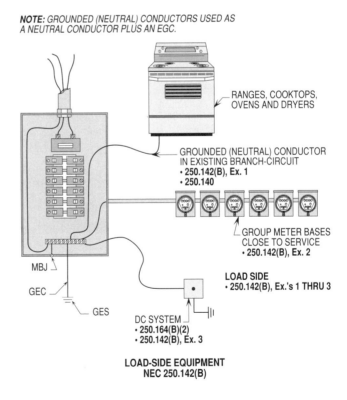

Grounding Tip 227: It is the responsibility of the electrician to ensure that each wiring method supplying equipment, etc., is equipped with an equipment grounding means.

MULTIPLE CIRCUIT CONNECTIONS
250.144

There will be cases where the equipment will be required to be grounded, that is, it is supplied by one or more circuits of grounded wiring systems on the premises, in this case, separate equipment grounding conductors shall be supplied from each system as specified in **250.134** and **250.138**. **(See Figure 7-15)**

Methods of Equipment Grounding

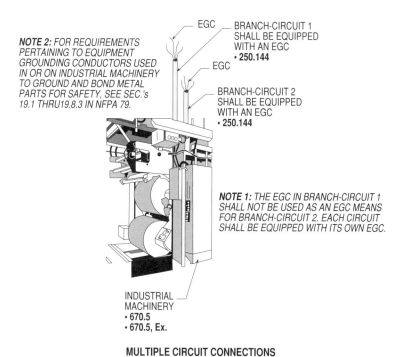

Figure 7-15. This illustration shows that branch-circuits 1 and 2 are required to be provided with an equipment grounding conductor.

CONNECTING RECEPTACLE GROUNDING TERMINAL TO BOX
250.146

A good connection to ground shall be made by connecting the receptacle grounding terminals to the box. A bonding jumper shall be permitted to be used to connect the grounding terminal of a grounding-type receptacle to a metal grounded box. The following are four exceptions when a connection shall be permitted to be made without using a bonding jumper:

- Surface-mounted boxes
- Contact devices or yokes
- Floor boxes
- Isolated receptacles

See Figure 7-16 for a detailed illustration when applying these requirements.

Grounding Tip 228: Receptacle yokes mounted to metal boxes shall be bonded to the metal per the NEC to protect personnel from electrical shock.

SURFACE-MOUNTED BOXES
250.146(A)

If a surface-mounted box such as a handy box is surface-mounted, then a bonding jumper providing metal-to-metal contact shall not be required. In other words, the 6-32 screws on the yoke of the device shall be considered an acceptable means of bonding the box and yoke together. Note that a bonding jumper shall be required if metal-to-metal contact is not provided by the connection above. **(See Figure 7-16(a))**

CONTACT DEVICES OR YOKES
250.146(B)

A flush-mounted box shall be permitted to be installed with self-grounding screws to ground receptacles with a bonding jumper. Note that these contact devices and yokes are specifically listed for proper grounding when used in conjunction with supporting screws to ensure adequate grounding between the yoke and flush-type mounted boxes, receptacles or switches. **(See Figure 7-16(b))**

Grounding Tip 229: If a 1 1/2 in. space is between box and receptacle yoke, an equipment bonding jumper shall be used to ground yoke to the metal box or self-bonding devices shall be used.

Figure 7-16. This illustration shows the accepted methods that are used to bond and ground the yoke of receptacles for safety.

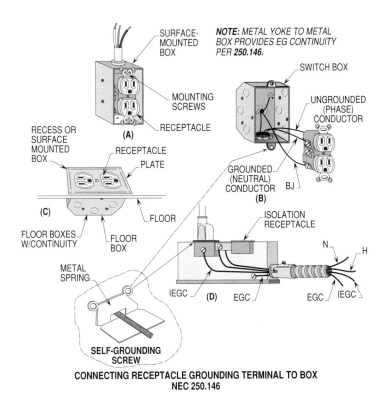

Figure 7-16(a). This illustration shows the surface box and the receptacle yoke making metal-to-metal contact.

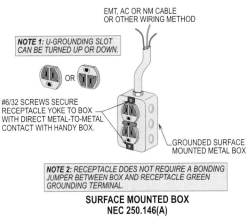

Figure 7-16(b). This illustration shows a self-bonding device used to make metal-to-metal contact between switch box and receptacle yoke.

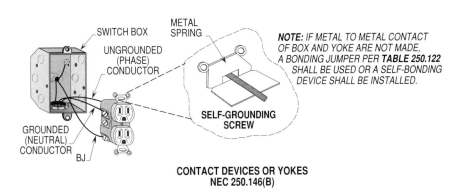

FLOOR BOXES
250.146(C)

The metal yoke of the receptacle requires no bonding jumper to be installed for floor boxes. Floor boxes are designed to provide a good contact between the metal box. However, such floor boxes shall be specifically designed and listed for such purpose. **(See Figure 7-16(c))**

Methods of Equipment Grounding

Figure 7-16(c). This illustration shows the technique for bonding and grounding a receptacle yoke to a metal floor box.

ISOLATED RECEPTACLES
250.146(D)

An isolated equipment grounding conductor shall be used to connect the receptacle isolated ground terminal back to the grounded connection at the service equipment or separately derived system to reduce electromagnetic interference.

This rule applies, when electrical noise known as electromagnetic interference occurs in the grounding circuit, an insulated equipment grounding conductor shall be permitted to be run to the isolated terminal with the circuit conductors. This insulated grounding conductor shall be permitted to pass through one or more panelboards in the same building without being connected thereto as permitted in **408.20, Ex.** Note that it still has to terminate at the equipment grounding terminal at a separately derived system or service panel. Also, see **250.96(B)**. **(See Figure 7-16(d))**

Grounding Tip 230: For hard-wiring (permanent) of sensitive electronic equipment, see Figure 5-16 in chapter 5 of this book.

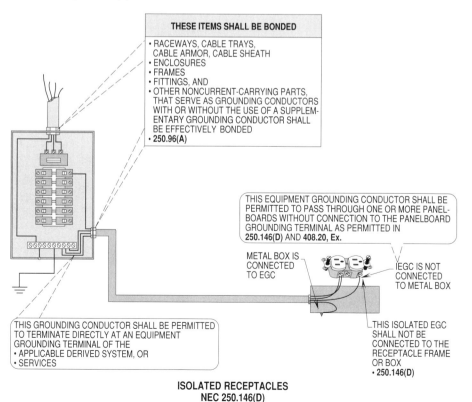

Figure 7-16(d). This illustration shows the techniques for bonding and grounding the yoke of an isolation receptacle to a metal box.

7-13

CONTINUITY AND ATTACHMENT OF EQUIPMENT GROUNDING CONDUCTORS TO BOXES
250.148

Where more than one equipment grounding conductor enters a box, all such conductors shall be spliced within the box with devices identified for the purpose. Equipment grounding conductors shall be connected and bonded to metal boxes to ensure grounding continuity. In other words, all grounding conductors entering the box shall be made electrically and mechanically secure, and the pigtail serves as the grounding connection to the device being installed. If the device is removed, the continuity of the equipment grounding conductors will not be disrupted. **(See Figure 7-17)**

Figure 7-17. This illustration shows that equipment grounding conductors shall be bonded and grounded to boxes, enclosures, etc. with a method that will ensure uninterrupted continuity.

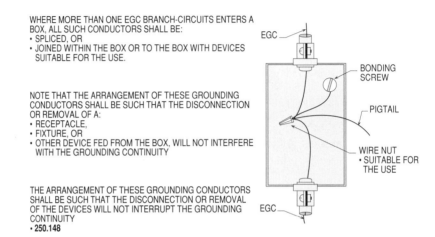

METAL BOXES
250.148(A)

A grounding screw or listed grounding device shall be used for no other purpose than providing a connection for one or more equipment grounding conductors to a metal box to ensure continuity. **(See Figure 7-18)**

Figure 7-18. This illustration shows a screw and bonding jumper used to bond and ground the metal of the electrical apparatus being supplied.

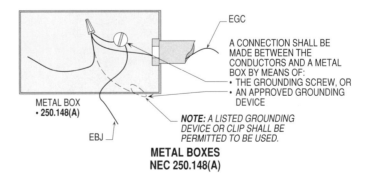

NONMETALLIC BOXES
250.148(B)

Where one or more equipment grounding conductors are installed in a nonmetallic box, the equipment grounding conductors shall be arranged in such a manner so as to provide connection to any fitting or device in the nonmetallic box requiring grounding. **(See Figure 7-19)**

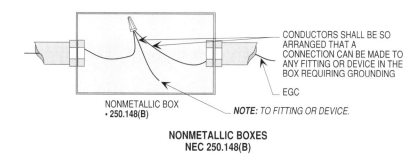

Figure 7-19. This illustration shows an equipment bonding jumper used to bond and ground the metal of fitting or device that the circuit is supplying.

METHODS OF GROUNDING MOBILE HOMES
550.16(A) THRU (C)

The distribution panel of a mobile home is not service-entrance equipment. It is, in essence, a feeder-circuit subpanel. The service-entrance equipment shall be located adjacent to the mobile home. When a mobile home is located and installed as a permanent home, then the necessary alterations shall be made in the wiring system to comply with the requirements of the NEC for a permanent home.

Grounding of electrical and noncurrent-carrying metal parts in a mobile home is through made connections to a grounding busbar in the mobile home distribution panelboard. The grounding busbar is grounded through the green colored equipment grounding conductor in the supply cord or the feeder-circuit wiring to the service grounding point in the service-entrance equipment located adjacent to the mobile home location. Note that neither the frame of the mobile home nor the frame of any electrical appliance shall be permitted to be connected to the grounded (neutral) conductor in the mobile home panelboard.

See Figures 7-20(a) through (d) for a detailed description of how to ground and bond the grounding busbars in the panelboard in the mobile home.

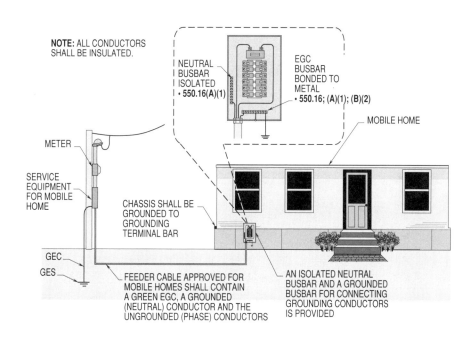

Figure 7-20(a). This illustration shows pole-mounted service-entrance equipment used for mobile home with the feeder-circuit cable routed above ground.

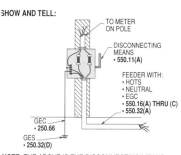

NOTE: THE ABOVE IS THE DISCONNECTING MEANS FOR THE SERVICE EQUIPMENT WHICH SUPPLIES POWER TO THE SUBPANEL IN THE MOBILE HOME. THE DISCONNECTING MEANS IS LOCATED ON THE POLE UNDER THE METER IN **FIGURES 7-20(a) AND 7-20(b).**

Figure 7-20(b). This illustration shows an underground feeder-circuit run in conduit to supply the subpanel in the mobile home.

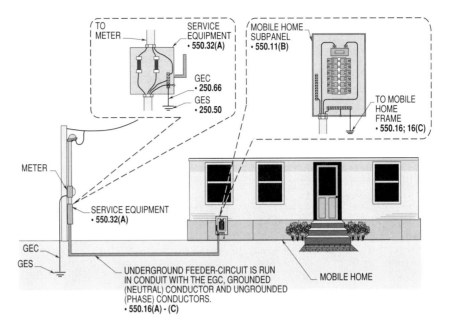

Figure 7-20(c). This illustration shows a pedestal with the service equipment grounded and a feeder-circuit used to supply the subpanel in the mobile home.

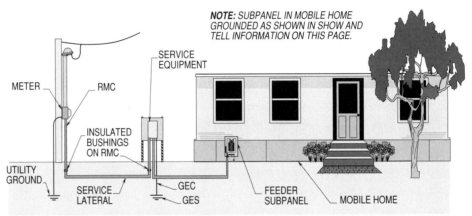

Figure 7-20(d). This illustration shows a feeder-circuit from the service equipment to a remote disconnecting means supplying the mobile homes subpanel. The requirements are also shown.

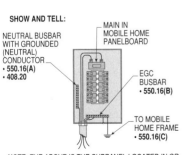

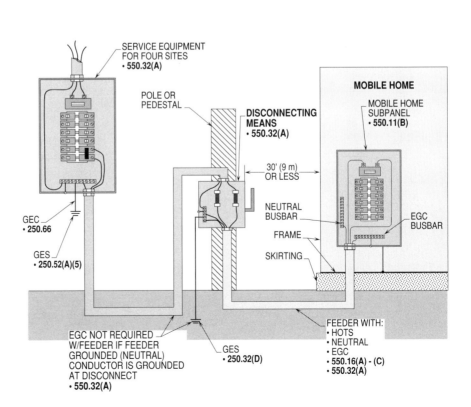

Methods of Equipment Grounding

GROUNDING AND BONDING CABLE TRAYS
392.7; TABLE 392.7(B); TABLE 250.122

Cable trays shall be permitted to have mechanically discontinuous segments between cable tray runs and/or between cable tray runs and equipment. Such discontinuous segments shall be bonded and grounded for safety to protect wiring methods, equipment and personnel from electrical shock and fire hazards.

For example, a cable tray having a cross sectional area of .40 will clear a 100 amp OCPD. The size equipment bonding jumper required to bond together discontinuous segments of a cable tray is 8 AWG copper per **Table 250.122** based on the 100 amp OCPD.

For conduits, cables, etc., that bonds to cable trays, a listed clamp or adapter shall be used. If a listed cable tray clamp or adapter is used, a nearby support, such as a support of 3 ft. (900 mm) shall not be required.

See Figures 7-21(a) through (c) for a detailed illustration of how to size equipment bonding jumpers to bond and ground wiring methods and equipment to cable trays.

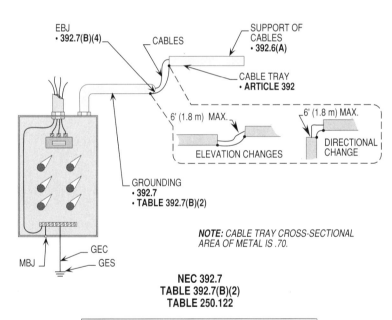

Figure 7-21(a). This illustration shows the procedure for sizing the equipment bonding jumper to bond and ground 2 cable trays together.

Figure 7-21(b). This illustration shows the procedure for sizing the equipment bonding jumper to bond and ground the electrical metallic tubing to the cable tray.

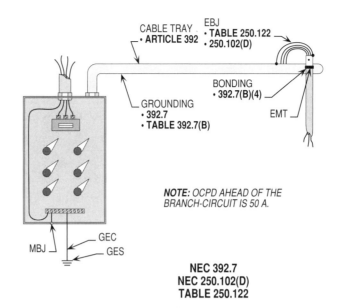

Problem: What size copper equipment bonding jumper is required to bond and ground the electrical metallic tubing to the cable tray?

Step 1: Select the OCPD rating
50 A is the rating

Step 2: Sizing EBJ
250.102(D); Table 250.122
50 A OCPD requires 10 AWG cu.

Solution: The size copper equipment bonding jumper is 10 AWG.

Figure 7-21(c). This illustration shows the procedure for sizing the equipment bonding jumper to bond and ground the motor control center to the cable tray.

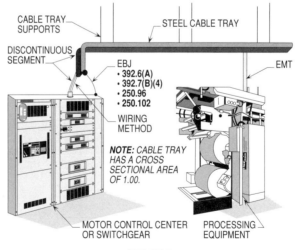

Problem: What size copper equipment bonding jumper is required to bond and ground the motor control center to the cable tray?

Step 1: Select the cross sectional area of cable tray
392.7; Table 392.7(B)
1.00 will clear 400 A OCPD

Step 2: Sizing EBJ
250.102(D); Table 250.122
400 A OCPD requires 3 AWG cu.

Solution: The size of the equipment bonding jumper is 3 AWG copper.

SWIMMING POOLS
ARTICLE 680, PART II

Swimming pools and decorative fountains shall be grounded and bonded to prevent electrical shock to people wading and swimming in the pool water. Proper size bonding and grounding conductors are essential to ensure the safety of persons having a good time in or around the wet areas of the swimming pool.

BONDING
680.26(A); (B); (C)

Grounding Tip 231: The 12 AWG equipment grounding conductor is used to clear a ground-fault while the 8 AWG bonding jumper is utilized to bond the metal parts of the pool to the grid system.

The bonding of all metal parts and grounding of the elements keeps the interconnected system and steel at the same potential which creates an equipotential plane. Stray currents moving in the water, metal and steel of the pool that are not cleared by the overcurrent protection device are kept at zero or at earth potential. This prevents electrical shock hazards to persons in the pool water or the area surrounding the pool, such as the deck where there may be stray currents in the steel or mesh in the concrete or gunite.

An 8 AWG or larger solid copper conductor shall be used to bond together all metal parts to the common bonding grid which is the foundation steel of the pool. The reinforcing bars in the pool and metal in the walls of the pool are used to form the common grid system. The bonding conductor shall not be permitted to be run to the service equipment and be connected to the grounded terminal bar in the service equipment enclosure per **680.22**. Note that this connection could allow more current from the service grounded (neutral) conductor to flow to the pool.

The equipment and metal parts of a swimming pool required to be bonded are as follows:

(1) All fixed parts located within 5 ft. (1.5 m) of the walls of the pool or they shall be separated by a permanent barrier.
(2) All metal parts and reinforcing steel in the pool or deck.
(3) All forming shells housing wet-niche or dry-niche luminaires (lighting fixtures).
(4) All metal parts and fittings of recirculating motor, etc.
(5) All fixed metal parts, such as conduit, pipes, cables, equipment, etc. located within 5 ft. (1.5 m) of the pool and not separated by a permanent barrier, and within 12 ft. (3.7 m) vertically of the maximum water level. **(See Figure 7-22)**

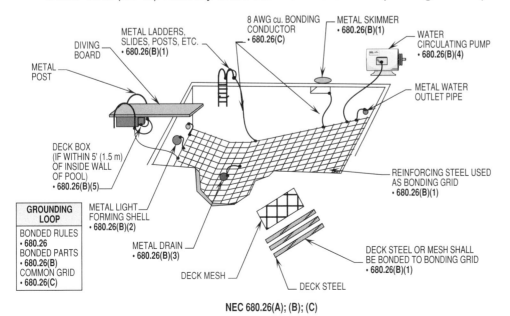

Figure 7-22. The bonding of all metal parts and grounding of the elements keeps the interconnected system and steel at the same potential which creates an equipotential plane.

Grounding Tip 232: The 8 AWG bonding jumper used to bond the metal of the equipment and steel in and around pools, spas and hot tubs shall be used for bonding and not as an equipment grounding conductor.

GROUNDING
680.6

All metal parts around the pool shall be bonded and grounded to ground potential and thus reduce the threat of shock hazard. The equipment and parts that shall be grounded includes the following items:

(1) Through-wall lighting assemblies and unders luminaires (lighting fixtures) other than those low-voltage systems listed for the application without a grounding conductor

(2) All electrical equipment located within 5 ft (1.5 m) of the inside walls of the pool

(3) All electrical equipment associated with the recirculating pump

(4) Junction boxes

(5) Transformer enclosures

(6) Ground-fault circuit-interrupters

(7) Panelboards that are not part of the service equipment and they supply electrical equipment associated with the pool

The equipment grounding conductor used to ground the metal parts of equipment, junction boxes, etc. shall be sized from **Table 250.122**, based on the rating of the overcurrent protection device protecting the branch-circuit conductors.

Grounding of the junction box and wet-niche luminaires (lighting fixtures) shall be accomplished by a 12 AWG copper insulated equipment grounding conductor. The grounding conductor shall be sized according to the rating of the overcurrent protection device per **Table 250.122**. It shall not be smaller than 12 AWG and shall be unspliced unless the exception is applied.

The grounding conductor shall be installed and run with the circuit conductors in rigid metal conduit, intermediate metal conduit or rigid nonmetallic conduit from the panelboard to the deck box. The main function of the 12 AWG equipment grounding conductor is to clear the overcurrent protection device in case of a ground-fault.

Flexible cords used to ground wet-niche luminaires (lighting fixtures) shall be provided with a 12 AWG or larger insulated copper equipment grounding conductors connected to a terminal in the supply deck box. The type of connection provides continuity when the luminaire (lighting fixture) is removed for servicing.

HOT TUBS, SPAS AND HYDROMASSAGE TUBS
ARTICLE 680, PART IV AND VII

Hot tubs, spas and hydromassage tubs shall be bonded and grounded for the same reasons as swimming pools for the safety of those using them. Such tubs shall be permitted to be located inside or outside of the dwelling unit or building.

BONDING
680.42(B)

Bonding by metal-to-metal mounting on a common frame or base shall be permitted. The metal band or hopps that are used to secure wooden staves shall not be required to be bonded, per **680.26**. (See Figure 7-23)

Methods of Equipment Grounding

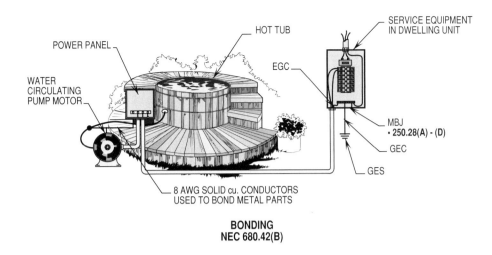

Figure 7-23. An 8 AWG solid copper conductor shall be used to ground all noncurrent-carrying parts associated with the hot tub, including metal piping, etc. that is located within 5 ft (1.5 m) of the tub.

GROUNDING
680.43(F)

The following electrical equipment shall be grounded:

(1) All electrical equipment located within 5 ft. (1.5 m) of the inside wall of the spa or hot tub.

(2) All electrical equipment associated with the circulating system of the spa or hot tub.

HYDROMASSAGE BATHTUBS
680.70; 680.71

Hydromassage bathtubs are treated in the same manner as conventional bathtubs. The wiring methods for hydromassage bathtubs shall comply with **Chapters 1 through 4** of the NEC. All elements for hydromassage bathtubs shall be supplied by GFCI-protected circuits. See **410.4(D)** for hanging luminaires (lighting fixtures) over and around the tub. Location of switches for luminaires (lighting fixtures) and receptacle outlets are treated in the same manner as a regular bathtub. **(See Figure 7-24)**

Grounding Tip 233: See **404.4** and **406.8(A)** for locating and installing switches and receptacles which are not associated with the hydromassage tub. Access to the circulating motor for servicing is required per **Article 100**, **430.14(A)** and **680.73**.

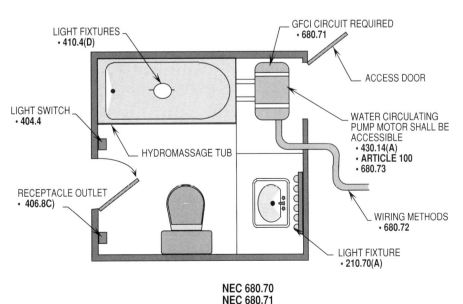

Figure 7-24. An 8 AWG solid copper conductor shall be used to ground all noncurrent-carrying parts associated with the hot tub including metal piping, etc. located within 5 ft. (1.5 m) of the tub. Note that this includes any and all metal water piping servicing the tub.

Name _____   Date _____

Chapter 7
Methods of Equipment Grounding

	Section	Answer

1. For ungrounded systems, the equipment grounding conductor shall be bonded to the grounded (neutral) conductor and the grounding electrode conductor at the service equipment neutral busbar terminal.   _____   T   F

2. For grounded systems, the equipment grounding conductor shall be bonded to the grounded (neutral) conductor and the grounding electrode conductor at the service equipment neutral busbar terminal.   _____   T   F

3. A nongrounding type receptacle shall be permitted to be replaced a GFCI receptacle protecting a single outlet and additional outlets downstream.   _____   T   F

4. The structural frame of a building shall be permitted to be used as an equipment grounding conductor for equipment secured to grounded metal supports.   _____   T   F

5. A separate flexible wire or strap shall be permitted to be used to ground the metal cases of portable cord-and-plug connected stationary equipment.   _____   T   F

6. The grounded (neutral) conductor in existing branch-circuit installations shall be permitted to bond and ground the metal frames of ranges, cooktops, ovens and clothes dryers.   _____   T   F

7. The grounded (neutral) conductor shall not be permitted to bond and ground separate building and structures under any circumstance.   _____   T   F

8. A flush-mounted box shall be permitted to be installed with self-grounding screws to ground and bond receptacles.   _____   T   F

9. The isolated equipment grounding shall be permitted to pass through one or more panelboards in the same building without being connected.   _____   T   F

10. Where more than one equipment grounding conductor enters a box, all such conductors shall be spliced within the box with devices identified for the purpose.   _____   T   F

7-23

8

DIRECT-CURRENT SYSTEMS

The grounding procedure for DC systems differs from those required for AC systems. This chapter is designed to familiarize the designer, installer and inspector with these requirements and how to apply such rules when designing and installing DC power systems.

GENERAL
250.160

This section requires DC systems to fully comply with Part VIII and other pertinent sections of **Article 250** which are not specifically intended for AC systems.

DIRECT-CURRENT CIRCUITS AND SYSTEMS TO BE GROUNDED
250.162

Under the following conditions of use, direct-current circuits and systems shall be required to be grounded:

- Two-wire, direct-current systems
- Three-wire, direct-current systems

TWO-WIRE, DIRECT-CURRENT SYSTEMS
250.162(A)

The main rule requires two-wire, DC systems of over 50 up to 300 volts supplying premises wiring to be grounded. However, there are three exceptions to those requirements listed below:

- Ground detectors
- Rectifier-derived DC systems
- DC fire alarm circuits

Grounding Tip 234. Ground-fault locators with built-in devices make it easy to locate ground-faults. These devices indicate the presence of leakage current or serious developing fault current conditions and sound alarms to take corrective action.

GROUND DETECTORS
250.162(A), Ex. 1

A system equipped with a ground detector and supplying only industrial equipment in limited areas shall not be required to be grounded.

Note that **Ex. 1** allows an ungrounded, two-wire, DC system to be used where such a system is installed with ground detectors and supplies only industrial equipment in a limited area. For example, an industrial plant might use a two-wire, ungrounded DC system for process line control of components. **(See Figure 8-1)**

Figure 8-1. This illustration shows ground detectors installed to sense a ground-fault and verify which line the ground-fault has occurred on.

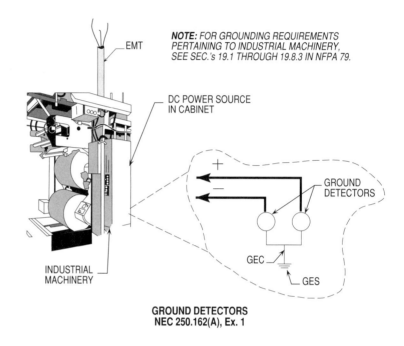

RECTIFIER-DERIVED DC SYSTEM
260.162(A), Ex. 2

A rectifier-derived DC system supplied from an AC system shall not be required to be grounded by the requirements that are listed in **250.20**.

Note that **Ex. 2** permits an ungrounded, two-wire, DC system to be used where the direct-current is derived from a rectifier. However, the AC system supplying the rectifier shall meet the requirements set down for AC systems. For example, such a system is the case where 120 volts AC is rectified for use as a 120 volt, DC control or signaling system. **(See Figure 8-2)**

Figure 8-2. This illustration shows the procedure for bonding and grounding the AC and DC side of a rectifier-derived DC system used for DC control and signaling circuits.

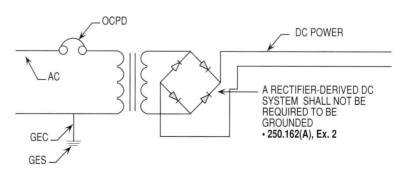

DC FIRE ALARM CIRCUITS
250.162(A), Ex. 3

DC fire-protective signaling circuits having a maximum control of .03 amps shall not be required to be grounded per **Article 760, Part III** in the NEC.

Note that **Ex. 3** permits the use of an ungrounded, two-wire, DC system where the power supply is limited to a maximum of 30 mA (.030 amps) when used for DC fire-protection signal circuits. The 30 mA comes from **Article 760, Part III** (Power-Limited Fire Alarm (PLFA) circuits). **(See Figure 8-3)**

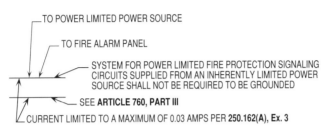

Figure 8-3. This illustration shows that under certain conditions of use, a DC fire alarm circuit shall be permitted to operate ungrounded.

THREE-WIRE, DIRECT-CURRENT SYSTEMS
250.162(B)

The neutral of all three-wire DC systems supplying premises wiring shall be grounded, regardless of voltage. Section **250.162(B)** requires the neutral conductor of all three-wire DC systems that supply premises wiring to be grounded. There are no minimum or maximum voltages mentioned, nor are there any exceptions listed. The grounding electrode conductor shall be installed at the source of supply, but located other than on the premises. Note that the grounded (neutral) conductor shall not be required to be bonded to a second grounding electrode conductor at the service as it is in the case of AC grounded systems. Note that **250.164** permits the grounding electrode conductor of DC systems to be installed at the direct-current generator, or at the first disconnecting means or overcurrent protection device. This rule applies when the DC source is located on the premises. **(See Figure 8-4)**

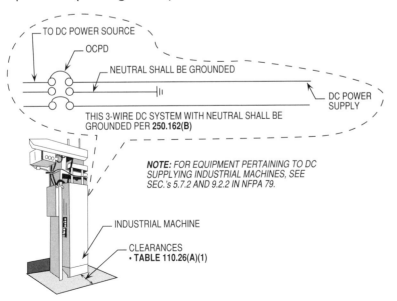

Figure 8-4. This illustration shows a three-wire, DC system supplying power to an industrial machine.

Grounding Tip 235: In three-wire DC distribution systems, the neutral shall be grounded at the off-premises generator site. Grounding of a two-wire DC system can be accomplished in the same manner. For an on-premises generator, a grounding connection shall be required and shall be located at the source of the first system disconnecting means or OCPD. Note that other equivalent means that utilize equipment listed and identified for such shall be permitted.

POINT OF CONNECTION FOR DIRECT-CURRENT SYSTEMS
250.164

On DC systems that are required to be grounded, the grounding shall be done at one or more supply stations, but not at the individual services. This pertains to the grounding of the supply conductors and not to the service equipment. The equipment shall be grounded at the service, but the neutral shall be isolated from the equipment at the service equipment. These rules and regulations are greatly different from those for AC systems because any passage of current between the ground at the supply station and at the service equipment may cause objectionable electrolysis, which must be avoided.

OFF-PREMISES SOURCE
250.164(A)

DC systems that are required to be grounded and supplied from an off-premises source shall have the grounding connection made at one or more supply stations. A grounding connection shall not be made at individual services or at any point on the premises wiring. **(See Figure 8-5)**

ON-PREMISES SOURCE
250.164(B)

Where the DC system source is located on the premises, a grounding connection shall be made at one of the following locations:

- The source
- The first system disconnection means or overcurrent device
- By other means that accomplish equivalent system protection and that utilize equipment listed and identified for the use

See Figure 8-6 for a detailed illustration when applying these requirements.

Figure 8-5. This illustration shows a three-wire DC system bonded and grounded at the off-premises DC power source.

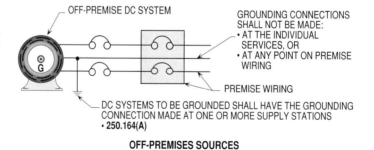

Figure 8-6. This illustration shows the procedure for bonding an on-premise DC power system at the source or the disconnect switch.

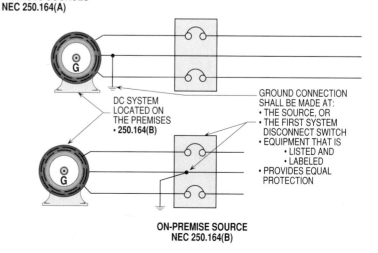

Direct-current Systems

SIZE OF DIRECT-CURRENT GROUNDING ELECTRODE CONDUCTOR 250.166

The grounding electrode conductor for DC systems shall be designed using different procedures than those for AC systems. **Table 250.66** is used for AC systems and the grounding electrode conductor is selected from this table based on the size of the largest ungrounded (phase) conductor supplying the service equipment. However, the grounding electrode conductor for DC systems shall be sized on the type of DC power system used and the type of grounding electrode that is used to ground the system. There are five ways of sizing the grounding electrode conductor to earth ground DC power systems and they are listed below:

- Not smaller than the neutral conductor
- Not smaller than the largest conductor
- Connected to rod, pipe or plate electrodes
- Connected to a concrete-encased electrode
- Connected to a ground ring

NOT SMALLER THAN THE NEUTRAL CONDUCTOR 250.166(A)

For a three-wire balancer set with overcurrent protection, such as in the case of an unbalanced condition, the grounding electrode conductor shall not be smaller than the neutral. Overcurrent protection shall be provided as required per **445.12(D)**. The grounding electrode conductor used to earth ground the DC system shall be at least 8 AWG copper or 6 AWG aluminum.

>**Problem 8-7.** What size grounding electrode conductor is required to ground a DC supply (three-wire balancer set) that has its ungrounded (phase) conductors rated at 250 KCMIL copper and a neutral rated at 1/0 AWG copper? (Building steel is used as GE)
>
>**Step 1:** Sizing GEC
>**250.166(A)**
>1/0 AWG cu. requires 1/0 AWG cu.
>
>**Solution: The size grounding electrode conductor required to earth ground the DC supply is 1/0 AWG copper.**

See **Figure 8-7** for an exercise problem when sizing the grounding electrode conductor for a DC supply with a three-wire balancer set.

Grounding Tip 236: To earth ground a DC power source to a metal water pipe and structural building steel, the grounding electrode conductor is sized from **250.166(A)** and **(B)**. When grounding a DC power source to other grounding electrodes, select the grounding electrode conductors as follows per **250.66(C) thru (E)**:

Made electrodes
- 6 AWG cu. or 4 AWG alu.
- **250.66(C)**

Concrete-encased electrodes
- 4 AWG cu.
- 250.66(D)

Ground Ring
- 2 AWG cu.
- 250.66(E)

Note: Review 250.166(A) thru (E) before sizing grounding electrode conductor to earth ground a DC power source.

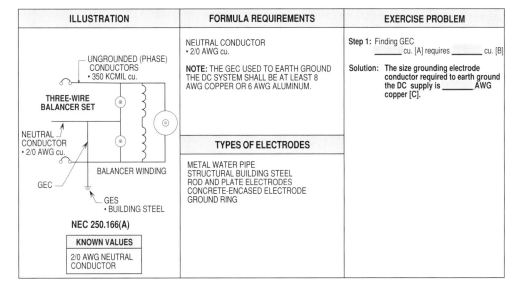

Figure 8-7. This illustration shows an exercise problem for sizing the grounding electrode conductors to earth ground the DC power source to structural building steel.

Grounding Tip 237: Because of the fault current that DC systems can produce, under certain conditions of use, the grounding electrode conductor shall be larger than for AC systems with the same size ungrounded (phase) conductors.

NOT SMALLER THAN THE LARGEST CONDUCTOR
250.166(B)

For DC systems other than the three-wire balancer, the grounding electrode conductor shall not be smaller than the largest conductor. The grounding electrode conductor used to earth ground the DC system shall be at least 8 AWG copper or 6 AWG aluminum.

Problem 8-8. What size grounding electrode conductor is required to ground a DC system that has its ungrounded (phase) conductors rated at 250 KCMIL copper and a neutral rated 1/0 AWG copper? (Building steel is used as GE)

    Step 1: Sizing GEC
              250.166(B)
              250 KCMIL cu. requires 250 KCMIL cu.

    Solution: **The size grounding electrode conductor required to earth ground a DC supply is 250 KCMIL copper.**

See Figure 8-8 for an exercise problem when sizing the grounding electrode conductor for a DC supply.

Figure 8-8. This illustration shows an exercise problem for sizing the grounding electrode conductor for a DC power source based on the size of the largest conductor.

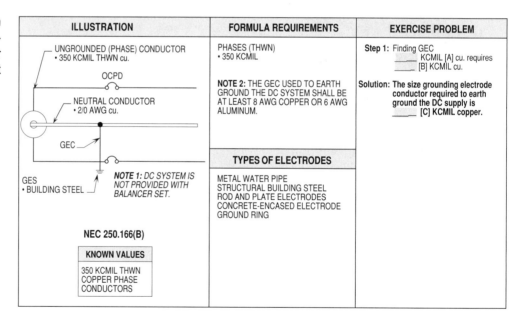

Grounding Tip 238: Since the 1996 Edition of the NEC, a grounding electrode conductor to made electrodes for DC systems has been the same size as for AC systems.

CONNECTED TO ROD, PIPE OR PLATE ELECTRODES
250.166(C)

Where the DC system is connected to made electrodes as in **250.52(A)(5)** or **(A)(6)**, such as driven rods, that portion of the grounding electrode conductor that is the sole connection to the grounding electrode shall not be required to be larger than 6 AWG copper or 4 AWG aluminum.

Problem 8-9. What size grounding electrode conductor is required to earth ground a DC supply to a made electrode (driven rod)?

Step 1: Sizing GEC
250.166(C)
Driven rod requires 6 AWG cu.

Solution: **The size grounding electrode conductor required to earth ground the DC supply to a driven rod electrode is 6 AWG copper.**

See Figure 8-9 for an exercise problem when sizing the grounding electrode conductor for connection to a made electrode.

ILLUSTRATION

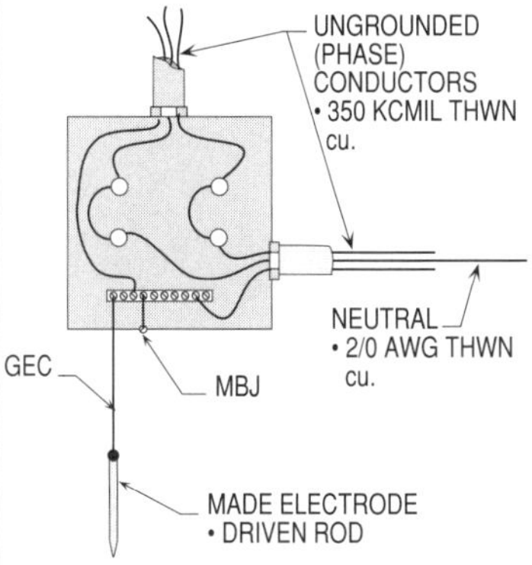

NEC 250.166(C)

KNOWN VALUES
350 KCMIL THWN cu. UNGROUNDED (PHASE) CONDUCTORS
2/0 AWG THWN cu. NEUTRAL CONDUCTOR

FORMULA REQUIREMENTS

PHASES (THWN)
• 350 KCMIL cu.

NEUTRAL (THWN)
• 2/0 AWG cu.

TYPES OF ELECTRODES

METAL WATER PIPE
STRUCTURAL BUILDING STEEL
ROD AND PLATE ELECTRODES
CONCRETE-ENCASED ELECTRODE
GROUND RING

EXERCISE PROBLEM

Step 1: Finding GEC
Made electrodes require ____ [A] cu.

Solution: **The size grounding electrode conductor required to earth ground the DC supply to a made electrode is ____ AWG [B] copper.**

Figure 8-9. This illustration shows an exercise problem for sizing the grounding electrode conductor to a driven rod electrode.

CONNECTED TO A CONCRETE-ENCASED ELECTRODE 260.166(D)

Where the DC system is connected to a concrete-encased electrode (such as 1/2 in. x 20 ft. (13 mm x 6 m) or greater lengths of rebar) as permitted in **250.52(A)(3)**, that portion of the grounding electrode conductor that is the sole connection to the grounding electrode shall not be required to be larger than 4 AWG copper.

Grounding Tip 239: Concrete-encased electrodes have been known in the electrical industry for many years as the UFER grounding system.

Problem 8-10. What size grounding electrode conductor is required to earth ground a DC supply to a concrete-encased electrode?

Step 1: Sizing GEC
260.166(D)
CEE requires 4 AWG cu.

Solution: **The size grounding electrode conductor required to earth ground the DC supply to concrete-encased electrode is 4 AWG copper.**

See Figure 8-10 for an exercise problem when sizing the grounding electrode conductor to a concrete-encased electrode.

Figure 8-10. This illustration shows an exercise problem for sizing the grounding electrode conductor to earth ground the DC power source to a concrete-encased electrode (UFER).

ILLUSTRATION

NEC 250.166(D)

KNOWN VALUES
350 KCMIL THWN cu. UNGROUNDED (PHASE) CONDUCTORS
2/0 AWG THWN cu. NEUTRAL CONDUCTOR

FORMULA REQUIREMENTS

PHASES (THWN)
• 350 KCMIL cu.

NEUTRAL (THWN)
• 2/0 AWG cu.

TYPES OF ELECTRODES

METAL WATER PIPE
STRUCTURAL BUILDING STEEL
ROD AND PLATE ELECTRODES
CONCRETE-ENCASED ELECTRODE
GROUND RING

EXERCISE PROBLEM

Step 1: Sizing GEC
CEE requires ____ cu.[A]

Solution: **The size grounding electrode conductor required to earth ground the DC supply to a concrete-encased electrode is ____AWG copper [B].**

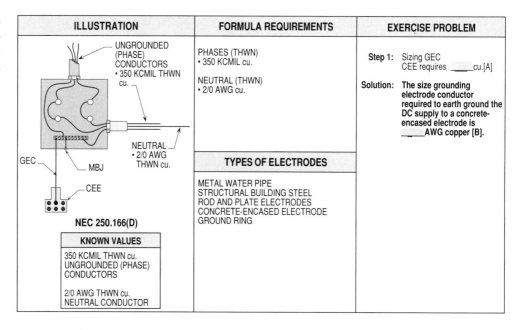

CONNECTED TO A GROUND RING 250.166(E)

Grounding Tip 240: The minimum earth resistance for an earth ground is 25 ohms per **250.56**. Any resistance lower than 25 ohms is considered a good resistance for an earth ground. A resistance of less than 5 ohms should be obtained if practical.

Where the DC system is connected to a ground ring as in **250.52(A)(4)**, that portion of the grounding electrode conductor that is the sole connection to the grounding electrode shall be as large as the conductor used for the ground ring.

Problem 8-11. What size grounding electrode conductor is required to earth ground a DC supply to a ground ring?

Step 1: Sizing GEC
250.166(E)
Ground ring requires 2 AWG cu.

Solution: **The size grounding electrode conductor required to earth ground the DC supply to a ground ring is 2 AWG copper.**

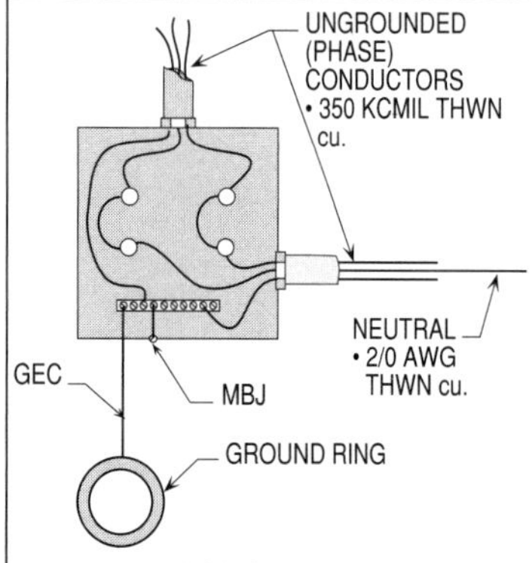

See Figure 8-11 for an exercise problem when sizing the grounding electrode conductor to a ground ring.

Figure 8-11. This illustration shows an exercise problem for sizing the grounding electrode conductor to earth ground the DC power source to a ground ring.

ILLUSTRATION

NEC 250.166(E)

KNOWN VALUES
350 KCMIL THWN cu. UNGROUNDED (PHASE) CONDUCTORS
2/0 AWG THWN cu. NEUTRAL CONDUCTOR

FORMULA REQUIREMENTS

PHASES (THWN)
• 350 KCMIL cu.

NEUTRAL (THWN)
• 2/0 AWG cu.

TYPES OF ELECTRODES

METAL WATER PIPE
STRUCTURAL BUILDING STEEL
ROD AND PLATE ELECTRODES
CONCRETE-ENCASED ELECTRODE
GROUND RING

EXERCISE PROBLEM

Step 1: Sizing GEC
Ground ring requires ____ cu. [A]

Solution: **The size grounding electrode conductor required to earth ground the DC supply to a ground ring is ____AWG copper [B].**

DIRECT-CURRENT BONDING JUMPER
250.168

For DC systems, the size of the bonding jumper shall not be smaller that the system grounding conductor specified in **250.166**. See Problems below for sizing the bonding jumper based on the size grounding electrode conductor required to earth ground the DC power source to its grounding electrodes.

Sizing bonding jumper based on the size of the neutral

Problem 8-12(a). What size bonding jumper is required to bond a DC power source with a three-wire balancer set. Ungrounded (phase) conductors are rated at 250 KCMIL copper with a neutral rated at 1/0 AWG copper?

Step 1: Sizing BJ
250.168; 250.166(A)
1/0 AWG cu. requires 1/0 AWG cu.

Solution: **The size bonding jumper required based on the DC supply is 1/0 AWG copper.**

See **Figure 8-12(a)** for an exercise problem when sizing the bonding jumper based on the size of the neutral.

Grounding Tip 241: The bonding jumper collects all the fault current in amps that can collect at this bonding connection and transfers it so that it flows through the windings of the generator and balancer set.

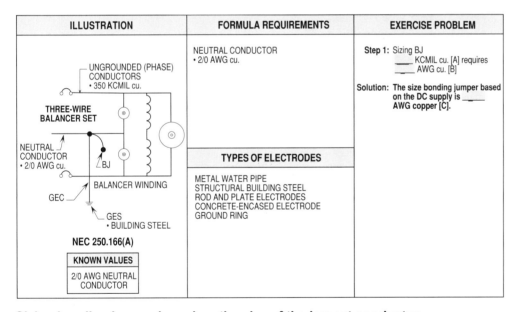

Figure 8-12(a). This illustration shows an exercise problem for sizing the bonding jumper to bond and ground the metal case of the DC power source to the grounding electrode conductor.

Sizing bonding jumper based on the size of the largest conductor

Problem 8-12(b). What size bonding jumper is required to bond a DC power source without a three-wire balancer set. Ungrounded (phase) conductors are rated at 250 KCMIL copper with a neutral rated at 1/0 AWG copper?

Step 1: Sizing BJ
250.168; 250.166(B)
250 KCMIL cu. requires 250 KCMIL cu.

Solution: **The size bonding jumper required is 250 KCMIL copper.**

See **Figure 8-12(b)** for an exercise problem when sizing the bonding jumper based on the size of the largest conductor.

Grounding Tip 242: Care must be exercised to ensure that the bonding jumper in **Problem 8-12(b)** is not under sized so that the full amount of fault current produced fails to transfer and trip open the OCPD's.

Figure 8-12(b). This illustration shows an exercise problem for sizing the bonding jumper to bond and ground the metal case of the DC power source to the grounding electrode conductor.

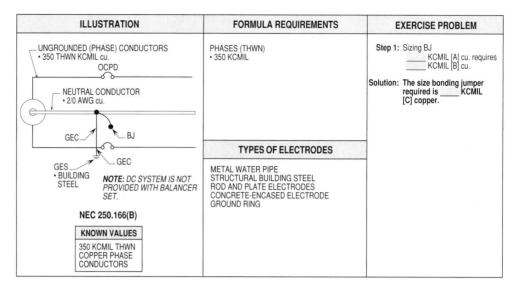

Sizing bonding jumper based on grounding to a rod, pipe or plate electrode

Grounding Tip 243: The types of made electrodes to be used and installed are listed in **250.52(A)(5)** and **(A)(6)** in the NEC.

Problem 8-12(c). What size bonding jumper is required based on the DC power supply grounded to a driven rod electrode. Ungrounded (phase) conductors are rated at 250 KCMIL copper with a neutral rated at 1/0 AWG copper? (Three-wire balancer set on DC supply)

Step 1: Sizing BJ
250.168; 250.166(C)
Driven rod electrode requires 6 AWG cu.

Solution: **The size bonding jumper required to a driven rod electrode is 6 AWG copper.**

See Figure 8-12(c) for an exercise problem when sizing the bonding jumper to driven rod electrodes.

Figure 8-12(c). This illustration shows an exercise problem for sizing the bonding jumper to bond and ground the metal case of the DC power supply to the grounding electrode conductor.

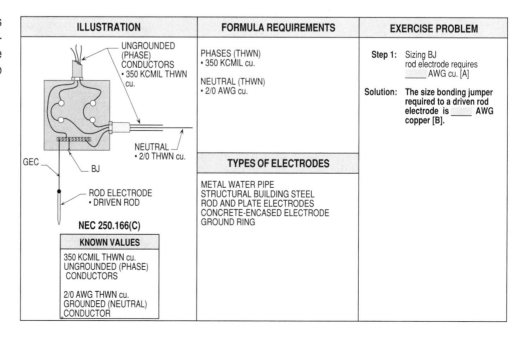

Direct-current Systems

Sizing bonding jumper based on grounding to a concrete-encased electrode

Problem 8-12(d). What size bonding jumper is required based on the DC power supply grounded to a concrete-encased electrode. Ungrounded (phase) conductors are rated at 250 KCMIL copper with a neutral rated at 1/0 AWG copper? (No three-wire balancer set on DC supply)

Step 1: Sizing BJ
250.168; 250.166(D)
CEE requires 4 AWG cu.

Solution: The size bonding jumper required to a concrete-encased electrode is 4 AWG copper.

See **Figure 8-12(d)** for an exercise problem when sizing the bonding jumper to a concrete-encased electrode.

Grounding Tip 244: The concrete-encased electrode shall be designed and installed by the requirements listed in **250.52(A)(3)** in the NEC.

ILLUSTRATION	FORMULA REQUIREMENTS	EXERCISE PROBLEM
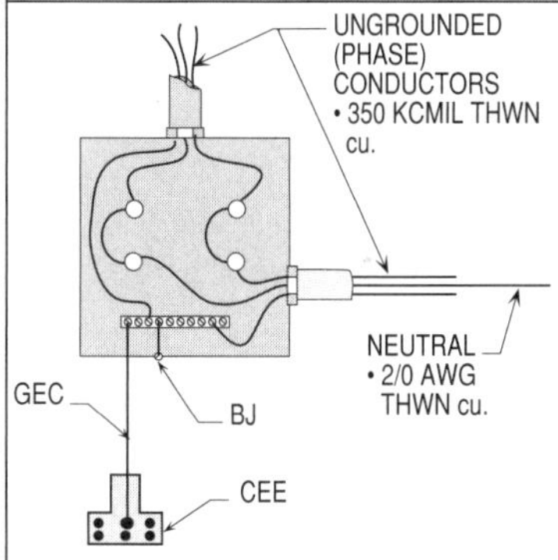 UNGROUNDED (PHASE) CONDUCTORS • 350 KCMIL THWN cu. — NEUTRAL • 2/0 AWG THWN cu. — GEC — BJ — CEE — NEC 250.166(D) **KNOWN VALUES** 350 KCMIL THWN cu. UNGROUNDED (PHASE) CONDUCTORS 2/0 AWG THWN cu. GROUNDED (NEUTRAL) CONDUCTOR	PHASES (THWN) • 350 KCMIL cu. NEUTRAL (THWN) • 2/0 AWG cu. **TYPES OF ELECTRODES** METAL WATER PIPE STRUCTURAL BUILDING STEEL ROD AND PLATE ELECTRODES CONCRETE-ENCASED ELECTRODE GROUND RING	Step 1: Sizing BJ CEE requires ___ AWG cu. [A] Solution: The size bonding jumper required to a concrete-encased electrode is ___ AWG copper [B].

Figure 8-12(d). This illustration shows an exercise problem for sizing the bonding jumper to bond and ground the metal case of the DC power supply to the grounding electrode conductor.

Sizing bonding jumper based on grounding to a ground ring

Problem 8-12(e). What size bonding jumper is required based on the DC power supply grounded to a ground ring. Ungrounded (phase) conductors are rated at 250 KCMIL copper with a neutral rated at 1/0 AWG copper? (Three-wire balancer set provided)

Step 1: Sizing BJ
250.168; 250.166(E)
Ground ring requires 2 AWG cu.

Solution: The size bonding jumper required to a ground ring is 2 AWG copper.

See **Figure 8-12(e)** for an exercise problem when sizing the bonding jumper to a ground ring.

Grounding Tip 245: The ground ring shall be installed by the requirements listed in **250.52(A)(4)** in the NEC.

Figure 8-12(e). This illustration shows an exercise problem for sizing the bonding jumper to bond and ground the metal of the DC power supply to the grounding electrode conductor.

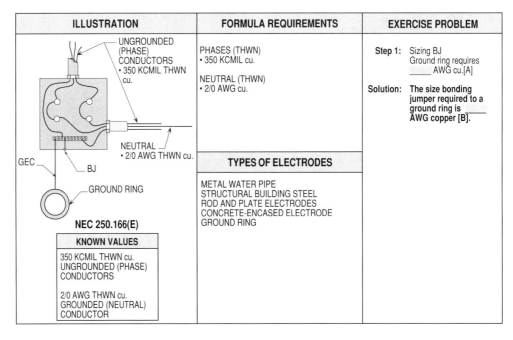

Grounding Note: DC motors are used in a great variety of industrial applications, particularly where a wide range of speeds are needed or where precise control of speed or position is required. At one time, DC applications were restricted by the availability of a DC power source. However, with the availability of reliable rectifiers and silicon-controlled rectifiers (Thyristors), DC machines can easily be DC from AC power lines. In fact, in many cases, it is easier to derive power from a rectified AC source than from a pure DC source. DC power sources are very reliable and safe if the proper grounding techniques in this chapter are applied.

UNGROUNDED DIRECT-CURRENT SEPARATELY DERIVED SYSTEMS
250.169

Except as otherwise permitted in **250.34** for portable and vehicle mounted generators, an ungrounded DC separately derived system supplied from a stand-alone power source (such as an engine-generator set) shall have a grounding electrode conductor connected to an electrode that complies with **Part III** in **Article 250** to provide for grounding of metal enclosures, raceways, cables and exposed noncurrent-carrying metal parts of equipment. The grounding electrode conductor connection shall be to the metal enclosure at any point on the separately derived system from the source to the first system disconnecting means or overcurrent device, or it shall be made at the source of a separately derived system that has no disconnecting means or overcurrent devices. Note that the size of the grounding electrode conductor shall comply with **260.166** in the NEC. **(See Figure 8-13)**

Direct-current Systems

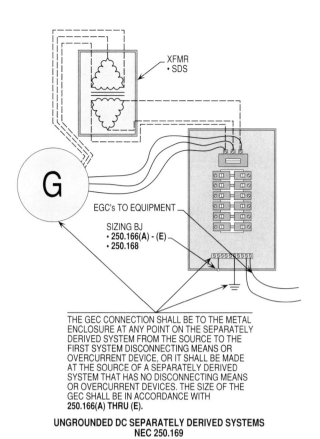

Figure 8-13. This illustration shows an ungrounded DC separately derived system supplying power to a panelboard with OCPD's used to protect DC circuits.

Name _____   Date _____

Chapter 8
Direct-Current Systems

	Section	Answer

1. A system equipped with a ground detector and supplying only industrial equipment in limited areas shall be required to be grounded.   _____   T   F

2. A rectifier-derived DC system supplied from an AC system shall be required to be grounded.   _____   T   F

3. The neutral of all three-wire DC systems supplying premises wiring shall be grounded, regardless of voltage.   _____   T   F

4. The grounding connection for a DC system shall be permitted to be made the first system disconnecting means or overcurrent protection device.   _____   T   F

5. For DC systems, the size of the bonding jumper shall not be smaller than the system grounding conductor.   _____   T   F

6. The main rule requires two-wire, DC systems of over 50 volts up to _____ volts supplying premises wiring to be grounded.   _____   _____

7. DC fire-protective signaling circuits having a maximum control of _____ amps shall not be required to be grounded.   _____   _____

8. The grounding electrode conductor for DC system other than the three-wire balancer used to earth ground the DC system shall be at least _____ copper.   _____   _____

9. Where the DC system is connected to rod electrodes, such as driven rods, that portion of the grounding electrode conductor that is the sole connection to the grounding electrode shall not be required to be larger than _____ AWG copper.   _____   _____

10. Where the DC system is connected to a concrete-encased electrode, that portion of the grounding electrode conductor that is the sole connection to the grounding electrode shall not be required to be larger than _____ AWG copper.   _____   _____

11. What size grounding electrode conductor is required to ground a DC supply (three-wire balancer set) that has its phase conductors rated at No. 4/0 copper and a neutral rated at No. 1 copper? (Building steel is used as GE)   _____   _____

12. What size grounding electrode conductor is required to ground a DC supply (three-wire balancer set) that has its phase conductors rated at 500 KCMIL copper and a neutral rated at 4/0 AWG copper? (Building steel is used as GE)   _____   _____

13. What size grounding electrode conductor is required to ground a DC system that has its phase conductors rated at 4/0 AWG copper and a neutral rated at 1 AWG copper? (Building steel is used as GE)   _____   _____

Section	Answer

14. What size grounding electrode conductor is required to ground a DC system that has its phase conductors rated at 500 KCMIL copper and a neutral rated at 3/0 AWG copper? (Building steel is used as GE)

15. What size grounding electrode conductor is required to ground a DC system that has its phase conductors rated at 4/0 AWG copper and a neutral rated at 1 AWG copper? (Driven rod is used as GE)

16. What size grounding electrode conductor is required to ground a DC system that has its phase conductors rated at 500 KCMIL copper and a neutral rated at 3/0 AWG copper? (Driven rod is used as GE)

17. What size grounding electrode conductor is required to ground a DC system that has its phase conductors rated at 4/0 AWG copper and a neutral rated at 1 AWG copper? (Concrete-encased electrode is used as GE)

18. What size grounding electrode conductor is required to ground a DC system that has its phase conductors rated at 500 KCMIL copper and a neutral rated at 3/0 AWG copper? (Concrete-encased electrode is used as GE)

19. What size grounding electrode conductor is required to ground a DC system that has its phase conductors rated at 4/0 AWG copper and a neutral rated at 1 AWG copper? (Ground ring is used as GE)

20. What size grounding electrode conductor is required to ground a DC system that has its phase conductors rated at 500 KCMIL copper and a neutral rated at 3/0 AWG copper? (Ground ring is used as GE)

21. What size bonding jumper is required to ground a DC supply (three-wire balancer set) that has its phase conductors rated at 4/0 AWG copper and a neutral rated at 1 AWG copper? (Building steel is used as GE)

22. What size bonding jumper is required to ground a DC supply (three-wire balancer set) that has its phase conductors rated at 500 KCMIL copper and a neutral rated at 4/0 AWG copper? (Building steel is used as GE)

23. What size bonding jumper is required to ground a DC system that has its phase conductors rated at 4/0 AWG copper and a neutral rated at 1 AWG copper? (Building steel is used as GE)

24. What size bonding jumper is required to ground a DC system that has its phase conductors rated at 500 KCMIL copper and a neutral rated at 3/0 AWG copper? (Building steel is used as GE)

25. What size bonding jumper is required to ground a DC system that has its phase conductors rated at 4/0 AWG copper and a neutral rated at 1 AWG copper? (Driven rod is used as GE)

26. What size bonding jumper is required to ground a DC system that has its phase conductors rated at 500 KCMIL copper and a neutral rated at 3/0 AWG copper? (Driven rod is used as GE)

	Section	Answer

27. What size bonding jumper is required to ground a DC system that has its phase conductors rated at 4/0 AWG copper and a neutral rated at 1 AWG copper? (Concrete-encased electrode is used as GE)

28. What size bonding jumper is required to ground a DC system that has its phase conductors rated at 500 KCMIL copper and a neutral rated at 3/0 AWG copper? (Concrete-encased electrode is used as GE)

29. What size bonding jumper is required to ground a DC system that has its phase conductors rated at 4/0 AWG copper and a neutral rated at 1 AWG copper? (Ground ring is used as GE)

30. What size bonding jumper is required to ground a DC system that has its phase conductors rated at 500 KCMIL copper and a neutral rated at 3/0 AWG copper? (Ground ring is used as GE)

8-17

9

INSTRUMENTS, METERS AND RELAYS

This chapter covers the grounding and bonding schemes permitted to ground special circuits and equipment.

If these grounding techniques outlined are designed, installed and maintained properly they will operate and provide the user with grounded systems that are reliable and dependable to ensure the safety of personnel working on, around or near such systems.

INSTRUMENT TRANSFORMER CIRCUITS
250.170

Current and potential instrument transformers are small transformers utilized to take a reduced current or voltage from a circuit. The lower current or voltage wired to the ammeter or voltmeter is scaled to read correct current or voltage for the circuit. The secondary circuit of such transformer is the circuit leading to the ammeter, voltmeter or instrument involved. The requirements for grounding the secondary circuits are as follows:

On switchboards:

- Primary of 1000 volts or greater shall be grounded.

- Primary of 0 to 999 volts shall be grounded if any live parts or wiring are exposed or accessible to "unauthorized" persons. Otherwise, there is no requirement for it to be grounded.

Not on switchboards:

- Primary of 1000 volts or greater shall not be required to be grounded.

- Primary of 300 to 999 volts shall be grounded if any live parts are exposed or accessible to "unauthorized" persons. Otherwise, there is no requirement for it to be grounded.

- For a primary less than 300 volts-to-ground, there is not a requirement to ground.

INSTRUMENT TRANSFORMER CASES
250.172

Standards to Review
NEC 250.170
NESC 151

The cases of instruments, meters and relays operating at 1000 volts or less shall be grounded. Instrument cases and relays which are not on switchboards and are accessible to other than qualified personnel shall have the cases or other exposed metal parts grounded where they operate at 300 volts or greater to ground per **250.174(A)**. **(See Figure 9-1)**

Figure 9-1. This Illustration shows that cases or frames enclosing the elements of instrument transformers shall be grounded where they are accessible to unqualified personnel.

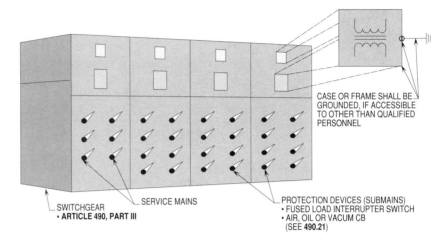

Instrument transformer cases or frames shall be required to be grounded if accessible to other than qualified personnel. Such cases or frames of current transformers shall not be required to be grounded where the primaries are not over 150 volts-to-ground and the current transformers are used exclusively to supply current to meters per **250.172, Ex. (See Figure 9-2)**

Figure 9-2. Under certain conditions of use, transformers used exclusively to supply current to meters shall not be required to have their cases or frames grounded.

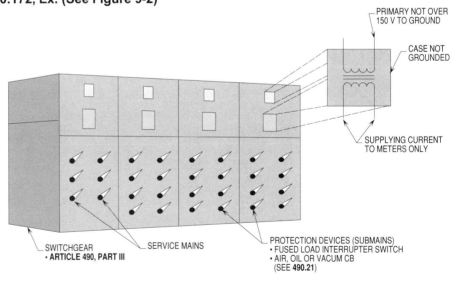

Grounding Tip 246: The NEC recommends that insulating mats for the protection of personnel to be used where the voltage-to-ground exceeds 150 volts. Insulating blankets and covers are also recommended if needed to provide additional safety per **250.174(C)**.

The cases of instruments, meters and relays operating at 1000 volts or less shall be grounded. Instrument cases and relays which are not on switchboards and are accessible to other than qualified personnel shall have the cases or other exposed metal parts grounded where they operate at 300 volts or more to ground per **250.174(A)**.

Dead-front switchboards having no live parts on the front of the panel, shall not be required to have the instrument cases grounded per **250.174(C)**.

Instruments, Meters and Relays

GROUNDING INSTRUMENT TRANSFORMER CASES

The grounding requirements of instrument transformer cases shall comply with the following:

- Potential transformer cases, if accessible to other than qualified persons, shall be grounded regardless of voltage. If not accessible, grounding shall not be required.

- Current transformer cases require grounding only if the primaries are over 150 volts, and then only if "accessible." If not accessible, grounding shall not be required.

Standard to review
NEC 250.172
NESC 151

CASES OF INSTRUMENTS, METERS AND RELAYS OPERATING AT LESS THAN 1000 VOLTS
250.174

Under the following conditions of use, instruments, meters and relays operating with windings or working parts at less than 1000 volts shall be grounded:

- Not on switchboards
- On dead-front switchboards
- On live-front switchboards

NOT ON SWITCHBOARDS
250.174(A)

Instruments, meters and relays not located on switchboards, operating with windings or working parts at 300 volts or greater to ground, and accessible to other than qualified persons, shall have the cases and other exposed metal parts grounded. **(See Figure 9-3)**

Grounding Tip 247: When testing for grounding problems, unless the various systems of indicators and wiring signals are working, all the personnel associated with such testing are in danger. High-voltage surges, heavy vibration, impact or just long use will burn out lamps. Frequent checks for burned out lamps should be made. Burn-outs should be replaced immediately. This routine also exposes problems not necessarily caused by grounding problems.

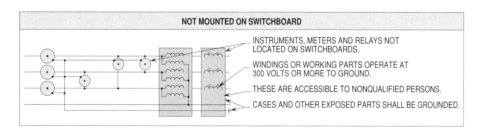

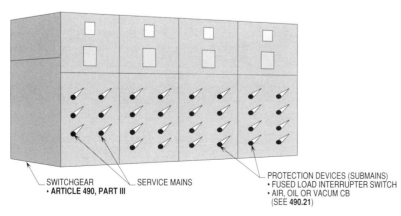

Figure 9-3. This illustration shows that exposed instrument devices, operating at 300 volts or more to ground, are not mounted on switchboard, but are accessible to unqualified personnel. Therefore, such instrument devices shall be grounded.

ON DEAD-FRONT SWITCHBOARDS
250.174(B)

Instruments, meters and relays whether operated from current and potential transformers or connected directly in the circuit on switchboards having no live parts on the front of the panels shall have their cases grounded. **(See Figure 9-4)**

Figure 9-4. These rules apply whether the equipment is operated from current or potential transformers or is directly connected to the circuit.

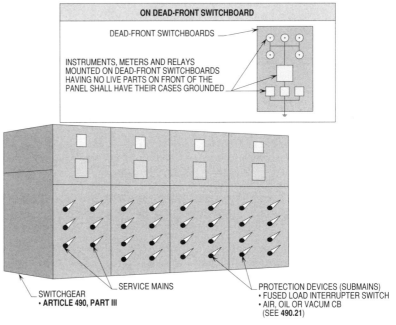

ON LIVE-FRONT SWITCHBOARDS
250.174(C)

Instruments, meters and relays operated from current and potential transformers or connected directly in the circuit on switchboards which have exposed live parts on the front of panels shall not be required to have their cases grounded. Mats of insulating rubber or other suitable floor insulation shall be provided for the operator where the voltage-to-ground exceeds 150 volts. Insulating mats will help isolate the operator from ground. **(See Figure 9-5)**

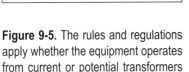

Standards to review
NEC 250.174(A) thru (C)
NESC 151

Figure 9-5. The rules and regulations apply whether the equipment operates from current or potential transformers or connects directly in the circuit.

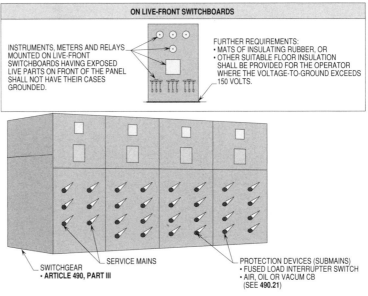

CASES OF INSTRUMENTS, METERS AND RELAYS - OPERATING VOLTAGE 1 kV AND OVER
250.176

For operating voltage that exceeds 1000 volts, the cases of instruments shall not be required to be grounded. However, they shall be isolated by elevation or protected by suitable barriers, grounded metal or insulating covers or guards. Electrostatic ground detectors shall be required to have the internal ground segments of the instruments connected to the instrument case and grounded properly. The ground detector is to be isolated by elevation per **250.176, Ex. (See Figures 9-6(a) and (b))**

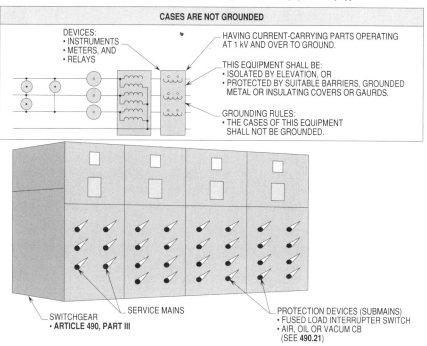

Figure 9-6(a). This illustration shows instrument devices operating at 1000 volts or greater with their cases ungrounded.

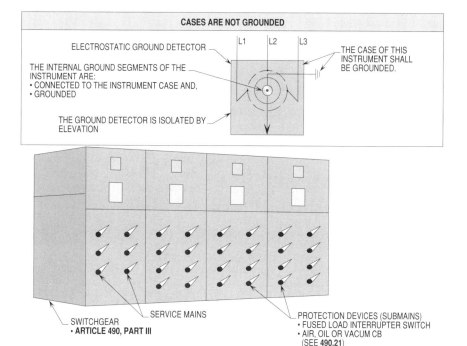

Figure 9-6(b). This illustration shows electrostatic ground detector operating at 1000 volts or greater shall have its case grounded.

The rules for enclosing and grounding the cases of ammeters, voltmeters, wattmeters, relays, etc. can be summed as follows:

- **Not on switchboards:** If voltage at instrument is less than 1000 volts-to-ground and the instrument is accessible to other than qualified persons, cases shall be grounded. Under 300 volts, they shall not be required to be grounded.

- **On switchboards:** Less than 1000 volts, on dead-front switchboards, cases shall be grounded. On live-front switchboards, cases shall not be required to be grounded.

- **1000 volts or greater to ground in any location:** Cases shall not be required to be grounded, but shall be isolated or suitably protected except for cases or electrostatic ground detectors having internal segments connected to the case, with the detector isolated by elevation. However, these cases shall be permitted to be grounded.

INSTRUMENT GROUNDING CONDUCTOR 250.178

The NEC specifies that the grounding conductor for circuits of instrument transformers and for instrument cases shall not be smaller than 12 AWG copper or 10 AWG aluminum. If the instruments are mounted on a metal surface, the instrument cases are considered properly grounded, and no additional grounding is required. **(See Figure 9-7)**

Standards to review
NEC 250.176
NESC 151

Standards to review
NEC 250.178
NESC 151
NESC 93.C(3)

Figure 9-7. This illustration shows the cases of instruments shall be permitted to be grounded by mounting them directly on metal surfaces or by using a grounding conductor.

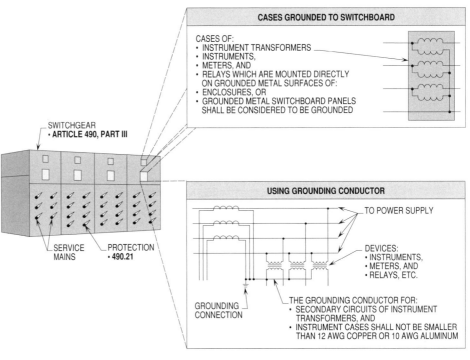

Name _____   Date _____

Chapter 9
Instruments, Meters and Relays

                                                                                                    Section   Answer

1. The cases of instruments, meters and relays operating at _____ volts or less shall be grounded.

2. Instrument transformer cases or frames shall be required to be grounded where the primaries are not over _____ volts-to-ground and the current transformers are used exclusively to supply current to meters.

3. Instruments, meters and relays not located on switchboards, operating with windings or working parts at _____ volts or greater to ground, and accessible to other than qualified persons, shall have the cases and other exposed metal parts grounded.

4. For operating voltage that exceeds _____ volts, the cases of instruments shall not be required to be grounded where isolated by elevation.

5. The grounding conductor for circuits of instrument transformers and instrument cases shall not be smaller than _____ AWG copper.

9-7

10

GROUNDING OF SYSTEMS AND CIRCUITS OF 1kV AND OVER (HIGH-VOLTAGE)

Special consideration must be given to grounding alternating-current (AC) systems and circuits of 1 kV (1000 volts) and over. The reason for grounding high-voltage systems (over 600 volts) is the same reasons as for grounding low-voltage systems (600 volts or less). System and circuits are solidly grounded for several reasons and they are as follows:

- To limit the voltage due to lightning
- To limit the voltage due to line surges
- To limit the voltage due to unintentional contact of the supply with higher voltage lines
- To stabilize the voltage-to-ground during normal operations
- To facilitate the operation of the overcurrent device in case of ground-fault conditions

Basically, it is up to the designer whether a high-voltage electrical system is grounded or not. Generally, it is a method in which the electrical system is utilized that really determines if it operates grounded or ungrounded.

DERIVED NEUTRAL SYSTEMS
250.182

A derived neutral from a grounding transformer shall be permitted to be used for grounding, rather than grounding directly to the neutral. However, there are basically three systems which are used for grounded neutral systems rated at 1000 volts and greater.

- Transformer-grounded neutral system
- Solidly grounded neutral system
- Impedance-grounded neutral system

See Figure 10-1 for the procedures used to ground high-voltage systems.

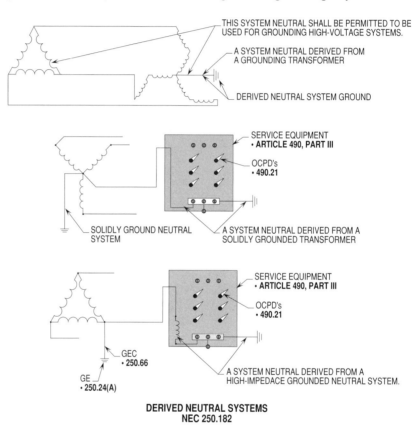

Figure 10-1. This illustration shows the procedures for deriving neutral systems to provide a grounded neutral for high-voltage systems.

Standards to review
NEC 250.182
NESC 215.B
NESC 93.C

The only condition where the NEC really requires high-voltage systems to be grounded is where the system supplies portable or mobile equipment per **250.20(C)**.

High-voltage systems supplying other than portable or mobile equipment are not required to be grounded but shall be permitted to be grounded if they are needed to be.

SOLIDLY GROUNDED NEUTRAL SYSTEMS
250.184

Standards to review
NEC 250.184
NESC 93.C
NESC 215.B

A solidly grounded neutral system is defined as grounded through an adequately grounded connection in which no impedance has been inserted intentionally. In other words, a solid metallic connection from system neutral to ground which has no impedance intentionally added in the grounding circuit.

Grounding of Systems and Circuits of 1 kV and over (High-voltage)

NEUTRAL CONDUCTOR
250.184(A)

Standards to review
NEC 250.184(A)
NESC 93.C
NESC 215.B

600 volt insulation shall be required for the neutral of high-voltage circuits. The neutral may be bare copper when used as a neutral of a service-entrance, or where used as the neutral of buried feeders. A bare neutral of copper or other metals may be run with overhead conductors outdoors.

MULTIPLE GROUNDING
250.184(B)

A high-voltage neutral shall be permitted to be grounded at more than one point when used in one of the following methods:

- Transformers supplying conductors to a building or other structure
- Buried feeders having a bare copper neutral
- Conductors outside routed overhead

Otherwise, only one ground shall be permitted if the installation does not comply with one of the methods listed above.

See Figure 10-2 for a detailed illustration of the grounding procedures used for solidly grounded neutral systems.

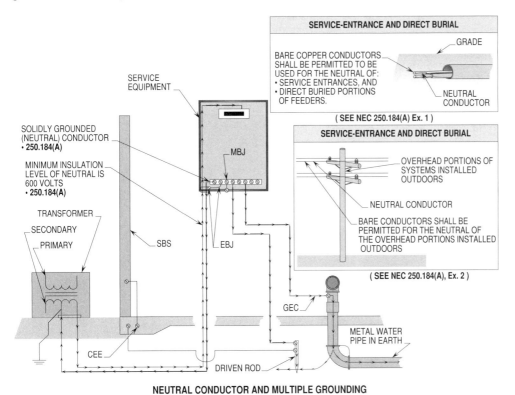

Figure 10-2. This illustration shows the grounding procedures involving a solidly grounded neutral system. Note that the grounded (neutral) conductor is used for fault-current return path.

Grounding Tip 248: In many existing installations, a ground grid of steel rebar is designed and installed with transformer, service equipment, feeder and branch-circuit fed equipment that is bonded and grounded to this low-resistance electrode. The grid system is used as a path by fault currents to travel over and clear OCPD's.

NEUTRAL GROUNDING CONDUCTOR
250.184(C)

Standards to review
NEC 250.184(C)
NESC 93.C
NESC 215.B

The neutral grounding conductor shall be permitted to be installed as a bare conductor if isolated from ungrounded (phase) conductors and protected from physical damage.

Standards to review
NEC 250.186
NESC 215.B

IMPEDANCE GROUNDED NEUTRAL SYSTEMS
250.186

Impedance grounding provides opposition to current flow by the impedance which is intentionally connected in series with the grounding electrode conductor and the system neutral. The impedance will intentionally limit the current flow to ground to a designed value. A resistor or impedance coil is utilized and both are known in the industry as "impedance grounding" due to the alternating current.

Grounding Tip 249: With reactance grounding, the fault current should be at least 60 percent of, but not more than the three-phase fault current. This is particularly important if generators are connected since such machines are generally braced only for the expected three-phase fault current.

The purpose of impedance grounding is to limit the current flow to ground if a ground-fault should occur, thereby creating a measure of protection for equipment from such ground-faults. By limiting the current flow, the arc at the point of fault can be eliminated or limited to a safe value. This is the main reason that the neutral grounded system is required for use with portable or mobile equipment operating at 1000 volts or greater to ensure a safe trip of the OCPD.

LOCATION
250.186(A)

The neutral of a high-voltage system shall be permitted to be grounded through an impedance coil placed in the grounding conductor.

Grounding Tip 250: With low-resistance grounding, the fault current will normally be in the range from 10 percent to 60 percent of the three-phase fault current. For high-resistance grounding, the fault current is intended to be limited to a smaller fraction of the three-phase fault current, generally less than 50 amps.

The impedance for an impedance-grounded neutral system has to be inserted in the grounding conductor between the grounding electrode of the supply system and the neutral point of the supply transformer or generator.

The impedance shall be connected in series with the following:

- The neutral
- The grounding electrode
- The grounding electrode conductor

See Figure 10-3 for the location of the impedance grounding unit.

Figure 10-3. This illustration shows the grounding impedance shall be installed in the grounding conductor between the grounding electrode of the supply side and the neutral point of the supplying transformer.

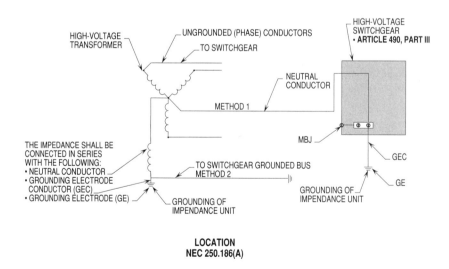

LOCATION
NEC 250.186(A)

Grounding Tip 251: The identification of the neutral shall comply with **200.6, 200.7** and **310.12(A)** and shall be insulated the same as the ungrounded (phase) conductors.

IDENTIFIED AND INSULATED
250.186(B)

When an impedance coil is used, no direct ground shall be permitted. The neutral of an impedance grounded neutral system shall be fully insulated with the same insulation as the ungrounded (phase) conductors and identified as such. **(See Figure 10-4)**

Grounding of Systems and Circuits of 1 kV and over (High-voltage)

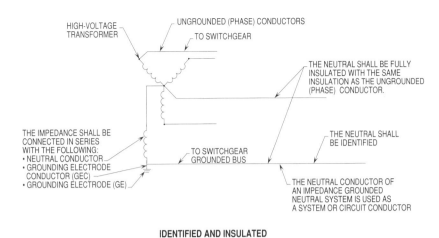

Figure 10-4. This illustration shows the neutral conductor of an impedance grounded system insulated and identified from the ungrounded (phase) conductors.

SYSTEM NEUTRAL CONNECTION
250.186(C)

The impedance neutral shall be identified and insulated. Insulation shall be equal to that of the ungrounded (phase) conductors. For an impedance neutral grounded system, the neutral originates at the wye point of the transformer and extends to the line side of the grounding resistor.

The line side of the resistor is the neutral. The grounded side of the resistor is the grounding electrode conductor. Therefore, there is a voltage between the line side and the grounded side of the system neutral impedance. This connection creates a voltage between the neutral and ground. The voltage from the neutral to ground is the voltage drop across the impedance, and varies with the impedance of the resistor and the current flowing through the impedance. **(See Figure 10-5)**

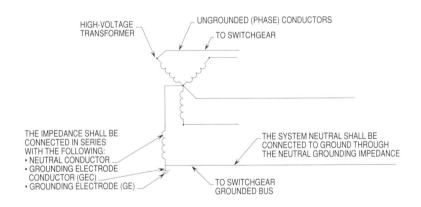

Figure 10-5. This illustration shows the system neutral shall be connected to ground through the neutral grounding impedance.

EQUIPMENT GROUNDING CONDUCTORS
250.186(D)

Equipment grounding conductors shall be permitted to be bare. They shall be connected to the ground bus and grounding electrode conductor and the service-entrance equipment and extended to the system ground. The neutral shall not be connected to the grounding electrode conductor. This conductor shall be connected to the grounded side of the grounding impedance that is connected to the grounding electrode conduc-

Grounding Tip 252: For requirements pertaining to grounding electric systems and equipment using the high-impedance technique to protect workers as well as preventing disorderly outage of power, see 2-6.1.5(5) in NFPA 70E.

tor. The equipment grounding conductor shall not be connected to the neutral at the service or ahead of the grounding impedance, but to the grounding electrode conductor on the grounded side of the grounding impedance. **(See Figure 10-6)**

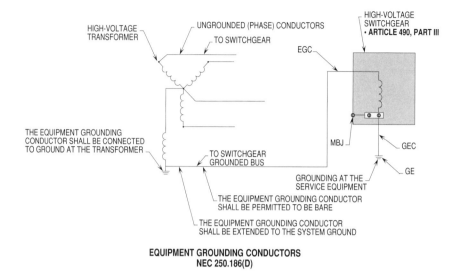

Figure 10-6. This illustration shows the equipment grounding conductor shall be connected to the grounding bus and grounding electrode conductor at the service-entrance equipment (switchgear).

Standards to review
NEC 250.188
NESC 93.C
NESC 215.B

GROUNDING OF SYSTEMS SUPPLYING PORTABLE OR MOBILE EQUIPMENT
250.188

High-voltage electrical systems supplying portable or mobile equipment shall be grounded and treated in a different manner than those used for substations installed on a temporary basis.

Standards to review
NEC 250.188(A)
NESC 93.C
NESC 215.B

PORTABLE OR MOBILE EQUIPMENT
250.188(A)

Portable equipment shall be supplied from a system having an impedance ground. In other words, the neutral shall be grounded through the impedance which is a special designed resistor or coil designed to limit current flow. **(See Figures 10-7(a) and (b))**

Figure 10-7(a). This illustration shows that portable or mobile equipment shall be supplied by a high-voltage supply which has its neutral grounded through an impedance unit.

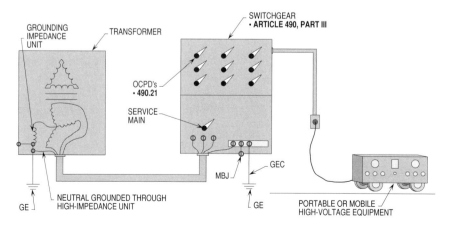

Grounding of Systems and Circuits of 1 kV and over (High-voltage)

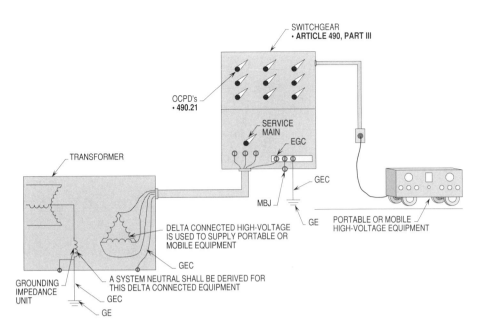

Figure 10-7(b). This illustration shows that delta-connected high-voltage systems are sometimes used to supply portable or mobile equipment.

EXPOSED NONCURRENT-CARRYING METAL PARTS
250.188(B)

Exposed metal parts of portable equipment shall be grounded to the point at which the neutral impedance is grounded. **(See Figure 10-8)**

Standards to review
NEC 250.188(B)
NESC 215.C
NESC 123.A

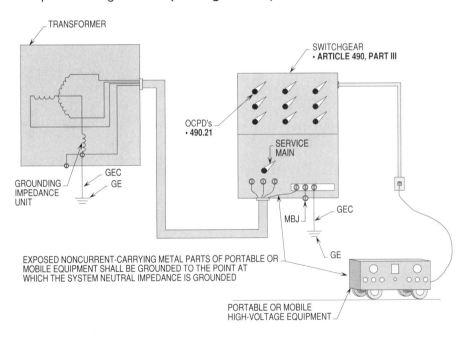

Figure 10-8. This illustration shows exposed metal parts of portable or mobile equipment shall be grounded by an equipment grounding conductor.

GROUND-FAULT CURRENT
250.188(C)

Maximum ground-fault current shall not develop a voltage between equipment frame and ground of over 100 volts. A properly designed impedance unit will limit the voltage to this value. **(See Figure 10-9)**

Standards to review
NEC 250.188(C)
NESC 123.A

Figure 10-9. This illustration shows that the maximum fault-current between portable and mobile high-voltage equipment shall be limited.

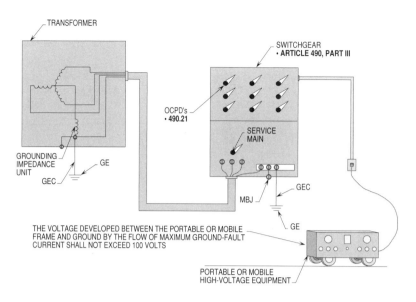

GROUND-FAULT DETECTION AND RELAYING
250.188(D)

Standard to review
NEC 250.188(D)
NESC 93.C(5)

A ground-fault detection system shall be provided that disconnects a high-voltage system that has developed a ground-fault. A detection system shall be provided that disconnects the high-voltage circuit to the portable equipment should a break in the equipment grounding conductor occur. **(See Figure 10-10)**

Figure 10-10. This illustration shows that ground-fault detection and relaying shall automatically deenergize any high-voltage element that develops a ground-fault.

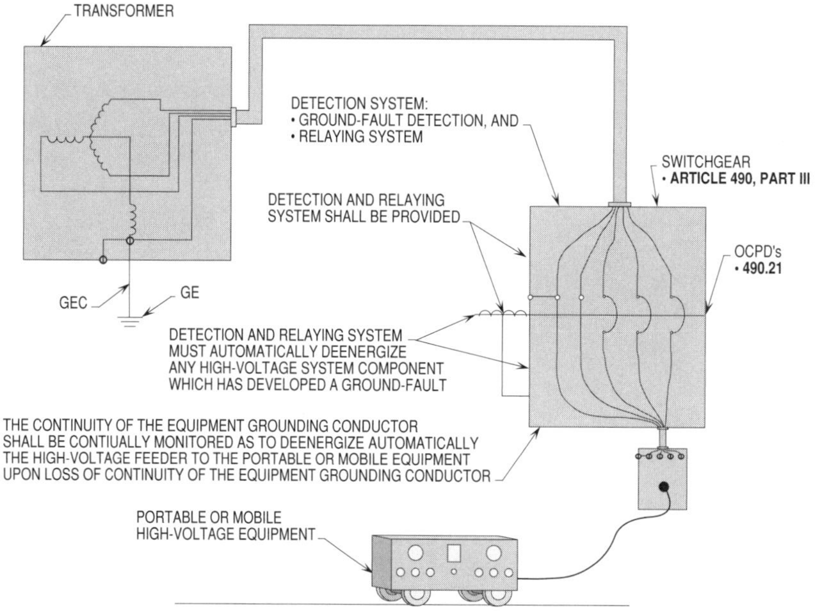

Grounding Tip 253: High-voltage system elements, developing a ground-fault shall be automatically deenergized by a dependable ground-fault detection and relaying system.

Standards to review
NEC 250-188(E)
NESC 123.A

ISOLATION
250.188(E)

The grounding electrode for the high-voltage circuit feeding portable equipment shall be separated at least 20 ft. (6 m) from other electrodes. There shall not be any connection between any buried pipe or fence electrode. **(See Figure 10-11)**

Grounding of Systems and Circuits of 1 kV and over (High-voltage)

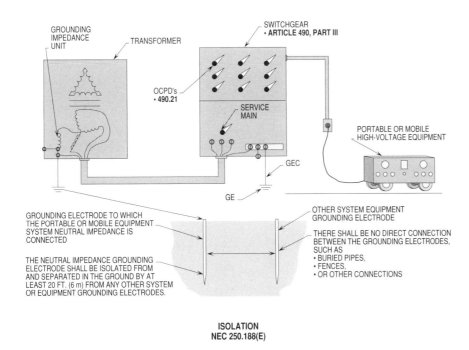

Figure 10-11. This illustration shows the grounding electrode, which grounds portable and mobile equipment, shall be isolated from other system electrodes.

TRAILING CABLE AND COUPLERS
250.188(F)

High-voltage trailing cables and couplers for interconnection of portable equipment shall be of a type approved for such use. **(See Figure 10-12)**

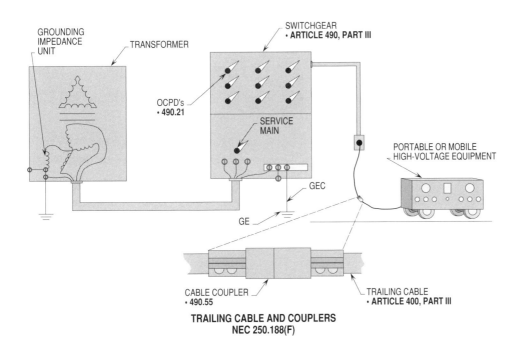

Figure 10-12. This illustration shows high-voltage trailing cables and couplers that are used for the interconnection of portable and mobile equipment and shall comply with the rules listed in **Part III** in **Article 400** for cables and **490.55** for couplers in the NEC.

Grounding Tip 254: The mandatory statement of **250.188(D)** to automatically deenergize does not allow the use of an audio visual alarm and then manual operation of the system disconnecting means.

GROUNDING OF EQUIPMENT
250.190

All noncurrent-carrying metal parts of fixed or portable equipment, including associated fences, housing, etc. shall be grounded. Such noncurrent-carrying metal parts that can be associated with fixed or portable equipment are as follows:

Standards to review
NEC 250.190
NESC 93.C(5)
NESC 123.A
NESC 215.C

- Fixed equipment
- Portable equipment
- Mobile equipment
- Associated fences
- Equipment housings
- Equipment enclosures
- Conductor enclosures
- Supporting structures
- 6 AWG copper
- 4 AWG aluminum

See Figure 10-13 for the rules concerning the grounding of noncurrent-carrying parts.

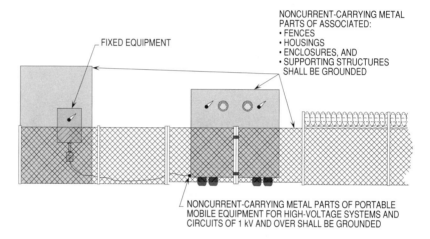

Figure 10-13. This illustration shows that all noncurrent-carrying metal parts, supplied by high-voltage systems, shall be grounded.

Standards to review
NEC 250.190, Ex. 1 and FPN
NESC 123.A
NESC 215.C

NOT REQUIRED TO BE GROUNDED
250.190, Ex.

Under certain conditions of use, the following equipment shall not have to be grounded:

- Where isolated from ground and located to prevent personnel who can make contact with ground from making contact with metal parts of equipment and ground when equipment is energized.
- Pole-mounted distribution apparatus per **250.110, Ex. 2**.

See Figures 10-14(a) and (b) for a detailed illustration of rules and regulations for equipment that shall not be required to be grounded.

Grounding of Systems and Circuits of 1 kV and over (High-voltage)

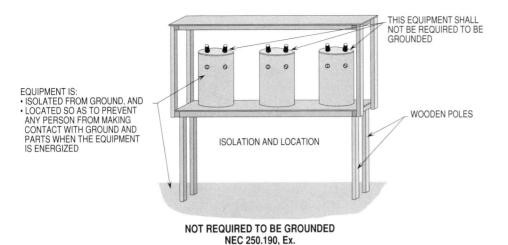

Figure 10-14(a). This illustration shows equipment that is isolated and shall not be required to be grounded, under certain conditions of use.

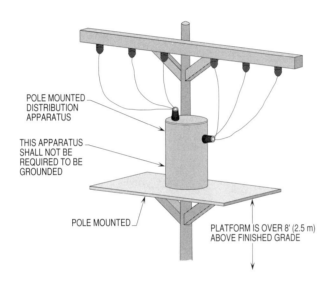

Figure 10-14(b). This illustration shows equipment that is isolated and shall not be required to be grounded, under certain conditions of use.

Name _____   Date _____

Chapter 10
Grounding of Systems and Circuits of 1 kV and Over (High-Voltage)

	Section	Answer

1. The neutral for high-voltage circuits shall be required to have an insulation of 600 volts. _____   T   F

2. The neutral for a solidly grounded system of 1 kV and over shall be permitted to be grounded at more than one point for services. _____   T   F

3. The neutral of an impedance grounded neutral system of 1 kV and over shall not required to be insulated. _____   T   F

4. Equipment grounding conductors for systems of 1 kV and over shall be required to be fully insulated and identified. _____   T   F

5. High-voltage portable equipment shall be required to be supplied from a system having an impedance ground. _____   T   F

6. Exposed metal parts of high-voltage portable equipment shall be required to be grounded to the point at which the neutral impedance is grounded. _____   T   F

7. Maximum ground-fault current shall not develop a voltage between equipment frame and ground of over 1000 volts for high-voltage portable equipment. _____   T   F

8. The grounding electrode for the high-voltage circuit feeding portable equipment shall be separated at least 10 ft. from other electrodes. _____   T   F

9. All noncurrent-carrying metal parts of high-voltage mobile equipment shall be grounded. _____   T   F

10. Pole-mounted distribution apparatus over 8 ft. 6 in. from finished grade shall be required to be grounded for high-voltage systems. _____   T   F

10-13

11

GROUNDING INFORMATION TECHNOLOGY EQUIPMENT

Grounding accomplishes multiple functions, all of which must be considered in the design and installation of information technology equipment (ITE) systems. Persons concerned with only one or two of these functions may violate the other in ignorance or because they are not his or her responsibility. Grounding is required both for safety reasons and because of the need for highly sensitive electronic circuits to operate reliably. Safety takes top priority, but the computer systems must be simultaneously safe and operationally reliable. There must be no compromise when it comes to safe grounding techniques.

GROUNDING FACTS
ARTICLE 250 OF THE NEC

For more than 100 years, attention to lightning protection and power transmission fault control, electricians and electrical engineers have been exhorted to create and use low-resistance ground connections to earth. This is appropriate for lightning and transmission line ground-faults since part of their paths is through earth. However, this in not the rationale for applying grounding principles to 120, 240, 208 and 480 volt utilization circuits. At these voltages, a system interconnected or bonded conductors acting as a voltage reference network can equalize voltage differences throughout the network much more effectively than multiple low-impedance earth contacts.

BONDED NETWORK
250.4(A)(1) thru (A)(5)

Such an interconnected, bonded network can serve as both a power and signal reference, regardless of its voltage with respect to earth ground. However, to avoid shock hazards in a building structure and to minimize voltage differences between individual reference networks, it is not only accepted practice but mandatory for safety purposes that these networks be connected to earth ground. Since the connection to earth is never expected to carry load or fault current, the National Electrical Code (NEC) permits this conductor to be sized per **Table 250.66**. **(See Figure 11-1)**

Figure 11-1. This illustration shows the supply grounded (neutral) conductor, equipment grounding conductor, isolated equipment grounding conductor, neutral conductor and the halo bonding jumper which are used together to form an interconnected bonded network.

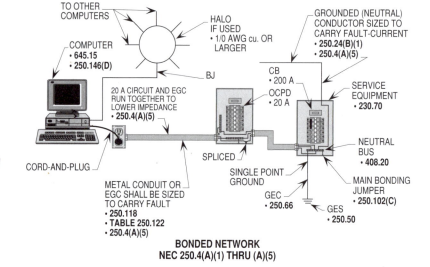

GROUND CURRENTS
250.53(G); 250.56

If there are ground currents in a driven earth ground electrode from any source, a low-resistance between earth and grounding conductor with its attached electrode will help prevent generation of unwanted electrical noise which sometimes appear in coincidence with a high-voltage gradient in dried-out soil. If ground currents appear to be exclusively high (more than about an ampere or two in a residence to more than 20 amperes in a large building), it is advisable to determine the source of such currents and reduce its magnitude if at all feasible to do so. High currents can cause deterioration of the electrode and become an increasing source of electrical noise. **(See Figure 11-2)**

Figure 11-2. This illustration shows the current in amps flowing through a driven rod to earth shall be limited in some cases.

Grounding Tip 255: High currents on a driven rod can bake-out and dry the soil around the rod and this can cause unwanted noise problems for computer systems.

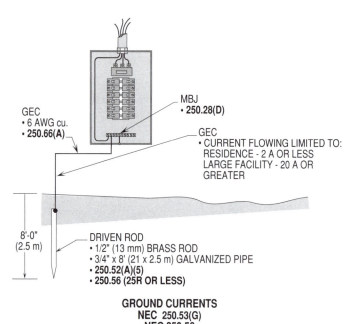

TECHNIQUES OF GROUNDING
ARTICLE 250 IN THE NEC

For ITE systems, minicomputers, microprocessors and office machines, a copper rod driven into moist soil is not a magic cure-all for grounding problems, or is it necessarily a requirement. Personnel shall be protected from electrical shock and facilities shall be protected from fire hazards. For this to be accomplished, sensitive electronic equipment shall be grounded with an equipment grounding conductor which provides a low-impedance path for a ground fault to return from the point of fault back to the common grounding bus of the service equipment or separately derived system supplying the power. Various functions and details of sensitive electronic system grounding may be summarized as follows:

- Touch voltage differences shall be limited by bonding and grounding to avoid a shock hazard.
- Ground-fault current return path to power source shall have low enough impedance to enable it to actuate OCPD's and disconnect the power source.
- Ground potential differences in the ITE shall be reduced to essentially a constant potential reference.
- Grounded conducting enclosures serve as electromagnetic shielding for sensitive circuits.
- Grounding in compliance with NEC and other safety codes is mandatory.
- ITE manufacturer's recommendations should be followed to the extent that they are consistent with the above bullets. Inconsistencies must be resolved and corrected properly.

Grounding Tip 256: All current must flow in loops and any current that enters the earth must also leave the earth. Each component of earth current results in an earth potential difference. Its these potential differences that are sources of trouble for users of electronic equipment and associated apparatus.

TOUCH VOLTAGE

Touch voltage is the voltage between any two conductive surfaces which can be simultaneously touched by an individual. The earth may at times be one of these surfaces. There should be no "floating" panels or enclosures in the vicinity of electric circuits. In the event of insulation failure or inadvertent application of moisture, any electric charge which appears on a panel, enclosure, metal cable or raceway shall be drained to "ground" or to an object which is reliably grounded. Note that a "safe" touch voltage of 30 volts rms or less will not necessarily satisfy system requirements for low noise signal differences (0.5 V or less) between parts of ITE grounded systems. **(See Figure 11-3)**

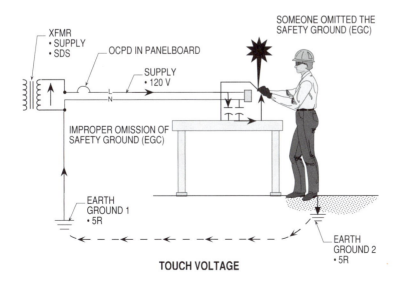

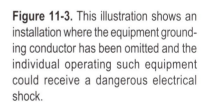

Figure 11-3. This illustration shows an installation where the equipment grounding conductor has been omitted and the individual operating such equipment could receive a dangerous electrical shock.

Grounding Tip 257: If there was a total resistance of 10 ohms at earth grounds 1 and 2, the current to the OCPD would be about 12 amps:

- A = 120 V ÷ 10R
- A = 12

Solution: About 12 amps will flow and would not trip open a 15 amp or 20 amp OCPD.

Grounding Tip 258: A driven rod and earth shall not be used as the sole (without the aid of an equipment grounding conductor) grounding means per **250.4(A)(5), 250.54** and **250.136(A)** in the NEC.

Figure 11-4. This illustration shows the proper path of a ground-fault as the current in amps flows from the faulted area back to the source.

Grounding Tip 259: An electrical pressure of 1 volt causes 1 amp to flow in 1 ohm. If the voltage increases, the current increases proportionately. Therefore, differences of potential between pieces of equipment can be a means of trouble for sensitive types of equipment.

Grounding Tip 260: The supplementary conductors are used to reduce unwanted noise problems, but are never used as equipment grounding conductors to clear ground-faults.

GROUND-FAULT RETURN PATH
250.4(A)(5)

A ground-fault return path to the point where the power source grounded (neutral) conductor is grounded is an essential safety feature. The National Electrical Code and some local wiring codes permit electrically continuous conduit and wiring device enclosures to serve as this ground-fault return path. Other codes may require the conduit to be supplemented with a bare or insulated conductor included with the other ungrounded (phase) conductors. In either case, the conduit, equipment grounding conductor (if used) and the power source's grounded (neutral) conductor are bonded together at the power source if it is located in the building or at the building entrance if the power source is located outside the building. If it is part of the premise's wiring, a transformer secondary winding which is isolated from the primary winding shall be considered to be a separately derived power source (SDS) and one of its conductors (neutral) shall be grounded.

If grounded properly, an insulation failure or other fault which allows an ungrounded (phase) conductor to make contact with the enclosure will find a low impedance path back to the power source neutral. This will be virtually a short-circuit by definition of **Article 100** in the NEC. The resulting overcurrent will cause the circuit breaker or fuse to disconnect the faulted circuit promptly. Driven ground rods used as the sole ground for such equipment are not part of this path and violate **250.54** in the NEC. **(See Figure 11-4)**

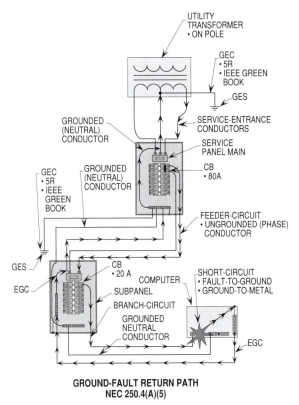

GROUND-FAULT RETURN PATH
NEC 250.4(A)(5)

GROUND VOLTAGE EQUALIZATION

Ground voltage equalization of voltage differences commonly found between parts of an ITE grounding system is accomplished in part when the equipment grounding conductors are connected to the grounding point of a single power source. If the equipment grounding conductors are long, and if the ground currents are significant, the impedance of the grounding conductors may be too high to achieve a constant potential throughout the grounding system. Supplementary conductors which may be needed are in addition to the equipment grounding conductors which are required for safety and not a replacement for them. These supplementary conductors control only the unwanted noise. **(See Figure 11-5)**

Grounding Information Technology Equipment

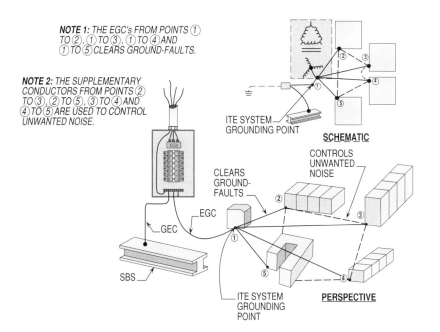

Figure 11-5. This illustration shows supplementary conductors used to control unwanted noise between the various ITE equipment enclosures.

SHIELDING REQUIREMENTS

Shielding may consist of metal barriers, enclosures or wrappings around one, the other, or both circuits. Such shielding, between sensitive circuits and sources of disturbance, must be connected to a signal reference such as ground in order to be effective. A floating shield can be worse than no shield at all. A long grounding conductor (5 ft. (1.5 m) or more) to a shield reduces the shield's effectiveness at high frequency since the inductance of the conductor obstructs the flow of current to the shield. To be effective at high frequencies, above 10 MHz, the shield must be grounded with short connections at multiple locations. Single point grounding at one end of a shield does not equalize the voltage along the length of the shield at high frequencies or during a pulse of noise current. Experience has proven that, shield voltage at the ungrounded end will couple to the inner conductor and will be partly converted to a normal mode signal as it reaches the grounded end. **(See Figure 11-6 and refer to Figure 11-54 and related text on page 11-38)**

Grounding Tip 261: A shield interposed between the primary and secondary coils can be used to control common-mode current flow. A common-mode signal simply stated is the average signal with respect to a reference conductor.

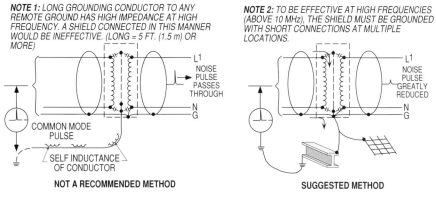

Figure 11-6. This illustration shows specific grounding techniques to either use or not use when dealing with shielding for unwanted noise.

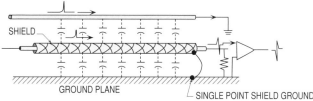

Grounding Tip 262: Noise and unwanted harmonics appear in the form of line-to-line and line-to-neutral voltage signals (normal mode), and also as noise voltages between the local reference point and each of the power conductors including neutral and sometimes equipment grounding conductors (common mode).

Grounding Tip 263: Safety code standards such as NFPA, UL, NEC and OSHA must be reviewed and understood before trying to apply their requirements to a computer installation.

COMPLIANCE WITH SAFETY CODES
90.7; 110.2; 110.3(B)

Compliance with safety codes is important, but it is also important to understand the basis for specific code requirements. Codes are often misinterpreted by individuals who lack understanding, but may be installing wiring, inspecting the job or interpreting specifications. A little knowledge of these basic principles can help one identify and resolve many problems before any harm is done to personnel and equipment.

TECHNIQUES OF GROUNDING SAFELY
110.3(B)

Grounding Tip 264: When in doubt that a computer installation complies with codes and standards, review one of the following:

- NEC
- OSHA 1910, Subpart S
- NFPA 50
- NFPA 75
- UL 467
- UL 1950
- UL 1363
- UL 1449

ITE manufacturer's grounding instructions and recommendations are important and should be followed. Occasionally, however, the grounding techniques specified for various units are inconsistent with the interpretations of wiring codes by contractors, electricians and inspectors. When there is doubt, there shall be no compromise with safety. The system shall be safe and shall be capable of operating reliably without compromise and maybe deleting safety grounds which will introduce new hazards.

Where the grounding instructions appear to conflict with NEC requirements and other codes and standards, the manufacturer who prepared the instructions should be consulted and asked to resolve the problem. Any arbitrary independent departure from the installation instructions or electrical codes can result in taking the responsibility and accepting part of the liability for safety hazards or failure of the sensitive electronic equipment to perform reliably. **(See Figure 11-7)**

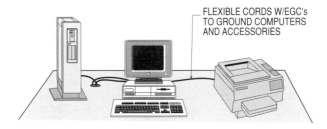

Figure 11-7. This illustration shows that grounding recommendations from the manufacturer must be considered when planning a grounding scheme for a computer system.

Grounding Tip 265: When codes place a specific requirement pertaining to a computer grounding scheme, such rules shall not be violated, unless the AHJ permits alteration of such rule.

REQUIREMENTS OF NEC AND OTHER CODES AND STANDARDS
ARTICLE 250; NFPA 50; NFPA 75; OSHA 1910, SUBPART S

Electrical safety and code requirements must always be given serious attention. Compliance is always recommended whenever it is possible to do so, especially if it removes a safety hazard. Occasionally, interpretations of codes and their applicability have to be questioned when they are believed to be wrong or unnecessarily restrictive. Codes are constantly being changed, amended or expanded because new situations are constantly arising which were not anticipated when the codes were first written.

APPROVAL OF EQUIPMENT
110.2; ARTICLE 100; 110.3(A); (B)

Sometimes an interpretation will depend upon whether the governing safety standard applies to building wiring or to a factory assembled product to be installed in a building. Underwriters Laboratories (UL) and other qualified testing laboratories examine products at the request and expense of manufacturers or purchasers, and will list products if the examination reveals that the product presents no significant safety hazard when properly installed.

Municipal, county and state inspectors generally accept UL and other qualified testing laboratories certification "listings" as evidence that a product is safe to install and be used. Without a listing, a test or examination by the safety department's own laboratory may be performed (if there is one), it may not be possible to obtain wiring permits and inspection sign-offs for connecting power to the loads until such an examination is performed and settled. Most sensitive electronic equipment manufacturers routinely seek and obtain UL listing of their products. Without it, there can be installation delays and problems with permits. **(See Figure 11-8)**

Grounding Tip 266: When approving equipment, it is important to review and understand the definitions of the following terms in **Article 100** in the NEC:

• Approved
• Identified
• Listed
• Labeled

Note: For acceptance of equipment, also review 1910.399 in OSHA, Subpart S.

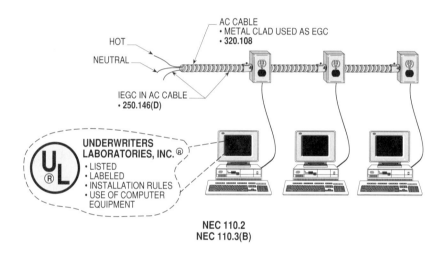

Figure 11-8. This illustration shows computer units which are listed and labeled. Such markings for equipment eliminates delays for obtaining permits and inspection sign-offs.

OSHA, NFPA, UL AND LOCAL CODES
110.2; OSHA 1910.399; NFPA 75; UL 478; UL 1950

On-site wiring must generally comply with local wiring codes as well as the NEC. Most of these are based upon the NEC. In addition, there is NFPA 75 which is the Standard for the "Protection of Electronic Computer/Data Processing Equipment". These codes specify wiring materials, wiring devices, circuit protection and wiring methods. There are sections devoted specifically to ITE installations. Such codes are primarily directed toward prevention of electrical shock and smoke or fire-related hazards to minimize loss of property.

Some communities, particularly large cities, will require site construction features which are considerably more restrictive and expensive to install than those specified by the NEC. In some instances, these features have been a factor for picking ITE sites outside of these areas. Sometimes it is not sufficient that the product to be installed and the wiring devices and methods are UL listed and covered by code. See OSHA 1910.399 and NEC **110.2** with **FPN** for equipment considered one of a kind (custom-made equipment). The cooperation and objective technical judgement of good electrical inspectors, electrical contractors and electricians can result in a safe and reliable ITE installations without arbitrary unnecessary restrictions and expense. **(See Figure 11-9)**

Grounding Tip 267: NFPA 75 and UL 1950 are two standards that designers, installers and inspectors should have in their library for fast and easy reference when designing, installing and inspecting a computer installation.

Figure 11-9. This illustration shows the standards that are sometimes referenced with the NEC to aid in designing, installing and inspecting computer installations.

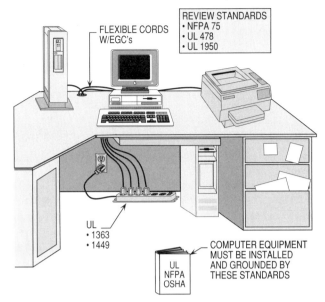

OSHA, NFPA, UL AND LOCAL CODES
NEC 110.2; OSHA 1910.399; NFPA 75; UL 478; UL 1950

Grounding Tip 268: UL 1363 is the standard used for "Relocatable Power Taps" and UL 1449 is the standard used for "Transient Voltage Surge Suppressors."

CENTRAL GROUNDING POINT
250.130(A)

The central grounding point of sensitive electronic system should be readily identifiable. It should be the point where the interconnected parts of the entire grounding system are connected to other grounding conductors which extend beyond and outside the ITE room. If there are two such points, each being interconnected to each other and to separate external grounds, the noise voltage difference between those separate external grounds will cause unwanted noise current to flow through the ITE grounding system through the ground loop which is formed. **(See Figure 11-10)**

Within very large systems, there may be subsystems, each with a central grounding point for connection to other central grounding points. However, separate connections to separate external grounds would create unwanted external ground loops. By such means, impulse ground currents can find paths in the grounded shields and grounded conductors of signal pairs and coaxial cables. Intercoupling with digital circuits and signal corruption can be the unwanted results. **(See Figure 11-11)**

Figure 11-10. This illustration shows a central grounding point at the UPS when all equipment grounding conductors from each ITE is connected.

Grounding Tip 269: The ITE at points ②, ③, ④ and ⑤ are grounded with an equipment grounding means to point ① which is the central grounding point used to reduce unwanted noise problems.

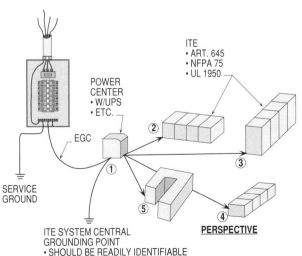

CENTRAL GROUNDING POINT
NEC 250.130(A)

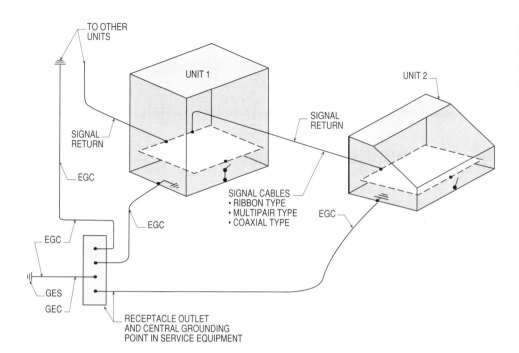

Figure 11-11. This illustration shows the equipment grounding conductors from each unit routed to a central grounding point in the service equipment.

FREQUENCY OF NOISE SIGNALS

The frequency of noise signals (any signal other than the desired signal) can vary from DC to MHz and even GHz frequencies. Few detectors will respond to the entire spectrum and fortunately the digital circuits will not either. However, the trend in devices used in sensitive electronic equipment circuits, today, is toward to ever increasing bandwidth and lower signal levels.

As signal frequencies in computers reach and exceed approximately 10 MHz, radiation of the signals and coupling to adjacent circuits become increasingly troublesome problems. It requires more costly wiring techniques such as twin lead, twisted pair, strip lines with ground planes or coaxial conductor construction to contain the electromagnetic fields associated with signals and noise. When computers operate at frequencies above 30 MHz, concern over line conducted signals and noise become overshadowed due to the importance of radiated signals and noise.

Computers with higher frequency performance circuits requires careful shielding of all circuits longer than a few inches. Coaxial conductors and other types of cables have become appropriate for higher frequencies. However, some systems are prime candidates for replacement with fiber optic signal transmission which is already being put into communications service to carry these digital signals. Fiber optics solves some of the common mode noise problems (refer to Grounding Tip 262 on page 11-5) that bother today's computer circuits. **(See Figure 11-12)**

Grounding Tip 270: Noise and interference can be carried to areas of sensitivity on powerlines, signal cables, building steel or on any utility conduit. The noise or interference can result from load switching, lightning, fault conditions, short-circuits or power factor corrections where capacitors are switched. This energy can also result from rf transmitters, radars, welders, electrostatic discharge or nearby electronic equipment and the problem is locating the point where such a noise is coming from. Note that this noise may also come from unsnubbed switched inductive loads such as relays, contactors, solenoids, etc.

GROUNDING HIGHER FREQUENCIES

In today's technologies with digital data and control signals, any DC and low AC frequency (100 kHz or less), signal circuits will follow the lowest resistance paths where conductors may be the largest and shortest. However, at high radio frequencies (above 100 kHz), stray capacitance and electromagnetic coupling become significant circuit paths. The path taken by DC currents can be tortuous and have too high an impedance to be a good high frequency path. At low frequencies where currents follow conductors,

Figure 11-12. This illustration shows the shields of signaling cables grounded at the chassis of the computers to help reduce unwanted noise.

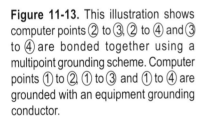

FREQUENCY OF NOISE SIGNALS

Grounding Tip 271: Single point grounding consist of grounding from the metal frames of computers to a single grounding point.

single point grounding is generally preferred. Note that the signal frequencies exceeding approximately 10 MHz and greater, the noise currents and voltage signals cannot easily be confined to conductors. In this realm, multipoint grounding of the system becomes necessary if its to be effective.

If the desired signal must be protected against both high and low frequency interference, a solid metallic galvanic grounding connection is needed for a single point ground, while at high frequencies one can use multiple ground paths through deliberate use of stray or discrete capacitors. A very effective technique is to have multipoint grounding connections to an outer shield over an inner insulated shield which has a single point ground. **(See Figure 11-13)**

Figure 11-13. This illustration shows computer points ② to ③, ② to ④ and ③ to ④ are bonded together using a multipoint grounding scheme. Computer points ① to ②, ① to ③ and ① to ④ are grounded with an equipment grounding conductor.

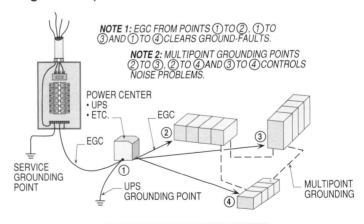

GROUNDING HIGHER FREQUENCIES

CONCEPT OF SINGLE POINT ENTRY

Grounding Tip 272: Single point entry consist of all conduits related to the computer system to enter from the same wall, ceiling, floor, etc. If such bonding of metal conduits and cables are performed properly, voltage differences can be reduced.

The single point of entry wiring strategy for ITE rooms is a practical approach to establishing a virtual single point ground for the ITE system, the communications system, the power source and life safety system. If all external conductors penetrating the ITE room were to enter at a single point rather than at multiple points around the room, their respective grounding conductors could all be interconnected at the point of penetration.

The expected benefit is that noisy ground currents flowing between the power, communications and other grounding conductors can flow through the short ground interconnections at the entry point rather than going through the grounding conductors within and among ITE system units. The remaining question is, "how short is short"? The answer is to limit the length of these interconnections with straps about 2 ft. (600 mm) to 4 ft. (1.2 m) in length. **(See Figure 11-14)**

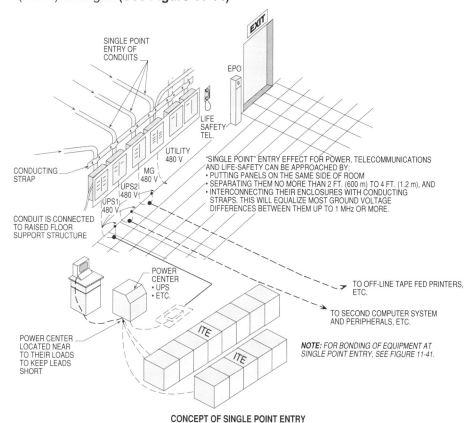

Figure 11-14. This illustration shows single point entry of conduits housing electric power, communications, life safety and other related systems.

Grounding Tip 273: Conducting strap (bonding strap) in interconnecting conduits and enclosures equalizes most ground voltage differences between them.

Grounding Tip 274: The conducting strap in Figure 11-14 can be formed by using wide straps of copper foil or from 22 gauge galvanized sheet steel. Copper straps of 0.02 in. thick (cut to the desired width) can also be used.

USING ISOLATION TRANSFORMERS
250.30

Placement of isolating transformers to achieve short connections to the loads they serve reestablishes a new ground reference point for the power source close to the ITE units they serve. If the isolation transformer is in an adjacent utility room, the ground reference point for that power source and its connection to the computer system ground will be outside the computer room. If it is on another floor or in the basement of a building, the ITE units which are served by that transformer will be connected to a remote grounding point. There may be considerable noise voltage differences between that remote grounding point and the ITE central grounding point within the computer room.

If the primary circuit is 480 or 600 volts rather than 208Y/120 volt, the feeder to the isolating transformer and wiring devices will cost less to install, will have lower losses and will provide better regulation. These are not insignificant for distances over 100 ft. (30 m) **(See Figure 11-15)**

Grounding Tip 275: Isolation transformers are equipped with shielded windings between primary and secondary windings which are used to attenuate transient noise. Such transformers are critical in applications such as computers, process controllers, micro processors, etc.

USING FILTERS

If grounded properly, the voltage differences between the grounding conductors of various services entering the single point of entry may be equalized at that point. However, any voltage difference between that equalizing point and the remotely located transformer's grounding point will be superimposed upon the line voltages as common mode noise. The power delivered to the ITE units with superimposed noise will be divested of the noise by the filters which will deliver clean power to the logic circuits within the ITE unit. However, the filters will shunt the noise current to the ITE frame and return to the source through the safety equipment grounding conductors and any metal conduit, raceway or cable which parallels that path.

Grounding Tip 276: An isolation transformer and panelboard can be installed on the 10th floor of a building and all computers can be wired and grounded to the transformers central grounding point. This will reduce and control unwanted noise.

Figure 11-15. This illustration shows the grounding procedures used when installing an isolation transformer to attenuate unwanted noise which affects ITE and associated apparatus.

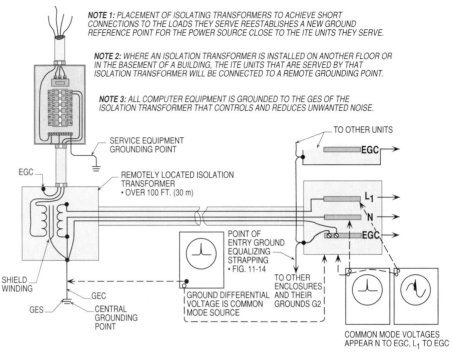

USING ISOLATION TRANSFORMERS
NEC 250.30

Grounding Tip 277: Filters shunt the noise current to the computer frame and if equipment isn't grounded properly, users are subjected to electrical shock. Review **300.3(B)** and **250.134(B)** in the NEC.

The noise currents flowing through these return paths create voltage drops in them, particularly at high frequencies or during impulses with fast rates of voltage change. The result is the creation of impulse noise voltage differences between various units of the ITE system, followed by equalizing current flowing in the grounded (neutral) conductors of the digital cables between ITE units. This is not a desirable condition. Typical noise impulses of up to 170 volt peak may occur regularly as a result of switching 120 volt AC circuits at random. This can result in ground current impulses of typically 1 to 4 amps peak which must be steered away from signal cables, since fractions of a volt and a few milliamperes or less can corrupt a data signal. **(See Figure 11-16)**

Figure 11-16. This illustration shows filters which are used to reduce unwanted noise that is capable of causing computer equipment to malfunction.

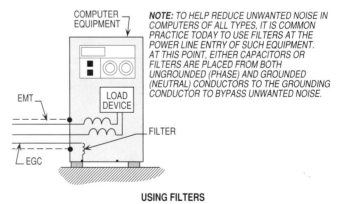

USING FILTERS

INSTALL ISOLATION TRANSFORMER CLOSE TO ITE
250.30

Grounding Tip 278: In present day, computer equipment manufacturers can provide filters which will limit such equipment distortion due to unwanted noise. Note that filters in most applications are used at the source of the nonlinear loads creating such problems. (See Note in Figure 11-16)

Installing the isolating transformer within the ITE room or computer locations eliminates one source of ground noise voltage. Putting the transformer close to the loads which it serves reduces the impedance of the return path for any residual noise, including that created by some of the ITE loads. This is the desired path for the noise current to flow, and because it is short, the noise voltage between ITE unit frames will be greatly reduced compared to installations with longer return paths to the transformer's neutral grounding point (central grounding point).

There are techniques for increasing the impedance of the interconnecting digital cables to common mode noise currents which will be described under the topic of "signaling cables create ground loops" on page 11-23. If putting the transformer close to the load does not provide sufficient reduction of noise interference, some of these techniques can offer additional attenuation of noise which would otherwise be transferred to sensitive digital signal circuit cables. **(See Figure 11-17)**

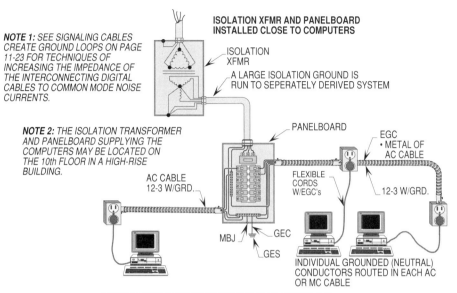

Figure 11-17. This illustration shows an isolation transformer and panelboard located near the computers that they supply.

Grounding Tip 279: There is an ungrounded (phase) conductor and grounded (neutral) conductor along with one equipment grounding conductor and one isolation equipment grounding conductor in the AC cables in Figure 11-17.

EARTH GROUND CONNECTION

The earth ground connection can be important to reliable performance of electronic circuits in ITE systems, minicomputers and word processors, but its role is frequently misunderstood. As a result, much effort and sometimes needless expense is incurred in achieving a "quiet, isolated ground" with a very low ground resistance. The quiet, low-resistance attributes are always desirable, but a misunderstanding of the term "isolated" can lead to dangerous grounding practices which violate the NEC and will not necessarily solve unwanted noise problems. **(See Figure 11-18)**

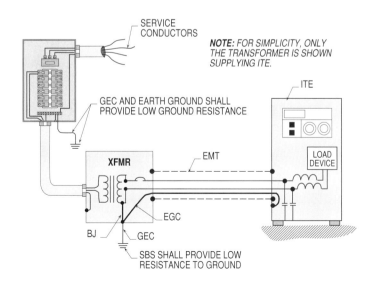

Figure 11-18. This illustration shows earth ground connections which shall be of low-resistance.

Grounding Tip 280: A reading of 1 to 5 ohms should be measured for the ITE systems to operate properly.

11-13

Grounding Tip 281: For grounding requirements pertaining to Relocatable Power Taps, see 20.1.1 through 20.2.8 in UL 1363.

STAND-ALONE UNITS

Stand-alone word processors, minicomputers, their peripherals and other electronic office machines are typically powered and grounded solely by their three-prong grounding type 120 volt plugs on their power cords. If a minisystem requires more than one separately powered unit, it is common practice to use a grounding type duplex receptacle, cube tap or portable receptacle strip provided that the total connected load does not exceed 80 percent of the 15 or 20 amp circuit protection rating for that circuit. Relocatable power taps and Transient Voltage Surge Suppressors (TVSS) units are used today by computer users to connect computers and their accessories. **(See Figure 11-19)**

Figure 11-19. This illustration shows relocatable power taps or Transient Voltage Surge Suppressors used to cord-and-plug equipment and provide protection from surges.

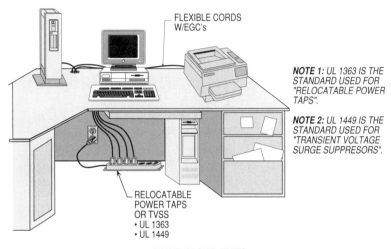

Grounding Tip 282: Before grounding techniques for computers and other sensitive electronic equipment are applied, review the requirements of the following NEC sections:

Routing EGC's together
- 300.3(B)
- 300.5(I)
- 300.20(A); (B)
- 250.134(B)

Installing rods or using structural building steel
- 250.4(A)(5)
- 250.54
- 250.136(A)

Installing isolation grounds
- 250.96(B)
- 250.146(D)

Installing Zero Signal Reference Grids
- 645.15

GROUNDING TIPS
250.4(A)(5)

In many instances, such an arrangement (minisystem grounding) has been used successfully. In its favor are having all minisystem grounds connected together at a common point as well as limiting noise voltage differences appearing between this minisystem central grounding point and the grounding point for the power source's grounded (neutral) conductor or any conducting structural member or surface in the vicinity of the load. Unfortunately, there are often reasons why this may be marginally successful or not work at all:

- The ground pin at the receptacle, the neutral grounding point at the building entrance service equipment and the transformer power source's secondary output winding grounding point are separated and may be grounded at two or three separate locations. This could permit noise voltages to develop between them and appear as common mode noise.

- Older buildings may contain wiring without an equipment grounding conductor and may even lack electrically continuous conduit to serve in its stead. If the receptacle outlet enclosure is grounded to a local metal water pipe, driven earth electrode or building structural steel and there is a continuous conductive path provided by the conduit back to the power source grounding point, such an installation could be unsafe. It also may inject excessive electrical common mode noise current. **(See Figure 11-20)**

- The equipment grounding conductor in the receptacle may be permanently connected to the conducting enclosure in which it is mounted. A connection which is integrally built into the receptacle normally creates this grounding path. Noise currents originating from a load plugged into an adjacent or nearby receptacle could reach the sensitive equipment through this path.

Grounding Information Technology Equipment

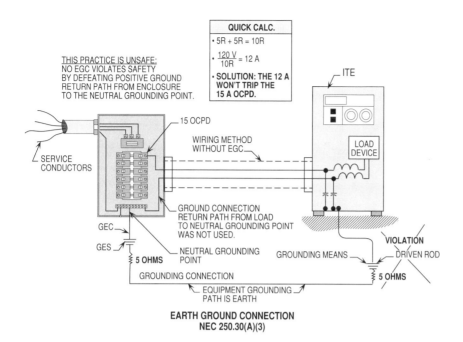

Figure 11-20. This illustration shows the earth being used as an equipment grounding path which is a violation of **250.4(D)** and **250.134(B)** in the NEC.

Grounding Tip 283: With earth used as a ground path (EGC), there is not enough current flow in amps to clear the OCPD rated at 15 amps. (See the Quick Calc in Figure 11-20)

Grounding Note: Electrical connections made to the earth with the intent to use the earth as a grounding path can be very dangerous. This connection is about 5 ohms at each grounding connection. For this reason, a fault condition that uses the earth as a conductor (grounding path) may not trip a protecting circuit breaker. The earth must be used for safety reasons, but it cannot be depended upon as a low-impedance conductor. If a fault current does use the earth, large potential differences can result and this can be very dangerous to the user. **(See Figure 11-20)**

ISOLATION GROUNDS
250.146(D); 408.20, Ex.

One solution is to install an "isolated ground" receptacle (sometimes defined by an orange color or orange triangle) in which the ground terminal is isolated from the mounting strap. An insulated equipment grounding conductor is then connected from the grounding terminal of the receptacle and passes through one or more panelboards without connecting to their grounding terminals. To reduce unwanted noise problems, this isolated equipment grounding conductor is then connected to the applicable derived system or service grounding terminal. **(See Figure 11-21(a) through (c))**

INDEPENDENT GROUNDING PROHIBITED
250.54; 250.136(A)

Warning, it is a violation for the equipment grounding conductor to be connected to an independent earth electrode without any other connections to the building ground, the equipment grounding conductor from the receptacle with the isolated equipment grounding conductor shall be connected directly to the neutral grounding point of the service equipment or separately derived system for the building. This is necessary for safety, compliance with the NEC and for low electrical noise at an ITE or computer unit. **(See Figures 11-22)**

Grounding Tip 284: The term "Isolated" refers to the method of equipment grounding not to whether it is grounded. The computer plugged into this form of receptacle is grounded at an upstream grounding point rather than locally at the outlet box. This requires a separate equipment grounding conductor (isolated) to connect to the receptacle grounding terminal to this remote upstream point. This upstream point shall be permitted to be any subpanel or the equipment grounding bus at the service equipment. This isolated equipment grounding conductor shall be permitted to pass through several subpanels before it is connected to the grounding electrode system. This point shall never be a separate or isolated grounding electrode. To be effective, the isolated equipment grounding conductor and equipment grounding conductor shall be insulated.

11-15

Figure 11-21(a). This illustration shows the isolated equipment grounding conductor and equipment grounding conductor pulled through the conduit and terminated at grounded busbar in the service equipment.

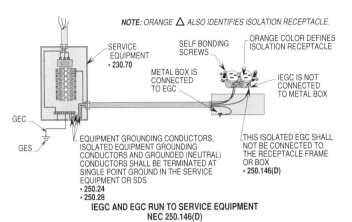

Figure 11-21(b). This illustration shows the isolated equipment grounding conductor passing through the subpanel and terminating at the single point grounding terminal in the service equipment. Note that an isolated equipment grounding conductor busbar can be installed in subpanel as shown in Figure 11-21(c).

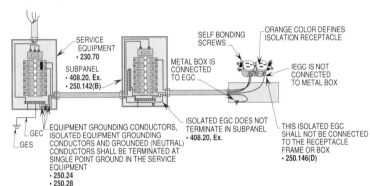

Figure 11-21(c). This illustration shows the isolated equipment grounding conductor passing through the subpanel and terminating at the grounding point in the separately derived system.

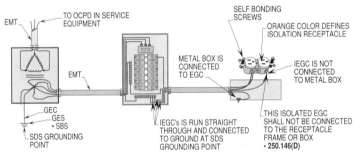

Figure 11-22. This illustration shows the computer grounded to and using the earth as a grounding path to clear a ground-fault if one should occur. This grounding installation is a violation of **250.4(D)** and **250.54** of the NEC.

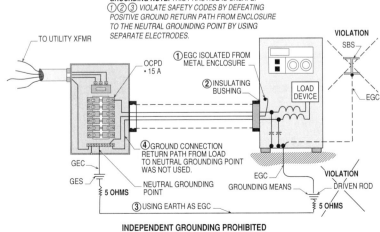

ALTERNATIVE TO REWIRING BUILDING
250.30; 250.146(D); 250.96(B)

An alternative to rewiring the building is to install equipment grounding conductors and use an isolating transformer with its own local ground near the load. An isolating transformer (k-rated) has separate primary and secondary windings and an interwinding shield. The secondary and shield can be referenced to the load's equipment grounding conductors simply by connecting short bonding straps between the shield, the secondary output neutral and the transformer frame which in turn is connected to a local ground (driven earth electrode, building steel, water pipe, raised floor structure, etc.). For safety, in the event of a primary circuit ground-fault, an equipment grounding conductor is also run with the input power to the service equipment ground or separately derived source ground. **(See Figure 11-23)**

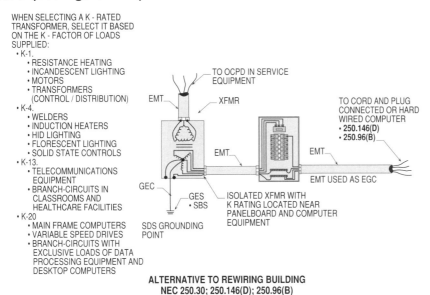

Figure 11-23. This illustration shows an isolation transformer used to establish a single grounding point for a panelboard to terminate isolated equipment grounding conductors and equipment grounding conductors to bond and ground computers where the existing wiring is without a grounding means.

DON'T BE CONFUSED BY THE TERM "ISOLATING TRANSFORMER"
250.30

The term "isolating transformer" may be confusing when used to describe this arrangement because it implies that no connection exists between primary and secondary, but this is not true. Primary and secondary are each connected to grounds or circuits which are grounded. However, these are not the same ground. Even though a grounding conductor may interconnect the two grounds, noise voltage can appear between them. Without the isolating transformer, this voltage difference would be superimposed upon the line voltage and delivered to the load. **(See Figure 11-24)**

Grounding Tip 285: Under no circumstances can the isolated equipment grounding conductor and equipment grounding conductor be routed externally to the conduit system. It shall run parallel and next to the circuit it is protecting. If the circuit is a separately derived AC system as from a transformer, the isolated equipment grounding conductor shall not go up stream beyond the transformer. There are no exceptions to this rule.

Figure 11-24. This illustration shows the connections of an isolation transformer to a panelboard with branch-circuits and isolated equipment grounding conductors and equipment grounding conductors routed to computers.

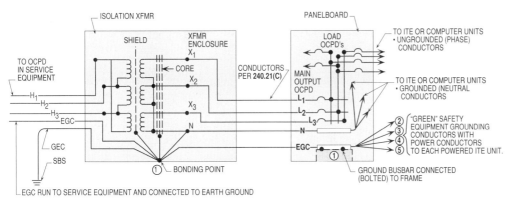

Grounding Tip 286: The earth is not a sump for electrical noise. A grounding conductor carrying noise to earth radiates just like an antenna and can cause noise problems for computers.

GROUNDING MYTHS RELATED TO ITE
250.52(A)(5); 250.56

One myth about earth grounds makes them to appear to be analogous to cesspools, allowing unwanted noise current to be drained into the earth and dissipated. If the ground current is small, one rod will do. If it is large, one must install a very long ground rod, install multiple interconnected rods or create a buried grid of conductors over a large area like a leach field for sewage effluent. Treatment of the soil with water and chemical salts further decreases the ground resistance and enables the ground connection to earth to carry more current with less voltage drop. **(See Figures 11-25(a) and (b))**

Figure 11-25(a). This illustration shows driven rods used as an earth grounding means.

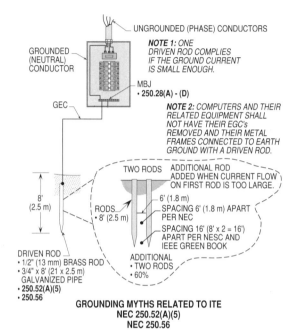

Figure 11-25(b). This illustration shows a driven rod that is treated to lower earth resistance.

Grounding Tip 287: The earth is not a sump for electrical noise. A grounding conductor carrying noise to earth radiates and serves as an antenna and can cause noise problems for computers. Currents flowing into the earth must reappear as a complete circuit somewhere. If reappearance of such current is in computer circuitry, there can be very serious unwanted noise problems.

LOW-RESISTANCE CONNECTION NOT A MYTH
250.4(A)(5); 250.54

This information on how to create a low-resistance earth connection is not a myth. However, the myth is that all unwanted noise current can be drained away and dissipated into the earth. This can be true for lightning induced noise where the earth is one of the terminals of the lightning current path. In most other instances, the earth is not or should not be one of the terminals of a noise source. Electricity flows in circuits. It follows Kirchhoff's law (what current goes in must come out, and generated voltages equal the voltage drops). If instruments indicate that an ampere of noise is flowing into the driven earth rod, it must be returning from some other earth connection. If the ITE system is part of that circuit, it is important to identify where it is entering as well as where it is leaving. **(See Figure 11-26)**

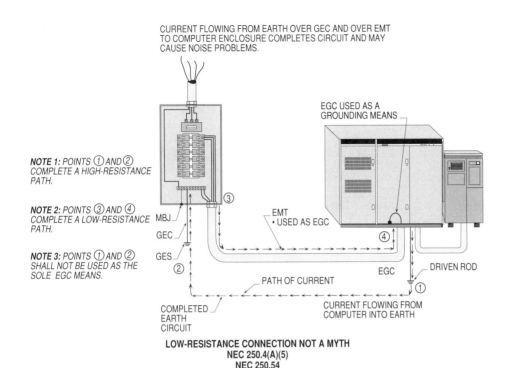

Figure 11-26. This illustration shows the electrical metallic tubing with an equipment grounding conductor used as the means to bond and ground the computer equipment.

Grounding Tip 288: Driven rods shall be permitted to be used as a grounding electrode, a supplementary electrode or for lightning protection. A driven rod shall never be used (without an equipment grounding conductor run with circuit conductors) to ground sensitive electronic equipment.

LOW-IMPEDANCE PATHS
250.4(A)(5)

The appropriate solutions are either to eliminate the source of the ground current or to provide short, low-impedance paths where that current can flow safely without creating significant voltage drops or unwanted coupling into signal circuits. A low-impedance connection to earth is difficult and sometimes expensive to achieve. A solution which requires two such grounds to provide a low-impedance path through earth cannot be expected to be better than a lower impedance conductor directly between the source and destination of that unwanted noise current. (See Note 2 in Figure 11-26 for such path)

DIRTY GROUNDS
250.4(A)(1) THRU (A)(5)

Another myth is that a computer requires a "clean, dedicated, isolated ground" with no electrical connection whatsoever between it and the "dirty ground". The term "dirty" is often used to describe the ground used for the utility power neutral ground where it enters the building or a downstream isolating or stepdown transformer within the building. This myth has been responsible for a number of unsafe installations and NEC violations, plus needless expense for additional grounding rods and long, heavy gauge copper grounding conductors to "clean ground" locations placed far away from the "dirty grounds".

Grounding Tip 289: When discussing grounding techniques, the term "dirty ground" that electrical personnel use is the ground used for the utility power ground where it terminates in the service equipment at the point it enters the facility.

If one were to believe this misinformation and act upon it, the power source conduit to the ITE enclosure could be broken at point "X" of **Figure 11-27** and replaced with an insulating bushing (a safety violation). Additionally, any equipment grounding conductor within that conduit would not connect the ITE or computer enclosure back to the power source grounding point, but instead would be connected to a "dedicated isolated, clean ground" (at point Y) which is another safety violation.

Figure 11-27. This illustration shows the electrical metallic tubing and equipment grounding conductor isolated and the ITE earth grounded to a driven rod.

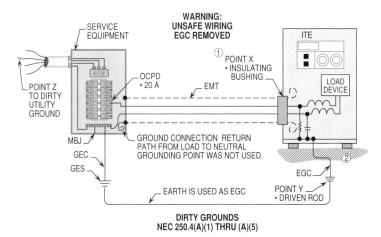

Grounding Tip 290: The method by which the ITE is grounded in Figure 11-27 is a shock and fire hazard and shall never be grounded in this manner.

RESULTS OF INDEPENDENT GROUNDS
250.4(A)(5); 250.54

Figure 11-28 illustrates the fault current path in the event of a short-circuit between a line conductor and the ITE unit's frame or a line-to-frame fault resulting from a line filter capacitor failing in a short-circuited mode. The return path of the fault current back to the grounded (neutral) conductor must flow through two driven earthing rods, the resistance of which is likely to be at least 5 ohms or more each. From a 120 volt power source, the maximum fault current would not exceed 12 amps. This is not enough to open circuit breakers which may range from 15 to 80 amps or more. A 15 amp rated circuit breaker needs 75 to 150 amps or more to cause it to open without delay. Larger circuit breaker ratings take proportionately more current to trip them and open the circuit.

If the circuit breaker does not remove the power promptly upon a fault, people could be injured or electrocuted, and small wire gauge conductors for signals together with electronic interface components could be literally fried and create a risk of fire. Ground-fault circuit interrupters are available which, if operational, would trip the power source circuit breaker with fault or leakage currents typically as low as 5 mA. However, a GFCI does not permit this to be a substitute for an equipment grounding conductor to be removed to the power source neutral grounding point. **(See Figure 11-28)**

Figure 11-28. This illustration shows an illegal grounding scheme used to ground the ITE system to help reduce and control unwanted noise problems to ITE.

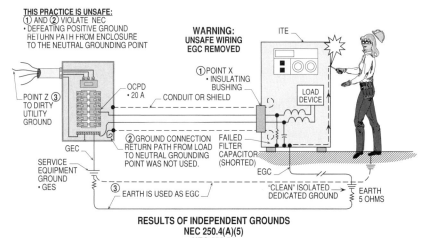

PROPER GROUNDING TECHNIQUES
250.4(A)(5); 250.54

Figure 11-29 shows methods of grounding which comply with the NEC, are safe and reduce the voltage differences between ITE enclosures and surrounding grounded objects. All of these, including the isolated grounding receptacle, have equipment grounding paths to the neutral grounding point. This general arrangement minimizes the noise currents which will flow in grounded signal circuits as well as in the safety equipment grounding conductors to each powered ITE enclosure. It complies with NEC requirements and has none of the hazards of the "isolated earth electrode grounding" connection. It recognizes that two earth rods in soil shall not be isolated from each other.

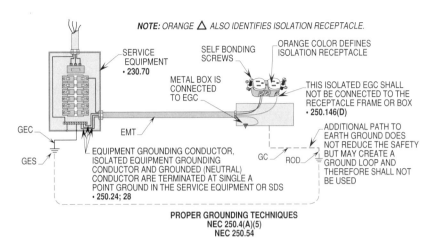

Figure 11-29. This illustration shows a legal installation where electrical metallic tubing with an isolated equipment grounding conductor is used to bond and ground safely the computer system that is plugged into the isolation receptacle. Note that the equipment grounding conductor is used to ground the metal outlet box and not the metal frame of the computer.

Grounding Note: Warning, if the driven rod in **Figure 11-29** is used and a lightning strike occurred close by, dangerous surges can enter through the rod and grounding conductors and cause serious damage to the components of the sensitive electronic equipment.

Grounding Tip 291: Sags and surges are short-term special cases of undervoltage and overvoltage conditions where the voltage fluctuation exceeds the allowable limit for at least some significant portion of one cycle (about 10 milliseconds) and reenters the limit band within 2.5 seconds (150 cycles). Because sags and surges are transitory conditions caused by momentary occurrences, they normally exhibit larger voltage excursions than longer term under voltages and overvoltages.

INSTALLING ADDITIONAL DRIVEN RODS
250.52(A)(5); 250.53(G)

It has been stressed that earth electrodes shall not be normal paths for ground-fault currents in ITE load utilization circuits. However, there may be some sources of ground current in driven earth electrodes associated with the power source. Lightning and switching transients are considered such sources. In addition, there may be fault sensing and relaying, plus the ground current resulting from multiple ground connections along a current-carrying grounded (neutral) conductor. Driven earth rods at various points along this conductor will have potential differences between them, thereby providing a source of ground current in these grounds and other grounds which are connected with them.

ROD-TO-EARTH RESISTANCE
250.56

Depending upon the amount of current flowing in a grounding rod, the need for low rod-to-earth effective resistance will vary. If only a few milliamperes are flowing, a 25 ohm ground would most likely be low enough. If many amps are flowing, it would be important that the heat loss about the grounding rod would not be so great so as to dry out the soil and cause the ground resistance to rise. If this occurs or if too much current causes a high-voltage drop in the soil around the rod, the back voltage is often rich in electrical noise which is unwanted in ITE computers and other sensitive electronic equipment sites. The solution is to find and reduce the source of the ground current, if feasible, or add more parallel grounding rods. These shall be at least 6 ft. (1.8 m) from the original driven earth rod and from each other. **(See Figure 11-30)**

Grounding Tip 292: Resistance to current through an earth electrode depends on the following:

- Resistance of the electrode itself and connections to it.
- Contact resistance between the electrode and the soil adjacent to it.
- Resistance of the surrounding earth.

Figure 11-30. This illustration shows multiple rods shall be permitted to be used to reduce the resistance to ground and help control electrical noise to ITE and computer systems.

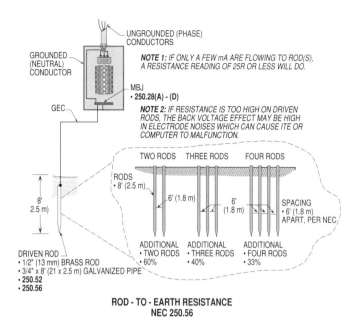

Grounding Tip 293: For frequency checks of equipment bonding and grounding techniques to ensure the reliability, see Sec. 27.7 and Appendix B in NFPA 70B.

FREQUENCY CHECKS
NFPA 780; NFPA 70B

Instrumentation specially designed for measuring the effective resistance of driven earth rods and buried or concrete-encased conducting grids is available. It should be used rather than ohmmeters or multimeters for measuring ground resistance of rods and other electrodes in soil or concrete. Where performance is dependent upon good, noise-free ground connections, the grounding rods and the connections to them shall be inspected and maintained frequently, typically at not less than 3 month intervals. If there is any DC component to the ground current, deterioration can appear rapidly. The deterioration rate with AC ground current is far less rapid. **(See Figure 11-31)**

Figure 11-31. This illustration shows a megger earth tester measuring the resistance to ground for a driven rod.

Grounding Tip 294: For ground measurement techniques using a megger earth tester, see Sec. 29.4 and Appendix I in NFPA 780.

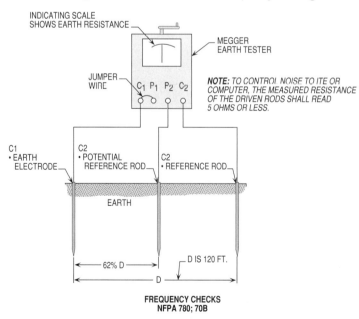

INTERCONNECTED GROUNDS BETWEEN SYSTEMS
250.54; 250.136(A)

Ground interconnections between separately grounded systems create ground loops. These are undesirable if loops create a path for noise currents to flow where they can enter signal cables in which a grounded (neutral) conductor is one of the signal conductors as well. This situation often exists in ITE and computers installations where data terminals and remotely located printers, etc. are located outside the ITE room.

Even if the remote terminals are located in the same building, they will not be likely operate from the same power circuits as those supplying the ITE units. Each of those terminals will be connected to power outlet receptacles on other branch-circuits. Each of their equipment grounding connections will most likely be attached to the metal enclosure into which each is mounted. The enclosures are connected to conduits which will be attached to building steel and other conduits all along the path to the branch-circuit breaker cabinet and to the power source within the building. **(See Figure 11-32)**

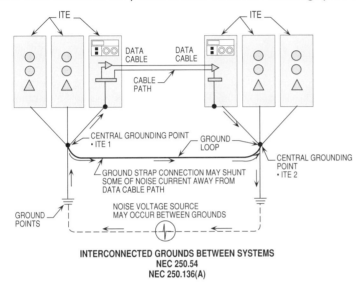

Figure 11-32. This illustration shows interconnections between separately grounded systems which create ground loops. This type of installation can cause unwanted noise problems for ITE systems.

SIGNALING CABLES CREATE GROUND LOOPS

Unless the terminal has been designed with double insulation so that its frame need not be grounded with the third "green" wire through the grounding pin, safety considerations demand that the frame of the terminal shall be grounded to the same point as the circuit's neutral grounding point. The fact that the terminal and the ITE system are each grounded to separated, but interconnected, grounding points, does not prevent noise voltage differences from appearing between them. The installation of a direct interconnection of signal cable, usually a twisted pair with a wrapped foil shield and drain wire, completes a ground loop path. Through the path, differences in ground potential will let noise current flow and intercouple with data signals. **(See Figures 11-33(a) through (d))**

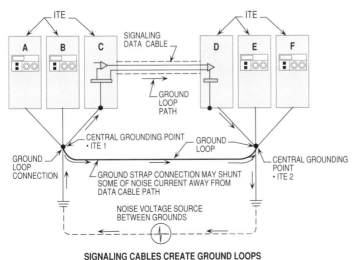

Figure 11-33(a). This illustration shows a ground strap used to shunt some of the unwanted noise current away from the path of the data cable in hope that malfunction of such equipment does not occur. Note that the strap helps control unwanted noise.

Figure 11-33(b). This illustration shows a ground loop between signaling cable and power supply which creates unwanted noise problems that causes terminals to malfunction.

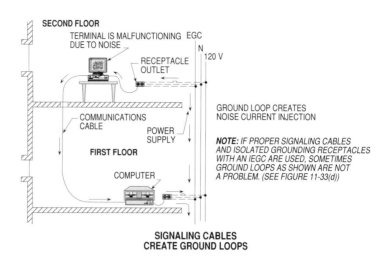

Figure 11-33(c). This illustration shows a temporary solution to reduce unwanted noise problems. Note that the terminal power supply is routed from the same receptacle power as the computer.

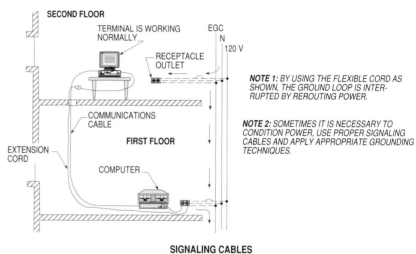

Figure 11-33(d). Today's computers and terminals are designed and manufactured with components to help reduce unwanted noise problems and by using isolated grounding receptacles, electrical noise is easier to control.

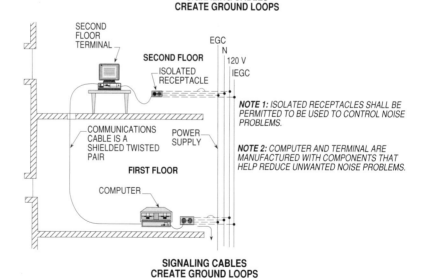

ZERO SIGNAL REFERENCE GRID
645.15

Grounding Tip 295: Zero signal reference grid is used to reduce voltage differences by creating a equipotential plane.

Zero signal reference grids (ZSRG) can be to ITE rooms what the metal chassis is to a radio or television set. Both serve as signal reference planes of constant potential over a very broad band of frequencies. A grid provides multiple parallel conducting paths between its parts. If one path is a high impedance path because of full or partial resonance, other paths of different lengths will be able to provide a lower impedance path.

A zero signal reference grid could be constructed of continuous sheet copper or aluminum, zinc-plated steel or any number of pure or composite metal with good surface conductivity. However, this type of construction would not only be expensive but also difficult to install in a computer room where other services have already been or are about to be installed. **(See Figure 11-34)**

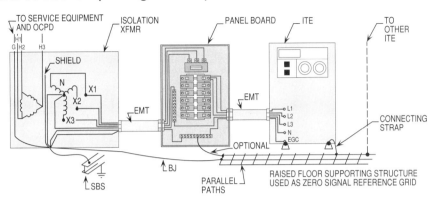

Figure 11-34. This illustration shows a zero signal reference grid with multiple parallel conducting paths between its parts. These paths are used to provide a lower impedance path to reduce unwanted noise problems that effect operation of the ITE system.

DESIGN OF GRID

Experience has demonstrated that a grid of conductors of approximately 2 ft. (600 mm) centers normally provides a satisfactory constant potential reference network over a very broad range of frequencies. Typically these have been formed of 4 AWG copper or aluminum conductors, which have been electrically joined at their intersections or by copper straps some 0.010 in. thick by 3 in. (75 mm) to 4 in. (100 mm) wide, also joint at their intersections. These grids typically lie directly upon the subfloor under the ITE room raised floor. Cables and conduits under the floor would normally lie below raised floor but above the grid. **Figure 11-35** shows a grid consisting of discrete conductors.

Grounding Tip 296: The zero signal reference grid provides an equipotential plane that reduces voltage differences between ITE which helps control unwanted noise.

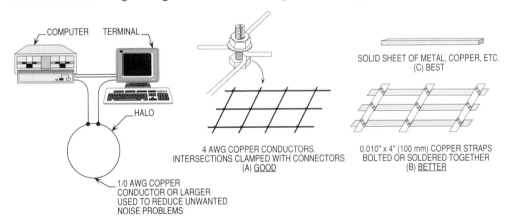

Figure 11-35. This illustration shows 4 AWG copper conductors used to build a grid system. Note that a solid sheet of metal, individual straps or a halo system are shown as grid systems.

ALTERNATE GRID

An alternative which appears to be equally effective and less expensive to install is the use of the raised floor supporting structure to serve as a zero signal reference grid. The two essential requirements are:

- Bolted-down stringers (the lateral supporting braces installed between supporting pedestals)

- Suitably plated (tin or zinc) members so that low-resistance pressure connections can be made

See Figure 11-36 for an alternate grid system.

Grounding Tip 297: Bolted-down stringers or raised floors can be used as an equipotential plane to reduce voltage differences between the metal frames of the computers.

Figure 11-36. This illustration shows the raised floor supporting structure of a computer room used as a zero signal reference grid. Note that the structure steel of the floor members provide the parallel paths to reduce unwanted noise.

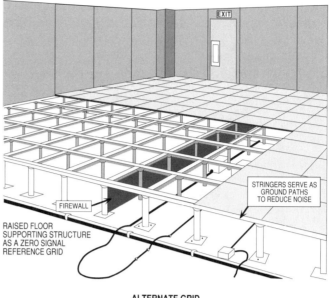

Grounding Tip 298: Before designing a zero signal reference grid system, consult with an electrical engineer who is familiar with and has experience installing one based on the operating frequency of the ITE system.

There are some stringerless supporting structures marketed in which the removable floor tiles lock onto the supports by gravity, but it has not been demonstrated that the contact has a low enough resistance and is free from intermittent contact as people or loads are moved across the floor. The bolted down horizontal stringers (braces) with pressure type spring washers or spring in the assembly have been shown to be highly suitable for the purpose.

CONNECTION OF ITE WITH STRAPS

Once a zero signal reference grid has been established in an ITE room, the various ITE units can each be connected to that grid by flexible flat braided copper straps. The connection should be made from each ITE unit, preferably at a point near its identified safety ground connection, to the nearest intersection of the reference grid. The strap should be no longer than necessary and should have few bends as possible and very little loop or sag to minimize the impedance at high frequency. Typical reduction in overall grounding impedance by use of a zero signal reference grid is illustrated in **Figure 11-37**.

Figure 11-37. This illustration shows the procedure for bonding and grounding the ITE system to the metal of the raised floor with straps to help control noise problems.

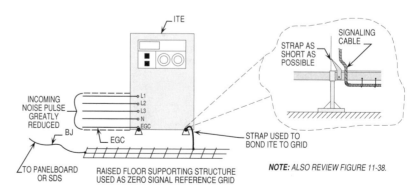

If the computer hardware manufacturer does not subscribe to the notion that ITE units should have these multiple connections to a reference grid, the stray capacitance between each ITE cabinet and the raised floor reference grid will still help reduce and control voltage differences between grounded parts of the ITE system.

Grounding Information Technology Equipment

LENGTH OF STRAP

The effectiveness of the zero signal reference grid is further enhanced if it is connected solidly to the ITE system's central grounding point by a very short strap. This can be accomplished by installing one or more isolating transformers, modular power center peripherals or UPS in the computer room, each being placed conveniently close to the loads which are to be served, particularly those ITE loads which transmit or receive the highest frequency data signals. Loads consisting primarily of small motors and no high performance digital circuitry (computer room air handlers, for example) need not be close nor have short leads. The isolating transformer in each load center will have a secondary output winding that is grounded to the zero signal reference grid as well as to all other grounding points required by the NEC. The AC voltage output to the ITE units will thereby be isolated from outside sources of disturbance and closely connected to the reference grid and to the loads by conductors too short to cause any resonance problems or to pick up other disturbances by induction.

If more than one power transformer, modular power center or UPS is used in the ITE room, the zero signal reference grid will be a sufficiently low impedance broadband interconnection to intertie their respective output circuit grounding conductors. The supporting structure for the floor with its multiple electrical paths prevents significant voltage differences from arising between them. Typical construction is illustrated in **Figure 11-38.**

Grounding Tip 299: Effective grounding straps can be formed by using wide strips of copper foil or from commonly available No. 22 gauge galvanized steel sheets of metal.

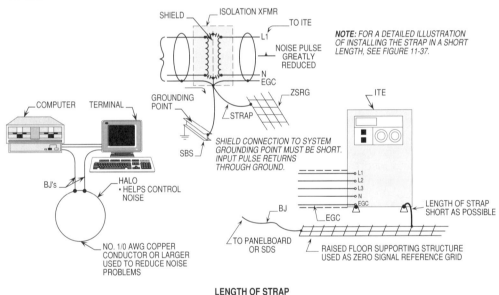

LENGTH OF STRAP

Figure 11-38. This illustration shows the strap must be as short as possible for connecting the ITE system to the metal of the raised floor to help reduce unwanted noise problems.

ZERO SIGNAL REFERENCE GRID NOT A SUBSTITUTE FOR EQUIPMENT GROUNDING CONDUCTOR

It is important, however, to adhere to the principle that the zero signal reference grid is not a substitute for any safety grounding prescribed by the NEC. It is used in addition as an overlay upon the equipment grounding conductors and their required routings and connections. The term "grounding grid" was used until someone pointed out that all equipment grounding conductors (safety grounds required by the NEC) were required to be put in conduits. The term zero signal reference grid is a more appropriate description of its function. The grid should be grounded for safety, just as all other conducting members near electrical conductors should be grounded. The conductor which grounds the grid at some convenient point to a ground bus in a branch-circuit distribution circuit breaker cabinet shall be a green wire, enclosed in a conduit and be of a conductor size appropriate to the largest ungrounded (phase) conductor ampacity (ampere capacity) to be brought into the ITE room. **(See Figure 11-39)**

Grounding Tip 300: Importance of equipment grounding conductor: If a 120 volt circuit (including short) has a total impedance of .25R and the equipment grounding conductor represents 50mR during a ground-fault, the temporary current will be as follows:

- $120 \text{ V} \div \left(.25\text{R} + \frac{50\text{mR}}{1000}\right)$
- 400 A rms

When the OCPD opens, the voltage on the computer chassis will be:

- $400 \text{ A} \times \left(\frac{50\text{mR}}{1000}\right)$
- 20 V

Figure 11-39. This illustration shows the equipment grounding conductor shall not be isolated from the chassis of the ITE system and the grid used as the equipment grounding means as well as the zero signal reference grid.

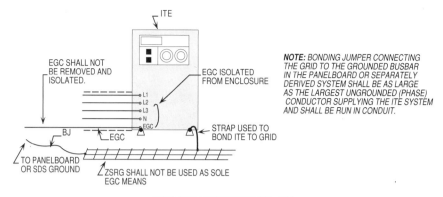

ZSRG NOT A SUBSTITUTE FOR EGC

WIRING SYSTEMS AND THEIR EFFECT ON ITE

Grounding Tip 301: The term "Ground Loop" does not have a real acceptable definition. Multiple ground connections can be formed to limit loop area and thus reduce common-mode coupling at High-Frequencies. **See Figure 11-40.**

As shown in **Figure 11-40**, the wiring system in a building with a large ITE installation will have many conductors and circuits, all of which will be in metal conduits, wireways, cables or in some cases in cable troughs. There may be freon pipes, chilled water pipes, sprinkler pipes, halon systems, compresses air tubing for thermostat controls and others in addition to air ducts and building steel structural members.

These may be hung by wires, straps or trapezes from the ceiling, sometimes above a drop room ceiling. Others may rest upon or be propped up slightly above the subfloor. They often penetrate the computer room and may be in walls and partitions around the ITE room. Many of the conductive conduits, pipes, ducts and structural members will be electrically interconnected at multiple points where they share the same hangers and clamp tie-down strips.

With these conditions, it is difficult to claim that the computer room has been designed as a "Faraday cage" (shielded room) because of the many foreign conductive members which penetrate it. Refrigerant cooling pipes or copper tubes are often connected to roof mounted condensers or heat exchangers. Another name for this arrangement might be "lightning rod" with conductors directly to the ITE room.

Figure 11-40. This illustration shows many ground loops in a facility that is created by wiring methods and other conductive materials.

Grounding Tip 302: The grounding loop between ground A and ground B can cause noise problems because of the voltage differences and amount of piping from the supplying transformer and the ITE.

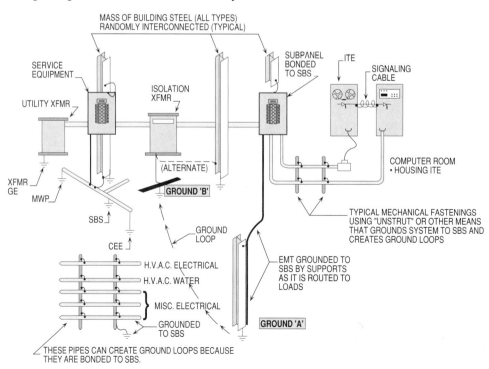

WIRING SYSTEMS AND PIPES AND THEIR EFFECT ON ITE

SINGLE POINT ENTRY BONDING

It is virtually impossible to enclose power, communications and other conductors in a conduit and have electrical noise voltages and currents on that conduit without coupling to the conductors contained within it. For this reason, it is possible to make a great improvement if some control can be exercised over all conducting members which penetrate the ITE room or which are run nearby. A single point of entry (previously illustrated in Figure 11-14) makes it possible to equalize the voltage differences without having the resulting current conducted throughout the ITE room.

Grounding Tip 303: The bonding together of metal conduits, cables and enclosures at a single point entry will help reduce unwanted noise problems flowing over and between these systems.

Finally, it is easier to isolate the incoming power if the isolating transformer is on the load side of the single entry point. Tests have shown that in many cases the zero signal reference grid of proper design are capable of equalizing voltage differences and handling considerable noise currents without injecting excessive noise into digital circuits of the ITE systems. However, this may be asking too much of the reference grid and should not be practiced as a planned approach unless carefully designed. **(See Figure 11-41)**

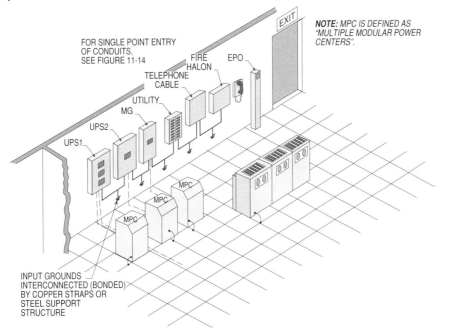

SINGLE POINT ENTRY BONDING

Figure 11-41. This illustration shows the procedure which can be used to bond together all metal enclosures that will provide an equipotential between the metal parts and help reduce unwanted noise problems.

Grounding Tip 304: The Multiple Modular Power Centers in **Figure 11-41** can be equipped with a UPS and isolation transformer to condition power, provide power during an outage and establish a single grounding point.

CIRCUITS LEAVING ITE ROOM

If single ground point strategy is to be followed, this may be possible within the ITE room, but as soon as circuits leave the room, they must be considered to be grounded at multiple points and be sources of ground loop noise currents.

It may be possible to retain the characteristics of a single point ground for the shielded conductors by using double shielded signal conductors or by running shielded cable with an outside insulating jacket through conduit. However, the outer shield or conduit may become a multipoint ground if not isolated properly.

Grounding Note: Single point vs multipoint grounding has been discussed in theory. However, grounding practice is seldom like the theory. Many unintentional ground connections not discussed in theory do appear in actual practice.

SHOW AND TELL:

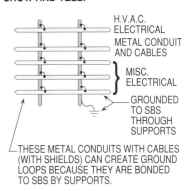

TRANSFORMERS - WHERE TO LOCATE THEM

Power available in many buildings suitable for ITE sites is often from transformers located in a basement utility room. In large buildings, the power available to each floor or to each of several widely separated areas may be supplied by dry type transformers closer to the loads which they serve. Some transformers supply 120/240 volt, single-phase. More frequently they will supply 208Y/120 volt, three-phase, four-wire system to panelboards (circuit breaker enclosures) from which single-phase or three-phase power may be distributed to branch-circuits. Most fluorescent lighting circuits in newer buildings operate at 277 volts which is supplied by 480Y/277 volt transformer outputs through separate panelboards.

Where the ITE installation is expected to be a significant load, these dry type transformers which serve local area may not be large enough in capacity to supply the additional ITE load. Even if they were, however, unwanted coupling of other load disturbances into the ITE system is reduced if it is powered from its own transformers and feeders which are dedicated to the ITE system only. This reduces unwanted coupling with other load circuits and reduces noise in many installations.

Grounding Tip 305: Where the grounded (neutral) conductor is shared between phases A, B and C for long feeder-circuit runs, a noise cross-coupling between sensitive electronic circuits can occur. Note that this coupling is most critical for load transients such as load switching or inrush current related loads.

LONG FEEDER-CIRCUIT RUNS

Installing long feeders to carry the 208Y/120 volt final utilization voltage over any significant distance, say, 150 ft. (45 m) or more, to an ITE room is not the best arrangement. Installing one or more feeders (and also, where applicable, branch-circuits) at a higher voltage such as three-phase, 480 volt to the ITE room and to specific load devices rated for 480 volts for large installations will be far more effective in performance and lower installation cost. At 480 volts, the line current will be 43 percent of a 208 volt system for the same transmitted power and will require three power carrying conductors rather than four. The resulting losses will be less and voltage regulation improved. Lower installation cost will also help offset the cost of the transformer in the ITE room which can be justified for reasons other than cost. **(See Figure 11-42)**

Figure 11-42. This illustration shows long feeder-circuit runs that can cause unwanted noise problems. All electrical loads (A/C units and ITE systems) in (a) are terminated in the same panelboard which is the worst condition for long feeder-circuit runs. Electrical loads in (b) are terminated in individual panelboards which is a fair condition to help reduce unwanted noise problems.

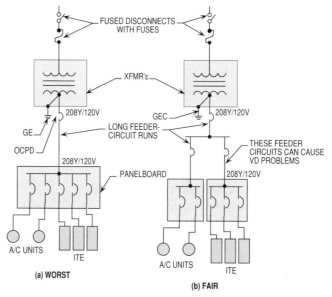

TRANSFORMERS CENTRALLY LOCATED

In general, transformers which provide power at the final utilization voltage to an ITE installation should be as close to the load as possible and not located in the basement with long feeder-circuit between ITE. There are advantages to putting the transformers directly in ITE rooms. Their effectiveness in isolating noise can be further strengthened by using transformers with interwinding Faraday shields. Other upstream transformers will have some advantage in achieving ground circuit noise reduction if they can be placed at some distance from the computer room transformers. **(See Figures 11-43(a) and (b))**

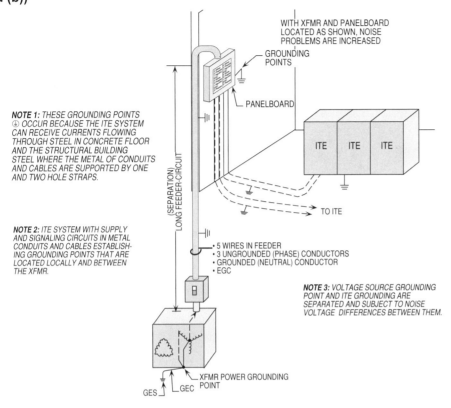

Figure 11-43(a). This illustration shows the grounding points locally, between and at transformer supplying panelboard and ITE system to be a considerable distance.

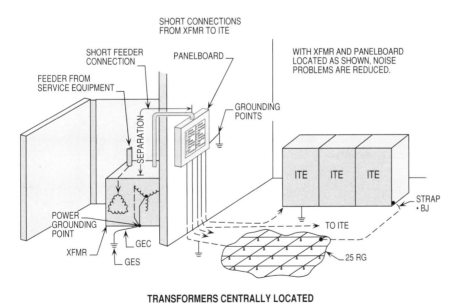

Figure 11-43(b). This illustration shows the distance of the feeder-circuit reduced due to the transformer being installed closer to the ITE system. This type of installation reduces the distance of the grounding points and therefore reduces and controls unwanted noise problems.

ISOLATING TRANSFORMERS

Isolating transformers have two basic functions:

- Transformation of voltage to the final utilization voltage
- Provisions of common mode shielding between primary and secondary circuits

Among secondary benefits, they limit output fault current to approximately 20 times the transformer's rated output current to protect smaller downstream circuit breakers from excessive fault current without the need for additional current limiting fuses. Three-phase, wye-connected transformers also improve power factor when they supply power to rectifier loads because third harmonics and their multiplies are not passed on to the upstream power source. **(See Figure 11-44)**

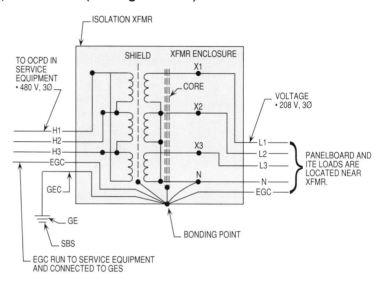

Figure 11-44. This illustration shows an isolation transformer installed locally near the panelboard and the ITE system supplied.

Grounding Tip 306: For isolation transformers to be capable of having ride-through power characteristics, a supplementary apparatus such as a UPS must be provided.

TRANSFORMER CHARACTERISTICS OF USE

Grounding Tip 307: Because most electrical loads in a building are nonlinear, the electrical systems are of low-impedance. Transformers can be used to supply a major portion of the required harmonic current. The transformer acts like a passive low-pass filter as the inductance begins to dominate. Note that when the harmonic current is high, the transformer can be designed for such loads.

While isolating transformers do not provide any voltage regulation, they do not create a significant internal voltage drop which would make it necessary to have an output voltage regulator in most applications. No useful energy is stored in an isolating transformer, so no extension of ride through is provided by a transformer during a power input interruption. Ride-through energy storage requires supplementary apparatus.

Transformers are among the more efficient of power conditioning devices, typically in the 95 to 98 percent range, even at partial loads of 25 percent or less. With no moving parts, they rank among the most reliable of electrical apparatus. Giving off little heat and creating little acoustic noise, isolating transformers in suitably packaged UL listed enclosures are completely suitable for installing in ITE rooms.

CONTROLLING HARMONICS

Transformers are available as single-phase or three-phase units. For all but the smallest ITE systems, however, three-phase, four-wire, 208Y/120 volt output is almost an industry standard in the United States and Canada. Transformers with a delta input winding configuration and a wye output configuration are preferred for their ability to reduce the apparent harmonic content of loads, improve power factor when the harmonic current content makes it low, and help balance the input currents when the individual output phase loads are unbalanced. Using a transformer with three-phase legs on one core will enable it to withstand some DC component and even harmonics in the load current. Three single-phase transformers on three separate cores with windings connected in delta on the primary and wye on the secondary would most likely saturate under similar load conditions and cause power source fuses to blow. **(See Figure 11-45)**

Grounding Information Technology Equipment

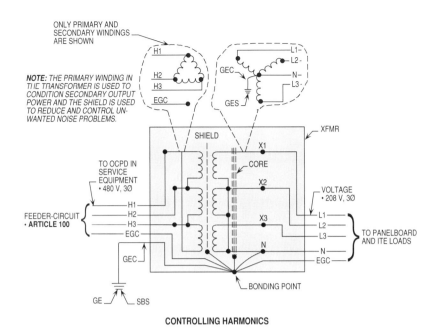

Figure 11-45. This illustration shows primary and secondary windings plus the shield in the transformer bonded to a central point and connected to ground.

TRANSFORMER WITH SHIELDS

With a shield between windings, the attenuation which it supplies is often specified in decibels (dB) or an equivalent voltage attenuation ratio. For a 1 to 1 ratio transformer, the dB attenuation is equal to 20 times the log to the base 10 of the voltage attenuation ratio. A voltage ratio of 30 becomes 29.5 dB, 100:1 becomes 40 dB, 1000:1 becomes 60 dB, etc. Super-isolation transformers with multiple shields are offered with claims of 140 dB or more or a voltage attenuation ratio exceeding 10,000,000:1. If these cost more than those in the 40 to 60 dB range, it may be a waste of money to purchase and install one unless equivalent shielding were applied to all other paths by which external electrical noise could reach the ITE equipment and the noise which the ITE equipment itself generates could be reduced as well.

Grounding Tip 308: Isolation transformers are equipped with an interposing electrostatic shield between the primary and secondary windings. This electrostatic shield is connected to the primary grounded (neutral) conductor which provides a localized path for reactive current flow. Note that reactive current will no longer flow in the entire building back to the service-entrance equipment.

TRANSFORMER WITHOUT A SHIELD

Most transformers without shields provide some common mode isolation, especially at low frequencies, primarily because of the electrical separation of primary and secondary windings. Adding an interwinding shield and connecting it to the secondary ground reference point with a short, direct conductor will improve the isolation. The ratio of the attenuation with the shield grounded to the attenuation with the shield floating with a 20 kHz signal is sometimes quoted as a shield effectiveness ratio, but is not directly comparable with the voltage attenuation ratio. Typical quoted values range from 20 to 30 or more. A second shield between primary and the grounded shield will reduce the conversion of common mode input to normal mode output noise voltage.

TRANSFORMER IMPEDANCE

The impedance of the transformer has been discussed previously under the heading of power source characteristics. To prevent load faults from drawing currents which would exceed the interrupting capacity (not the trip current rating) of circuit breakers, transformers with excessively low internal impedance are not desired. Typical impedance values are a compromise between a desire for a good regulation (low impedance indicated) and fault current limiting (high impedance indicated). Dry type transformers with internal impedances between 3 and 6 percent are usually acceptable. A 5 percent value is very common. In a ground-fault on the 208Y/120 volt output side of a 100 kVA transformer, the symmetrical RMS value would not exceed 5560 amps if the transformer impedance were 5 percent. This is usually within the capability of a standard 10,000 amp interrupting capacity circuit breaker. **(See Figure 11-46)**

Grounding Tip 309: Secondary conductor connections from a separately derived system shall be permitted to be increased up to 100 ft. (30 m) under certain installation requirements per **240.92(B)** and **250.92(D)** in the NEC. In some cases, this increased conductor length will reduce the available ground-fault current to equipment.

11-33

Figure 11-46. This illustration shows the procedure for calculating the fault-current which is available at the secondary terminals of the transformer.

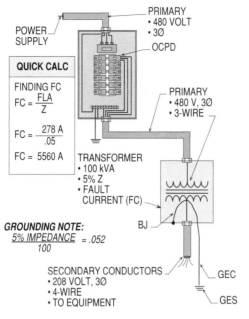

TRANSFORMER IMPEDANCE

Grounding Tip 310: Separate isolating transformers may be used to separate easily disturbed loads (memory, for example) from loads which create disturbances (document sorters, for example). Air handler cooling units in the ITE room which use multiple continuously running three-phase fan or blower motors of not more than, say, 2 HP each can be powered from the ITE isolating transformers. This will help stabilize the load circuits against unbalance, harmonics and external sources of line voltage disturbances.

Quick Calc

Use the following procedure for sizing a modular power center:

- Divide total power in watts by the ITE room floor space devoted to ITE use.
- Typical large systems use 50 W to 60 W per square foot.
- A 50 ft. x 100 ft. ITE room full of equipment might be expected to use approximately 250 kW of power, exclusive of HVAC power requirements.

Quick Calc

- 50 W x 50' x 100' = 250,000 W

- $\dfrac{250{,}000\text{ W}}{50' \times 100'} = 50\text{ W}$

- $\dfrac{250{,}000\text{ W}}{1000} = 250\text{ kW}$

Grounding Tip 311: Branch-circuits are those conductors between the final OCPD's and the loads supplied. Feeder-circuits are those conductors between the service equipment and panelboard or switchgear.

ISOLATING TRANSFORMERS NOT ALWAYS REQUIRED

Isolating transformers are not always listed as a "requirement" by ITE manufacturers. There are circumstances where existing electrical systems may be adequate. Additionally, some ITE users are more tolerant of occasional computer malfunctions than others. However, ITE manufacturers who do not insist upon isolating transformers generally recommend their use. Some ITE manufacturers offer them as optional "power peripherals" as part of their ITE systems, contained in enclosures which harmonizes with the rest of the system. Such systems can be listed by Underwriters Laboratories under UL 478, "Standard for Electronic Data-Processing Units and Systems".

MODULAR POWER CENTERS ARE SOMETIMES USED

Whether isolating transformers are installed as part of the building wiring or as power peripherals, they can perform the same functions. If in the form of modular power center peripherals, they are easier to position near the loads which they serve, are infinitely easier to move, and make it possible to reconfigure the output circuits in accordance with needed additions or changes to the ITE system. Having these transformers in the computer room typically adds about 2 percent to the air-conditioning load in the computer room, but putting them in a nearby utility closet does not entirely eliminate the need to cool them. As part of the computer system, the power centers can be moved with the system if a move is required. As part of the building wiring, this becomes more difficult. The financial treatment may also be affected, depending upon whether the isolating transformer is part of the computer system of a leasehold improvement to the building wiring. **(See Figure 11-47)**

FEEDERS AND BRANCH-CIRCUITS
ARTICLE 100; UL 478

The definitions of "branch-circuit" and "feeders" in the NEC and UL 478 affect the names applied to an isolating transformer's input and output circuits. If the transformer is installed as part of the site's "premises" wiring, its input and output circuits are "feeders" and the individual circuits from the circuit breaker panelboard which distributes power supply to individual ITE units are considered the "branch-circuits" per NEC.

Grounding Information Technology Equipment

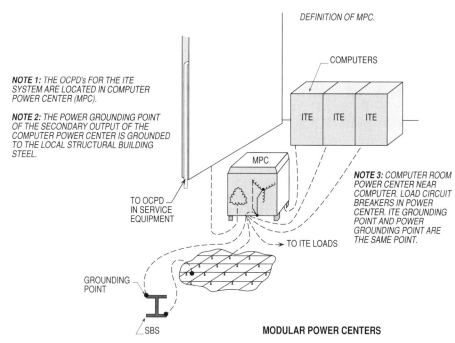

Figure 11-47. This illustration shows a computer power center with OCPD's located near the ITE system and grounded to the local structural building steel. This illustration will reduce unwanted noise problems created by long feeder-circuit runs. **(See Figure 11-48)**

Grounding Tip 312: The computer power center can be equipped with an isolation transformer and UPS to condition power, provide ride-through power during an power outage and establish a central grounding point.

If a power peripheral is used, the final overcurrent protection device and premises wiring to the outlet receptacle or permanent connection terminal block becomes a "branch-circuit," even if it is 480 volt and is not the final utilization voltage in the ITE system. The circuit between the power peripheral and other ITE system units become "interconnecting cables" rather than "branch-circuits" since they are not the circuit conductors between the final (site wiring) overcurrent protecting device and the output(s). These definitions can affect what is subject to inspection, who inspects and inspection fees. **(See Figure 11-48)**

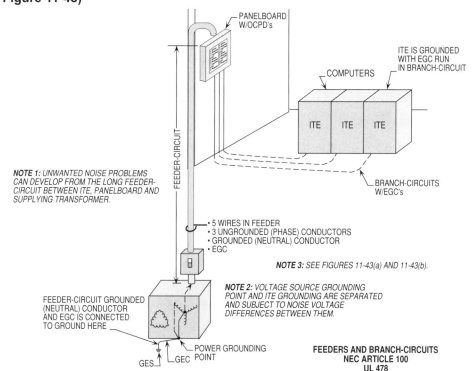

Figure 11-48. This illustration shows noise problems that can develop in the ITE system where long feeder-circuits are used for supplying power to panelboards housing OCPD's that connect branch-circuits to ITE systems.

NO MODIFICATIONS OF EQUIPMENT
90.7; 110.2; 110.3(B)

Ordinarily the premise's wiring is subject to the NEC. The appliances to be connected to the premise's wiring by pluggable or permanent connection are also subject to examination for safety if they have not been found acceptable by a recognized testing

Grounding Tip 313: All ITE systems and their accessories shall be listed or approved by UL or other acceptable safety testing laboratories which are recognized by the AHJ for the city, state, etc., that such units are installed.

laboratory such as UL. In most municipalities, electrical safety inspectors accept UL and other recognized testing laboratory "listings" as evidence of suitability for safe connection to premise's electrical wiring. ITE users should be aware, however, that any modification to a product which has been UL listed voids that listing until the product has been reexamined and found safe. The modification may be trivial and entirely safe, but the implications of liability and the trouble, effort and expense of reexamination should discourage ITE users from modifying a UL listed piece of equipment without being prepared to go all the way in reinspection. **(See Figure 11-49)**

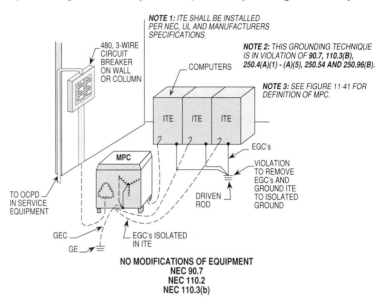

Figure 11-49. This illustration shows it is a violation to modify ITE systems if the listings of the equipment does not permit such modification.

Grounding Tip 314: A UPS installation with emergency standby diesel power can extend ride-through almost indefinitely, depending upon fuel available and the rated running time of the diesel between overhauls or downtime for service. This may range from 200 hours to more than 2000 hours. Turbines generally run longer between service attentions.

UNINTERRUPTIBLE POWER SYSTEMS (UPS)
645.11; 645.15

Uninterruptible power sources (UPS) have become a virtual necessity for powering large or small systems where the application serves a critical need, as when interruption of the function would risk personal safety or monetary loss.

Computers and data communications can be no more reliable than the power from which they operate. The difference between UPS and Emergency Standby Power is that the UPS is always in operation. It reverts to an alternative power source such as the public utility only when the UPS fails or needs to be deenergized for service. Even then, the transfer of power source occurs so quickly (milliseconds) that it does not interrupt the proper operation of the ITE system. **(See Figure 11-50)**

Figure 11-50. This illustration shows UPS with OCPD's for the protection of the branch-circuits supplying the ITE systems. Note that the grounding point of the UPS is near the ITE systems.

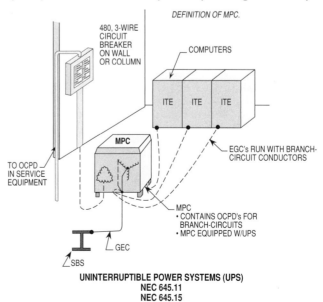

Grounding Information Technology Equipment

DISADVANTAGE OF STANDBY POWER SYSTEMS

Emergency standby power is normally off and does not start (manually or automatically) until the public utility power fails. Diesels can regularly be started within 10 to 30 seconds if their condition for rapid starting has been continuously maintained. However, during this interruption, ITE systems would have shut down. Most ITE systems cannot ride through more than 8 to 22 milliseconds of power interruption. Those which ride through very short duration power interruptions so far as energy continuity is concerned may be interrupted because of the electrical noise created by the interruption. The success of UPS systems which permit ITE operations to continue without interruption require power switching to be done in substantially less than a half cycle and without substantial transient noise during the switching. **(See Figure 11-51)**

Grounding Tip 315: Emergency, legally required and optional standby systems will have a delay before they power-up. Therefore, battery banks, UPS or other power sources can be utilized until such systems are powered-up.

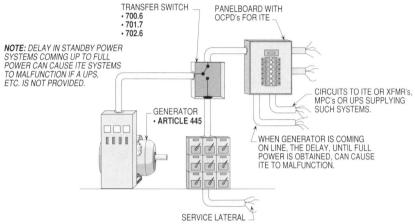

DISADVANTAGE OF STANDBY POWER SYSTEMS

Figure 11-51. This illustration shows a standby power system such as an emergency, legally required or optional type used to supply power to ITE system if utility power is lost. UPS is usually installed to provide ride through power if utility power is temporarily lost.

RIDE THROUGH PERIODS

There are some ITE and data communications units which will ride through much longer power interruptions of 0.5 seconds or longer. Furthermore, the nature of most communications equipment is to be tolerant of momentary interruptions in the communication line and of extraneous noise. Received signals are automatically checked for errors and missing bits regardless of cause. The signals are retransmitted until received correctly. In such systems, momentary interruption of power is not disastrous. Consequently, simple power transfer switching is feasible where one source may be disconnected before the other source is connected. There can be 50 to 100 milliseconds or longer between these events using a simple break-before-make transfer switch as shown in **Figure 11-52**.

Grounding Tip 316: UPS can extend ride-through to typically 5 min. to 30 min. or longer. When requesting quotes on UPS installations, ask for incremental cost of increasing ride-through times, say in 5 min. or 10 min. increments. The incremental cost of extra ride-through may be worthwhile, especially if there is a back-up diesel which will not start because of water in the fuel line or need to change a cranking battery. Not much can be done in 5 min., more can be done in 30 min. or longer to restore emergency power.

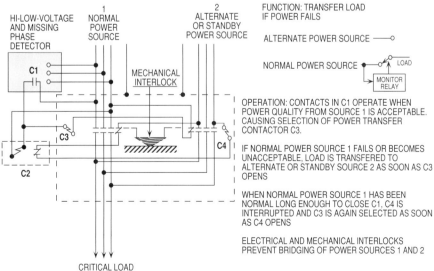

RIDE THROUGH PERIODS

Figure 11-52. This illustration shows a simple break-before-make transfer switch used to connect standby power sources, if needed for temporary power to ITE or computer systems.

Grounding Tip 317: Engineers can design and install filters on a power line that will stop most interference from entering or leaving sensitive circuits.

FILTERS

Filters which pass basic power frequency and reject electrical noise and unwanted harmonics in the line voltage are feasible, and are very popular with some designers. However, if they are not designed and installed correctly, they may create more problems than they solve. They require application skill.

Noise and unwanted harmonics appear in the form of line-to-line and line-to-neutral voltage signals (normal mode) and also as noise voltages between the local grounding reference point and each of the power conductors including the grounded (neutral) conductor and sometimes the equipment grounding conductor (common mode) as illustrated in **Figure 11-53**. Typically, the unwanted noise includes radio frequency signals and fast rise-time impulses. A low-pass filter will therefore pass the 60 Hz power and its low order harmonics needed to operate an ITE unit's power conversion equipment, but will attenuate the high frequencies which interfere with logic circuits and do little if anything to provide power. Such filters typically have individual inductors in series with each power wire and shunt capacitors between power conductors and from each power conductor (including grounded (neutral) conductor) and the ITE unit's conducting enclosure.

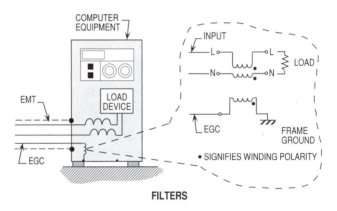

Figure 11-53. This illustration shows a simplified hook-up of filters used to attenuate noise and cause such to by-pass the sensitive circuits of the ITE system.

Grounding Note: Filters are a network of capacitors and inductors designed by engineers to provide attenuation to a band of signals. Such filters are normally packaged in metal cans which are usually installed on a metal bulkhead. Their operation shunts unwanted signals to the surface of the can, allowing them to return over the metal conduits and equipment grounding conductors.

Grounding Tip 318: There are many ground loops in a facility that are benign under certain conditions of use and not harmful to the systems operation and they are as follows:
- Structural building steel
- Ground rings
- Reinforcing steel rods
- Metal conduits and cables
- Other metal piping
- Etc.

MULTIPLE SEPARATED GROUNDS
250.54; 250.136(A); 250.6

The conducting enclosure shall be grounded using an equipment grounding conductor for reasons of safety. However, this conductor often creates a path for common noise currents to flow in a loop circuit, starting with equipment grounding conductor's connection to a grounded conduit which becomes a source or a sink for noise currents which may be generated in the total electrical premises wiring or in the ITE load. Part of that ground loop may be the shield or the grounded wire of a conductor pair used for digital signals. Any noise current flowing in that ground loop can create noise in logic signal conductors by electromagnetic coupling or direct connection and stray capacitance. Such noise voltages and currents are associated with the voltage differences which appear throughout a large interconnected system of grounding conductors. **(See Figure 11-54 and also reference Figure 11-40)**

Grounding Information Technology Equipment

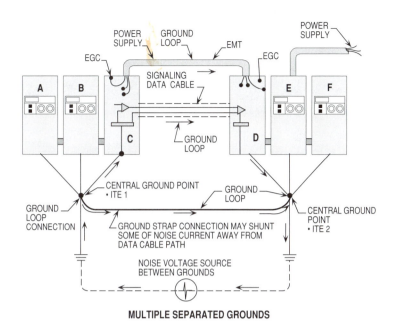

Figure 11-54. This illustration shows multiple separated grounds located at ITE systems A, B, C, D, E and F.

Grounding Tip 319: When OCPD's are opening due to power line disturbances such as slow over or under voltage problems or very sharp narrow transients, one of the following or combination of such is recommended:
- Measure EGC's at source for noise
- Power-line filters
- DC distribution filtering
- Filter system with power elements
- Shielded type power cords
- Provide equipment with another branch-circuit
- Isolation transformer
- UPS if needed

BALUNS AND FILTERS

It may be impossible to eliminate common mode noise voltages, but it is possible to exert some control on where they will occur and the noise current paths which might result. **Figure 11-55** illustrates a form of a common mode filter, also known in the past as a "balun". It passes the flow of normal mode current without significant voltage drop because the power conductors, wound on a common core, are wound in opposition. Load current does not saturate the core. These components can be used with computers to reduce noise problems.

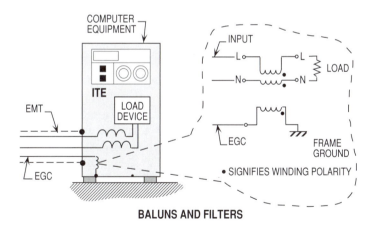

Figure 11-55. This illustration shows a simplified installation of filters used to reduce unwanted noise problems by utilizing their winding polarities.

On the other hand, any high frequency noise or impulse appearing between the equipment grounding conductor on the left of **Figure 11-56** and the ITE enclosure or frame on the right would find the connection through the filter to be high impedance connection. The ground voltage offset would appear across this coil where it would do no harm. Almost identical voltages would also appear across the grounded (neutral) conductor and line conductor coils, thereby offsetting their voltage to equipment ground by the same amount as the frame to ground offset.

11-39

Figure 11-56. This illustration shows that almost the same voltages appear at the grounded (neutral) conductor and line conductor coils. Thereby offsetting their voltage to equipment grounding conductor by the same amount as the frame to ground which will reduce unwanted noise problems.

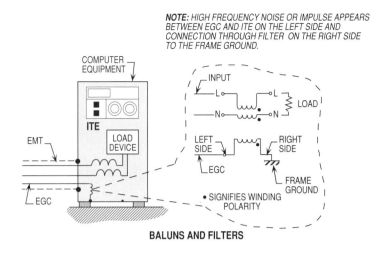

Grounding Tip 320: To help reduce unwanted noise in computers of all types, it is common practice today to use filters at the power line entry of such equipment. At this point, either capacitors or filters are placed from both ungrounded (phase) conductors and grounded (neutral) conductors to the grounding conductor to bypass unwanted noise.

HOW TO CONTROL NOISE WITH FILTERS

ITE units susceptible to electrical noise or which produces it almost always contain line filters. When the ITE unit must have direct connected digital signal cable connections with other units, it is sometimes supplied with a composite line filter which will attenuate both normal and common mode noise voltages and currents. When users of ITE attempt to assemble and interconnect units supplied by different manufacturers and operate them as an interconnected system, they may find that none of the units have the total amount of filtering needed to prevent common mode ground loop current problems. In such cases, supplementary filters may be needed.

MATCHED FILTERS

To operate properly without interaction, filters are typically designed to operate effectively when the input and output impedances are matched to the respective impedances of the power source and the load. These are not always constant, so practical filters must often be a compromise. The result is often unwanted "ringing" or line voltage distortion at some frequency other than the disturbance frequency which the filter is intended to correct.

Grounding Tip 321: Three-phase power systems generate dominant odd triplen harmonics rated at 3, 9, 15 and so on. They affect many three-phase systems when they are 7, 13, 19, 25 and greater harmonics. These harmonic currents arise from nonlinear loads such as solid state inverters and capacitor input rectifier systems.

SEVERE HARMONIC DISTORTION

Filters are sometimes necessary to correct severe harmonic distortion from a small power source such as an uninterruptible power source inverter circuit. The more usual and generally effective filters are low-pass filters which remove high frequency normal mode electrical noise. However, equally effective in most cases is the use of a transformer having suitable leakage reactance and capacitors across output windings to perform several functions simultaneously. These are voltage transformation, tap adjustment, common mode isolation and substantial high frequency noise attenuation. This is described under transformers - **where to locate them on page 11-30**.

LINE VOLTAGE REGULATORS

Line voltage regulators are a solution to line voltage regulation exceeding the capability of the self-contained voltage regulators in most ITE units. They are available in units which use various operating principles, each having some advantages and disadvantages. Unless combined with a transformer with separate input and output windings, many regulators do not establish a new output voltage with its own ground reference. Instead, most regulators act as autotransformers to buck or boost the input line voltage, but the ground reference of the output is to the same input grounded (neutral) conductor and its upstream grounding point. **(See Figure 11-57)**

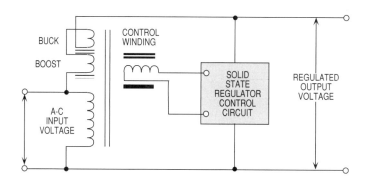

Figure 11-57. This illustration shows the line voltage regulator that is used to regulate the line voltage for an ITE circuit.

LINE VOLTAGE REGULATORS

MAGNETIC COUPLING CONTROLLED VOLTAGE REGULATORS

These regulators are available in which a direct current control winding controls the saturation in parts of a special transformer structure. This, in turn, controls the AC flux path through boost or buck coils to raise or lower output voltage in response to an electronic regulator circuit's DC output current.

Since the regulators are all solid state, there is no mechanical maintenance. Changes are smooth. Although the typical response time of 5 to 10 cycles is too slow to prevent surges and sags from reaching the load, it is much faster than electromechanical regulators. As with any other regulator placed ahead of another regulator, there is the danger of unstable interaction if the regulator in the load and the regulator in the line have comparable response time delays. The internal impedance of these regulators is greater than isolating transformers which are commonly used.

There are several type of voltage regulators available and a few or listed below:

- Motor operated variable ratio transformer
- Rotorol or induction
- Ferroresonant transformers
- Magnetic coupling controlled voltage regulators
- Tag switching regulators
- Capacitor bank switching for voltage control

Note that only type magnetic coupling controlled voltage regulators are called-out, explained and illustrated in **Figure 11-57.**

TECHNIQUES OF GROUNDING SIGNALING CABLES

There are basically three causes of noise problems in sensitive electronic signaling circuits which are capacitive, inductive and radioactive coupling. Such noise problems are usually corrected by the methods in which the shields of the signaling cable are terminated to the enclosure of the equipment.

COAXIAL CABLES

Coax can be used for signal circuits having high-frequency signals. Note that the fields used in transmission of signals are contained inside the cable and this has nothing to do with the way they are grounded at each end. To reduce high-frequency noise, the shield shall be grounded at both ends. **(See Figure 11-58)**

Grounding Tip 322: To control noise problems from capacitive, inductive and radioactive couplings, the appropriate cable system must be selected and the shields grounded properly. Review the construction of the cables in Figures 11-58 through 11-63 in this book on how such cables can be used to reduce and control unwanted noise.

Figure 11-58. This illustration shows the shields of the coaxial cable grounded at both ends to reduce and control high-frequency noise problems.

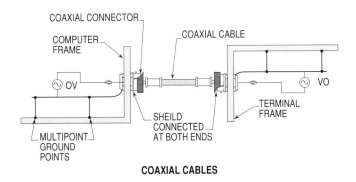

Grounding Tip 323: Shields used to reduce radiative noise shall be bonded to the computers chassis at each end. Review Figures 11-60(a) and (b) for shields connected at both ends.

SHIELDS CONNECTED AT ONE END

Shields that are terminated at one end, but do not carry signal current are used as electrostatic shields or guard shields. Such shields are connected to a zero potential reference point for the signal. The grounding of the shield in this manner is used to protect against electric fields which can cause noise problems in the signal. However, a single point ground will protect the signal from magnetic fields. This form of shielding is most effective at frequencies below 100 kHz. Shields connected together and grounded to a single point assumes no ground potential differences exist in the system.

Single point shield grounding can be used to control analog instrumentation signals. Coax and multiple grounding is used to control high-frequency energy related signals. Note that at low-frequencies, a shield grounded at both ends assumes a voltage gradient which is the same on the outside and inside surfaces of the shield.

To correct capacitive noise problems, connect the shield at one point so the capacitively coupled noise currents intercepted by the shield bypass the signal circuit as it returns to the noise surface. **(See Figure 11-59)**

Figure 11-59. This illustration shows a shield grounded at one end only to help reduce capacitive coupled noise problems.

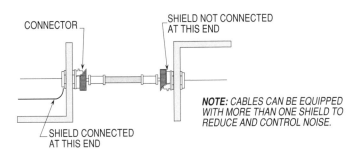

Grounding Tip 324: Radiative shields shall never be connected inside at chassis and inductive shields should be connected on the other side of the power source.

SHIELDS CONNECTED AT BOTH ENDS

To correct radiative noise problems, connect the shield at both ends of the signal circuit (at the chassis). Shield should be grounded at the driven source and at the receiver getting the signal. Note that a radiative shield shall never be terminated inside a chassis.

To correct inductive noise problems, connect shield on the other side of power source. **See Figures 11-60(a) and (b)** for grounding the shield for radiative and inductive noise problems.

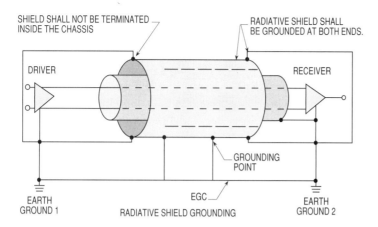

Figure 11-60(a). This illustration shows the shields of cables grounded at each end to reduce radiative noise problems. Note that shield shall terminate inside of the chassis frame.

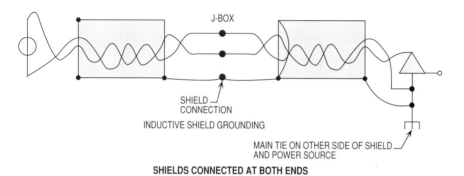

Figure 11-60(b). This illustration shows the main tie on the other side of the shield and power source.

RIBBON CABLE

Ribbon cables are used to control noise for logic signals. The radiation processes can be minimized by providing a grounding conductor next to each signal conductor. In design, this is done by making every third conductor a ground or by providing a grounding plane or shield. All grounding conductors shall be connected together at the ribbon termination and to the ground planes associated with the signal send. Note that these cables are ideal for many high-frequency installations. However, for them to be effective, proper connectors shall be used to terminate the shields individually at the chassis and the chassis frame shall be clean or of conductive material.

FOIL SHIELD CABLES

Aluminum foil is frequently used as shield material an shielded cables. The foil makes an excellent electrostatic shield. Cables using a foil wrap are not intended for the control of high-frequency energy. The foil is sometimes difficult to terminate because it can not be soldered and it tears very easily. To avoid these termination problems, cable manufacturers have added a drain wire to the cable. This conductor allows the shield to be terminated without contact problems. Designers usually prefer foil shields with external drain wires over drain wires installed inside the cable due to the noise that can be coupled into the cable. Due to the termination problems involved with the aluminum foil, these cables are usually avoided by most designers and installers.

Grounding Tip 325: Because of the terminating problems involved with foil shielded cables, they are not usually used by designers and installers.

BRAIDED CABLES

Braided copper is the most commonly used sheath for shielded cables. In many installations, single braided cable is adequate for the control of high-frequency signals. However, the state-of-the-art is the double braid-type cable.

At low-frequencies skin effect is considered negligible and the entire shield is used by the flow of current. Sheaths that are grounded at both ends provides a voltage gradient along the sheath and helps reduce and control noise problems.

Grounding Tip 326: Braided cables in many installations are fully capable of reducing and controlling high-frequency signals. The sheaths of such cables shall be grounded at each end.

Grounding Tip 327: A shield incorrectly connected or left floating is worse than no shield at all.

USING MULTIPLE SHIELDS TO CONTROL NOISE

Quard shields used in analog instrumentation does not provide shielding for high-frequency noise. Another external shield is needed if these fields are to be controlled. The inner shield is grounded where the signal grounds and the external shield is grounded to the source ground. Sometimes it is beneficial to ground the outer shield at more than one point because this reduces the loop areas that can cause noise problems. Note that this concept is not used in low-frequency shielding techniques.

See Figures 11-61(a) and **(b)** for using multiple shields to control noise related to capacitive, radiative and inductive coupling.

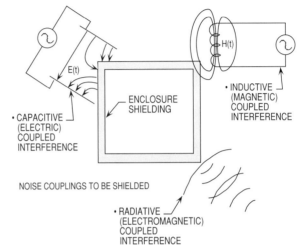

Figure 11-61(a). This illustration shows enclosure shielding used to control and reduce the noise associated with different types of couplings.

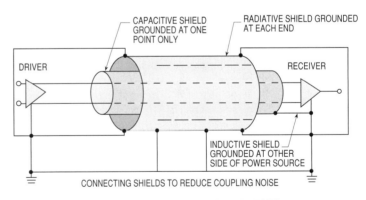

Figure 11-61(b). This illustration shows how the shields shall be connected to control capacitive, radiative and inductive noise producing couplings.

Grounding Tip 328: If shields are not terminated to the chassis properly by connectors, potential differences can develop between shield and chassis. Note that potential differences can cause noise problems and sensitive circuits to malfunction.

TERMINATING SHIELDS TO CLEAN SURFACES

If shields are not terminated properly, the quality of shielding is usually compromised at such points. A pinched down shield to a connecting conductor can cause current flowing on the outside surface of the shield to enter inside the cable system and create noise. Such a condition develops a potential difference between shield and chassis termination. This voltage can couple to all internal conductors and cause noise which creates malfunctions of the signaling circuits.

Note that the connector should never be mounted to a painted surface. Plated metal surfaces should be used and the surfaces should be protected against oxidation. Be careful with metal for it can look good, but circumstances can destroy the integrity of its bonding ability. It is very important for installers to ensure that shielding is terminated in a manner that provides bonding to the plated metal of the chassis. **(See Figure 11-62)**

Grounding Information Technology Equipment

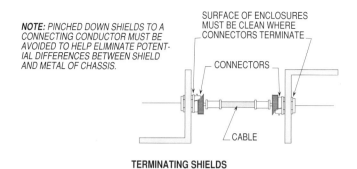

Figure 11-62. This illustration shows that installers shall ensure that terminating surfaces of enclosures are clean before connecting cable with connectors.

OPTICAL FIBER CABLES

Optical fibers are hair-thin structures capable of transmitting light signals with extremely low signal loss and at very high rates. Fibers are available in a variety of sizes and material compositions and with a wide range of optical performance. They are solid structures with fibers which function to guide the rays of light from source to destination with astonishing results.

An optical fiber cable must confine and guide light without losing the light when it arrives at the ITE system or terminal, etc. Care must be exercised when terminating the cable to the chassis frame of the ITE and associated equipment. Note that fiber optical cables provide a simple way to breach a shielded enclosure without providing a conductive path for radiation. Fiber optical cables is an effective way of achieving ground noise problems immunity. **(See Figure 11-63)**

Grounding Tip 329: There is an aperture (opening or hole) problem when routing optical fiber cables through walls and floors. One solution is to run the optics through a thin metal tube and bond parallel to the wall of the room. Such tubes serves as a wave guide and attenuates most of the radiation. Sometimes insulating tapes are used to enclose the optics instead of metal tubes.

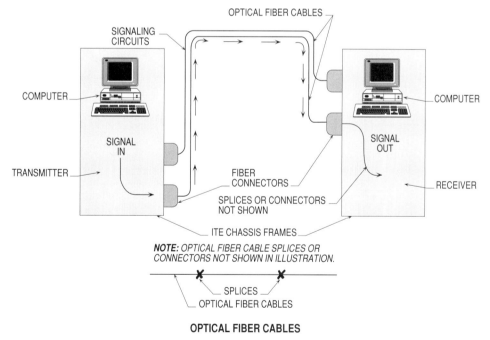

Figure 11-63. This illustration shows optical fiber cables used to carry signals to sensitive electronic apparatus.

Name _____   Date _____

Chapter 11
Grounding Information Technology Equipment

	Section	Answer

1. A bonded interconnected network can serve as both a power and signal reference grid. _____ T   F

2. For dwelling units, ground currents of an amp or two flowing over a grounding electrode conductor run to a driven rod is considered to be exclusively high and cause computer circuits to malfunction. _____ T   F

3. Ground-fault return paths are not required to have low-impedance paths. _____ T   F

4. Grounded conducting enclosures can serve as electromagnetic shielding for sensitive circuits. _____ T   F

5. Touch voltage is the voltage between any two conductive surfaces. _____ T   F

6. Under certain conditions, a driven rod can be used as the sole equipment grounding means for a computer. _____ T   F

7. To reduce and control voltage differences between the metal frames of computers, supplementary conductors are sometimes used. _____ T   F

8. Isolation transformers can be used to reduce and control electrical noise. _____ T   F

9. Manufacturer's grounding recommendations must always be adhered to even where they conflict with NEC requirements. _____ T   F

10. UL testing laboratory is a laboratory recognized by electrical inspectors. _____ T   F

11. The central grounding point of a sensitive electronic system should be readily identifiable. _____ T   F

12. Fiber optic solves some of the common mode noise problems that interrupt computer operation. _____ T   F

13. Coaxial cable can be used to help prevent noise problems from reaching the sensitive circuits in computers. _____ T   F

14. A single point entry is where all conduits enter the computer room from different locations. _____ T   F

15. A No. 22 gauge galvanized sheet steel can be formed and used to bond all conduits together where they enter a single point in a computer room. _____ T   F

16. Filters can be used to reduce and control unwanted harmonic noise. _____ T   F

17. Filters shunt the noise current to the ITE frame and such current returns over the equipment grounding conductor for the computer circuit. _____ T   F

_____ T F 18. It's not necessary to located the isolation transformer near the computer system served.

_____ T F 19. The isolation grounds must always terminate to the metal frame of a subpanel.

_____ T F 20. The grounding electrode used to earth ground the metal frames of computers should read 1 to 5 ohms.

_____ T F 21. Relocatable power taps can be used to cord-and-plug a computer and its accessories.

_____ T F 22. The earth must never serve as the sole equipment grounding conductor path for a computer.

_____ T F 23. An isolation ground cannot be used to ground the metal frame of the computer independently and separately from the equipment grounding conductor of the supply circuit.

_____ T F 24. The isolation ground can be routed through panelboards all the way back to the service equipment.

_____ T F 25. Multiple driven rods can be used to lower resistance to ground.

_____ T F 26. Long feeder-circuits can cause noise problems in computer's sensitive circuits.

_____ T F 27. ITE connected to the same panelboard feeding equipment with high-inrush current related loads can cause sensitive circuits in computers to malfunction.

_____ T F 28. Computer equipment must never be modified from its manufactured wiring scheme.

_____ T F 29. UPS systems must never be used to provide ride-through power for computers.

_____ T F 30. To control radiative noise problems, the shield in a signaling cable must be connected at both ends.

_____ T F 31. Foil shield cables are not usually used by designers because the foil tears so easily.

_____ _____ 32. Ground _____ can cause noise problems for computers.

    (a) isolation   (c) none of the above
    (b) loop   (d) all of the above

_____ _____ 33. A sag occurs for about _____ seconds.

    (a) 2.5   (c) 4
    (b) 3.5   (d) 5

	Section	Answer

34. For noise-free ground connections, such connections should be checked not less than _____ month intervals.

(a) 1 (c) 3
(b) 2 (d) 6

35. A ground _____ can be used to bond metal frames of computers together to control noise problems.

(a) conductor (c) all of the above
(b) strap (d) none of the above

36. Isolation transformers that are k-rated should be located _____ the ITE, if noise is to be controlled.

(a) far (c) none of the above
(b) near (d) all of the above

37. Shields that do not carry signal current and are connected at one end are used as _____ shields.

(a) electrostatic (c) all of the above
(b) guard (d) none of the above

38. To control capacitive noise problems, the shield is connected at _____ point only.

(a) one (c) all of the above
(b) two (d) none of the above

39. A zero _____ reference grid can be used to reduce and control noise problems when computers are connected to it with short ground straps.

40. _____ AWG copper conductors can be installed in 2 ft. centers to form a zero signal reference grid.

41. A _____ AWG copper conductor can be used to form a halo to ground computers for noise.

42. Stringers of a _____ floor can be used to form a zero signal reference grid.

43. A zero signal reference grid has to be _____ into the grounding electrode system.

44. A zero signal reference grid cannot be used as a _____ for the equipment grounding conductor run to a computer.

45. Ground straps used to ground the metal frames of computers to the zero signal reference grid must be as _____ as possible.

11-49

Section	Answer
_____	_____
_____	_____
_____	_____
_____	_____
_____	_____

46. Single point entry bonding of equipment enclosure will help _____ noise problems.

47. Isolation transformers have a grounded _____ that help control noise problems for sensitive circuits.

48. Ribbon cables are usually used to control noise for _____ signals.

49. Braided cables can be used to control circuits having high _____ signals.

50. To control high-frequency noise, another _____ shield is necessary.

12

LIGHTNING PROTECTION ON OUTSIDE OVERHEAD LINES

Lightning strikes are currents flowing from one cloud to another or to earth. Such currents will take the path of least resistance. High buildings, towers, trees, high lines, and other tall objects provide these striking points of lower resistance.

Wooden poles and towers supporting transmission and distribution lines are paths of less resistance due to their connections to ground. They are often struck by lightning strokes and these less resistance paths provide circuits in which these destructive currents use to flow to ground safely. Properly designed and installed grounding systems and associated equipment will aid in diverting such currents to ground without damaging the elements of overhead lines.

See **NFPA 780** and **NFPA 77** for further rules pertaining to the design and installation of lightning protection systems.

SURGES

Surges or transients are defined as large overvoltages which can develop suddenly on transmission and distribution lines due to lightning strokes or induced charges. Lightning strokes are surges as described above while other induced charges are developed by circuit-switching operations. These large overvoltages may damage the electrical elements connected to the lines, even though they disappear rapidly into the earth.

Grounding Tip 330: In areas where lightning is a real problem, the best protection that can be provided for aerial cables against lightning strikes consist of grounding the messenger and sheath at every pole and securing as low a ground resistance as possible.

SHIELD AND GROUNDED (NEUTRAL) CONDUCTOR

The surge protection of a transmission line is usually designed to either prevent flashover by using overhead shield conductors or allow flashover to occur in a controlled manner and use automatic reclosing to prevent disorderly outages.

Distribution lines employs a grounded (neutral) conductor which is sometimes installed above the ungrounded (phase) line conductors to serve as a shield conductor. These grounded (neutral) conductors intercepts the surge currents and diverts them to ground through grounding earth connections.

Grounding Tip 331: A shield conductor is highly effective in diverting lightning surges to earth, only if they are connected to a low ground resistance of 25 ohms or less.

SHIELD STATIC CONDUCTOR

A ground conductor or static conductor, installed above the other line conductors and strung along with such conductors, provides protection or shielding to the lines from direct lightning strokes.

The shield conductor on an overhead line may be a galvanized steel, copper covered steel or aluminum covered steel cable.

The shield conductor on lines supported by metal towers is grounded at each tower, with clamps connecting it to the tower grounding conductors and driven ground rods installed at the base of such towers.

This shield conductor on lines supported by wooden poles is connected to ground at different intervals by a grounding conductor from the clamp supporting the shield conductor at the top of the pole, to a driven rod below at the base of the pole.

To ensure a low-resistance path to ground, the grounds should be installed at each pole or at intervals not exceeding every fifth pole along the line. **(See Figure 12-1)**

Figure 12-1. This illustration shows the towers with a static ground conductor mounted above the ungrounded (phase) line conductors that shields them from lightning strikes and switching surges.

Grounding Tip 332: Shielding grounds shall be bonded to surge arrester grounds, where provided, at points where underground cables are connected to overhead lines per NESC Rule 92.A and B.

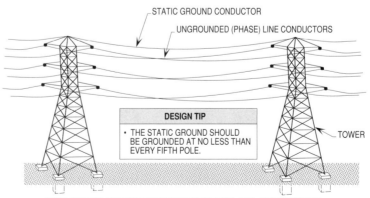

NOTE: TOWERS ARE USUALLY GROUNDED TO RODS, ETC., AT EACH TOWER.

SHIELD STATIC CONDUCTOR
NFPA 70B - CH. 19-3.13
NFPA 77 - CH. 3-3.1

GROUNDED (NEUTRAL) CONDUCTOR

On distribution systems, the grounded (neutral) conductor may be installed above the ungrounded (phase) line conductors to serve in the place of the shield conductor.

The grounded (neutral) conductor shields the ungrounded (phase) line conductor from direct lightning strokes by providing a zone of protection.

These grounded (neutral) conductors are designed to intercept strokes and divert them to ground prior to over voltage build up in the underlying ungrounded (phase) line conductors to flashover potentials.

The NESC requires grounding of such conductors used for primary and secondary lines. Additionally, circuits shall be grounded at each piece of electrical equipment and terminated to such lines with at least four grounding connections per mile along the line. This grounding procedure provides an effective grounding path. **(See Figure 12-2)**

Grounding Tip 333: When grounding the primary grounded (neutral) conductor at numerous points along its length, these connections to ground are known as multi-grounded techniques and they are used to derive lightning strikes, voltage surges, etc.

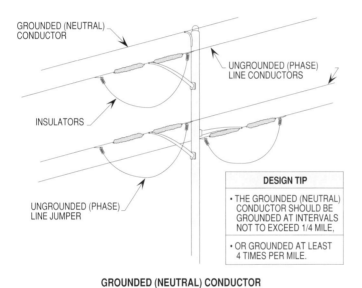

Figure 12-2. This illustration shows that the utility pole has a grounded (neutral) conductor that is utilized to take lightning strikes and provide a cone of protection for the ungrounded (phase) line conductors located below.

Grounding Tip 334: For requirements on the four-ground connection on a common neutral, see NESC Rule 97.C.

GROUNDING SYSTEMS

Shielded grounded (neutral) conductors if installed properly provide a highly effective grounding scheme. However, its effectiveness depends primarily on a very low ground resistance.

When shielded grounded (neutral) conductors are connected to a low resistance to ground (grounding electrode), they bring the ground (earth) potential above the transmission and or distribution lines. Hence, the stress on the insulation of the line and equipment due to surges is greatly reduced.

Grounding Tip 335: For grounded (neutral) conductors and shielded conductors to be effective in diverting lightning strikes, their connections to ground shall be verified on a time frame that ensures safety.

DRIVEN RODS

The most used ground in distribution systems is the driven rod. For lowest resistance it should be driven until the top end of the rod is at a burial depth of 2 ft. (600 mm) from finished grade. After the rod has been driven, the connection of the ground conductor from the pole is made and an effective ground path is provided.

The size of the grounding conductor should be at least a 6 AWG copper conductor. For physical protection, a 4 AWG copper conductor is normally used.

A cover is sometimes installed over the grounding conductor to protect persons or animals from electrical shock in case the ground around the rod bakes out and soil contact is lost. When a ground rod bakes out or the ground freezes, the rod loses contact with moist earth and the resistance on the rod is very high. This can cause current to flow over the grounding conductor to the rod which may create very hazardous conditions.

If a lightning arrester breaks down, a high current can flow over the grounding conductor to the rod and cause a high resistance earth ground. Until the failed arrester is found and fixed the grounding conductor cover protects personnel from touching the conductor and being shocked. **(See Figure 12-3)**

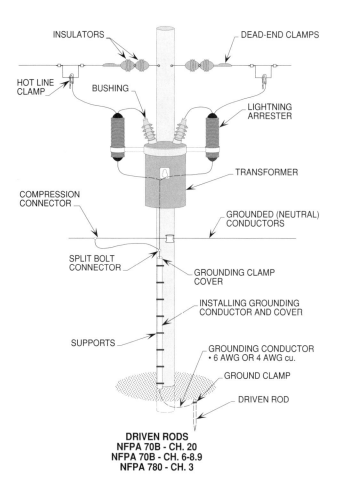

Figure 12-3. This illustration shows the proper way to install a grounding conductor and protect it from physical damage using a cover mounted to the pole.

Grounding Tip 336: For requirements pertaining to driven rods and their connections in the ground, See Sec.'s 3.13.1 through 3.13.1.2 in NFPA 780.

If one rod fails to lower the resistance to ground to the designed level needed, additional rods can be driven and connected in parallel. Rods in parallel and properly spaced will lower the resistance based upon number or a counterpoise ground can be added and used.

For example, two rods connected in parallel usually have a resistance of about 50 to 60 percent of that of one rod. Three rods have about 33 to 40 percent of one rod and four rods have about 25 to 33 percent of one rod. **(See Figure 12-4)**

Lightning Protection on Outside Overhead Lines

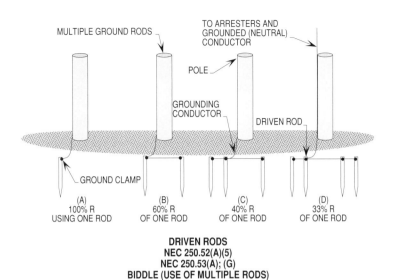

Figure 12-4. This illustration shows the installation of multiple rods used to reduce the resistance to ground.

Grounding Tip 337: For requirements pertaining to the installation of driven rods, See Rules 94.B1 and B2 in the NESC and Sec. 3.13.1 and appendix I of NFPA 780.

BUTT WRAP GROUND

A pole-butt grounding plate or butt coil is sometimes used on wooden poles. A butt coil consists of a copper conductor installed at the bottom of a pole. Butt grounds shall have enough turns of wire to make good contact with the soil of the earth. At least seven turns of wire with a length of 14 ft. (4.2 m) or greater provides a satisfactory grounding contact with the soil of the earth.

The end of the grounding conductor must be shorted across the turns of the butt ground conductor to prevent a choke coil effect. **(See Figure 12-5)**

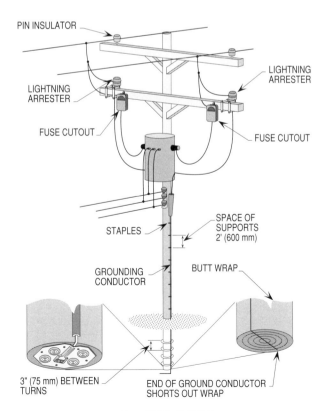

Figure 12-5. This illustration shows a butt wrap grounding system that diverts surge current into the ground (earth).

Grounding Tip 338: For requirements pertaining to the installation of butt wrap grounds, see Rules 94.Ba thru c in the NESC.

12-5

Grounding Tip 339: For requirements explaining the use of counterpoise grounding techniques, See Rules 94.3a and 93.E5 in the NESC.

COUNTERPOISE GROUNDS

A counterpoise grounding scheme is installed where the soil has a high-resistance. Counterpoise systems consists of one or more buried conductors connected in a paralleled (continuous) arrangement identified as radial or crowfoots type.

The counterpoise conductors are connected to the overhead grounded (neutral) conductor or static ground at all supporting towers and wooden pole structures.

The action of a counterpoise ground reduces the surge impedance of the ground connection which increases the coupling between the grounded (neutral) conductor and conductors, therefore lowering resistance.

Grounding Note: The conductors of the counterpoise ground only need to be buried deep enough to prevent theft or disturbance by cultivation. The recommended burial depth is to be installed at least 18 in. (450 mm) or more below grade. **(See Figure 12-6)**

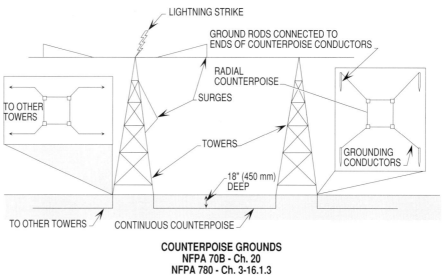

Figure 12-6. This illustration shows that counterpoise grounds can be used to lower the resistance to ground which provides a better path to ground to divert surges. **Note:** Counterpoise can be continuous or noncontinuous from tower to tower.

TOWER BASE GROUND

Steel structure tower bases are either grounded to extended buried steel footers or by means of driven ground rods. If driven rods are used, one rod is driven for each leg of the tower to ensure lower resistance to divert surges due to lightning strikes and switching. **(See Figure 12-7)**

GROUND GRID

Grounding Tip 340: For requirements pertaining to designing and installing a low-resistance ground grid, See Rules 96.A through C in the NESC.

A grid grounding scheme consists of buried conductors and driven ground rods which are interconnected to form a continuous grid network.

Ground grids are usually used for the grounding of lines and equipment located in the substation. The equipment and apparatus shall be bonded and grounded to the ground grid using large conductors to minimize the grounding resistance which limits the potential between equipment and ground surface.

Fences shall be constructed inside the ground grid and connected as needed to provide a safe bonding scheme. Sometimes large conductors are run underneath substations and spaced properly to create a very low resistance to ground. **(See Figure 12-8)**

Lightning Protection on Outside Overhead Lines

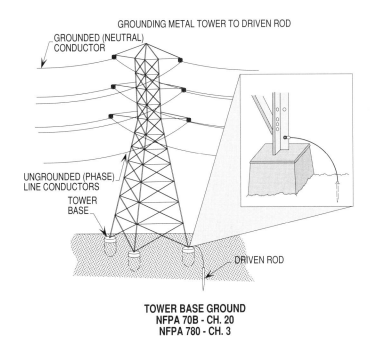

Figure 12-7. This illustration shows the base of the metal tower grounded to a driven rod. Note that each tower leg can be grounded to a driven rod or counterpoise per Figure 12-6.

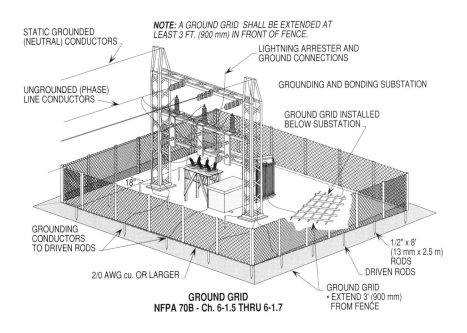

Figure 12-8. To establish equipotential planes between ungrounded (phase) line conductors and equipment, elements are grounded to a counterpoise ground and driven rods or ground grid designed for the purpose.

Grounding Tip 341: Ground grids can be designed and installed to provide a resistance of 5 ohms or less. It is recommended for large substations to have a resistance of 1 ohm or less and 5 ohms or less for small substations. For more information on how to bond and ground a substation, see IEEE 80.

LIGHTNING ARRESTERS

A lightning (surge) arrester is a device that protects line transformers and other electrical elements from high-voltage surges. These surges can occur either because of lightning strikes or abnormal switching in the circuit. The lightning arrester provides a path over which the surge can pass to ground, before such surge attacks and damages the lines, transformer or other electrical equipment.

The elementary lightning arrester consists of an air gap in series with a resistive element. The voltage surge causes a spark to jump across the air gap, which passes through the resistive element to ground.

Such resistive element is a material that provides a low-resistance path for the high-voltage surge, however has a high resistance to the flow of line current. This material is usually called the "valve" element.

Lightning arresters usually have an air gap in series with a resistive element, and the resistive (valve) element must act as a conductor for high-energy surges and also as an insulator toward the flow of line current. In other words, the lightning arrester bleads off only the surge current, and afterwards the air gap reopens and insulates the line voltage and current to ground through the arrester. **(See Figure 12-9)**

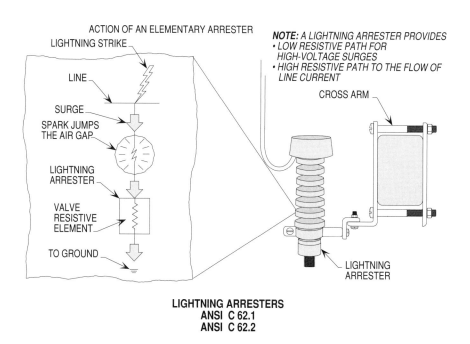

Figure 12-9. This illustration shows the elementary action that takes place when a lightning arrester diverts a power surge to ground.

Grounding Tip 342: For rules explaining the requirements for designing and installing lightning arresters, See Rules 97.A1 through A3 and 97.B1 and B2 in the NESC.

TYPES OF ARRESTERS

There are many different types of lightning arresters available in the industry today. The type chosen is based upon its characteristics and use in the protection of the lines, transformers and other equipment from surges.

EXPULSION-TYPE ARRESTER

Expulsion arresters consists primarily of a fiber expulsion chamber and a series of external and internal series gaps between lines and ground. Expulsion chamber and interior gaps are contained in a porcelain housing.

They are bridged by an high-voltage surge. In other words, when lightning occurs the gaps are bridged and the high surge current flows harmlessly to ground.

VALVE TYPE ARRESTERS

Valve arresters are by far the most common types in use today and consist essentially of a series gap structure and a valve element(s) made of nonlinear resistance material which are all sealed in a porcelain housing.

The series gap assembly is usually located at the top and when there is excessive voltage surge on the line, the spark jumps across the air gap and the surge energy flows through such material to ground. As the surge current decreases, the resistive power of the material increases and no line current flows to ground.

GAPLESS TYPE ARRESTERS

Gapless arresters are a variation of the valve arrester and have just recently been introduced to the industry. This basically does away with the requirement for the series gap assembly. Such arresters are known as "gapless" or "MOV" representing the term metal oxide varistor.

APPLICATION OF ARRESTERS

There are three classes of arresters and they are distribution, intermediate and station. Their selection and use is based on the terms of voltage rating, protective values, durability and pressure relief, including venting methods.

DISTRIBUTION CLASS ARRESTERS

Distribution class arresters are widely used on electric systems especially for the protection of oil insulated equipment. For protection of dry type equipment or rotating machines special low sparkover arresters are usually employed.

Distribution arresters are utilized on primary distribution systems to protect insulators, distribution transformers, lines and other associated equipment. **(See Figure 12-10)**

Grounding Tip 343: For the general requirements pertaining to the location of lightning arresters, See Rule 190 in the NESC.

Grounding Tip 344: For lightning arresters installed inside of buildings, See Rule 191 in the NESC.

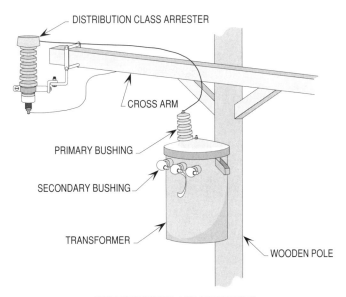

Figure 12-10. This illustration shows a distribution class arrester tapped from a distribution line and protecting the primary side of a transformer and other associated equipment.

DISTRIBUTION CLASS ARRESTERS
ANSI C 62.1
ANSI C 62.2

INTERMEDIATE CLASS ARRESTERS

Intermediate class arresters normally offer lower sparkover and discharge voltage characteristics and greater surge current discharge capability than distribution arresters.

Intermediate arresters are often used on substation exit lines and other areas in the distribution system needing a greater level of lightning and surge protection. **(See Figure 12-11)**

Figure 12-11. This illustration shows intermediate class arresters protecting a 34.5 kV, three-phase supply line from the source.

Grounding Tip 345: For requirements pertaining to the design and installation of the grounding conductors, See Rules 192 and 93.C4 in the NESC.

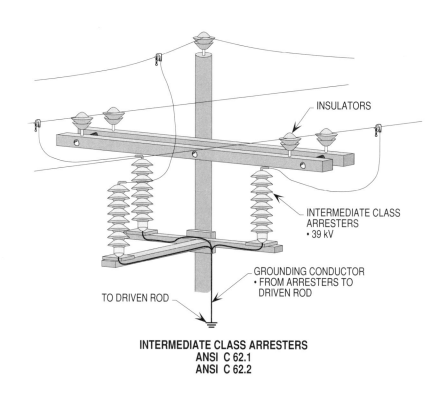

STATION CLASS ARRESTERS

Grounding Tip 346: When the value of equipment protection and uninterrupted power is desired, the use of station class arresters are usually used.

Station class arresters have the greater protective characteristics and thermal capability than most other arresters. Station class arresters are used in substations and generating plants to provide the best level of protection possible. **(See Figure 12-12)**

Grounding Note: Selecting and installing arresters

- The first point to consider is the rating of the arrester which is sized greater than the transformer rating, but not much greater. If it were sized much greater, it becomes a path of greater resistance and forces the high-voltage surges into the transformer, which damages the elements.

- The second point is to ensure that the lightning arrester provides the shorter path to the ground and diverts to ground the excess voltage as quickly as possible.

- The third point is to size the grounding conductor correctly. Both the line lead and the grounding conductor should be as large as the primary drop leading from the power main to the transformer.

- The fourth point is to make sure that all connections are very tight and the arrester is properly grounded.

Lightning Protection on Outside Overhead Lines

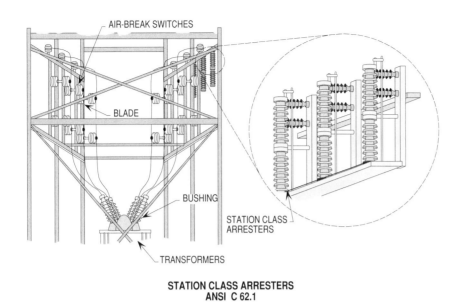

Figure 12-12. This illustration shows station class arresters installed at a substation location.

CONE OF PROTECTION

The area of protection for a well grounded object is considered to be a conical zone below and around such object that is based upon a 45° angle and 30° vertical respectively.

In other words, the grounded object throws a protective "shadow" over and below things located within such shadow and lightning strikes will not normally enter this shadow zone.

Grounding Tip 347: For a better understanding of the cone of protection technique, see Sec.'s 6.3.3.1 through 6.3.3.4 in NFPA 780.

OBJECTS UNDER CONE

All objects that are located within this "shadow" of protection is considered protected. Things touched by the "shadow" or outside the shadow are susceptible to direct lightning strikes and are not considered protected. **(See Figure 10-13)**

Grounding Tip 348: For requirements pertaining to objects and structures protected under the zone of protection, see Sec.'s 3.7.3.1 through 3.7.3.4 in NFPA 780.

OVERHEAD LINES BELOW CONE

An overhead shield conductor may be used as a cone of protection for ungrounded (phase) line conductors located below and within the zone of protection.

Lightning strikes are directed to the uppermost line conductor and diverted to ground through the grounding conductor connected to a driven rod or butt ground. If constructed and installed correctly with a resistance of less than 5 ohms to ground, this grounding scheme will normally perform as it is designed to do. **(See Figure 12-14)**

12-11

Figure 12-13. This illustration shows that facilities are considered protected from lightning strikes where they are located within the protective shadow provided by the cone of protection.

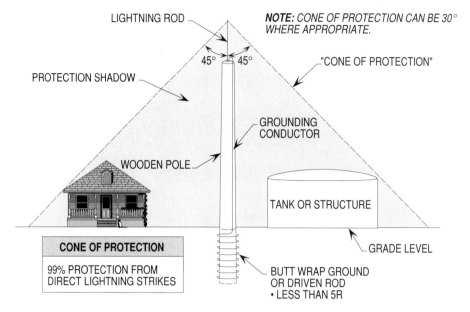

Figure 12-14. This illustration shows the cone of protection for three-phase conductors located within the protective shadow. The static ground protects the ungrounded (phase) line conductors from direct lightning strikes.

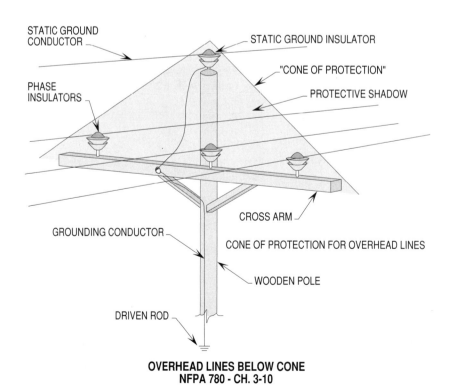

TYPES OF CONSTRUCTION

There are basically three types of overhead power line construction schemes used in the industry today to protect lines and equipment and they are as follows:

- Flat-top
- Shielded
- Hi-impulse

Grounding Note: These types of construction techniques are normally employed for distribution lines utilized to route overhead lines to areas serving special types of electrical equipment.

FLAT-TOP CONSTRUCTION

Flat-top construction is the most common used design for overhead power line installations. It provides satisfactory protection in regions with low lightning incidents.

Basically phase-to-phase flashover occurs through the crossarm wood through the steel insulator pins and through pole mounting bolts. **(See Figure 12-15)**

Grounding Tip 349: For rules pertaining to multi-grounding at poles supporting overhead lines, See Rule 96.C in the NESC.

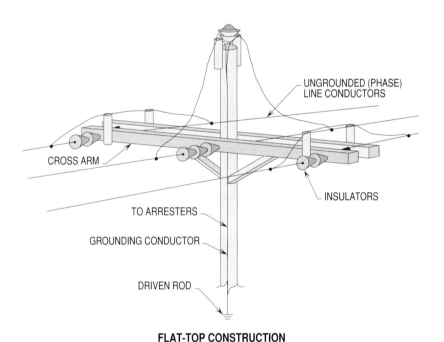

Figure 12-15. This illustration shows flat-top construction which is the most common type of overhead construction used.

Grounding Tip 350: Flat-top construction used to support overhead lines are recommended in low lightning incident regions. For rules to determine such regions, See Appendix H in NFPA 780.

SHIELDED CONSTRUCTION

Shielded construction consists of all ungrounded (phase) line conductors located within the "cone of protection" that is provided by an overhead shield conductor.

Lightning strikes are directed to this higher line conductor and diverted to ground through the grounding conductor that is connected to ground at the base of the wooden pole. This grounding conductor used to carry the surge of current to ground is connected to a driven rod, butt wrap or a counterpoise grounding scheme. **(See Figure 12-16)**

Figure 12-16. This illustration shows an overhead static grounding conductor, which is used to shield the ungrounded (phase) line conductors, located below, from lightning strikes.

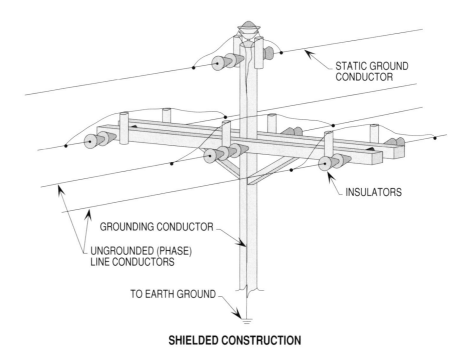

SHIELDED CONSTRUCTION

Grounding Tip 351: The basic insulation level (BIL) is defined as the coordination of insulation which requires the insulation of all elements of a system to be above a minimum level and that a selected OCPD operate satisfactorily below such minimum level.

Grounding Note: All three-phase conductors are within the "cone of protection" of the overhead shield conductor. Lightning strikes are directed to this uppermost conductor then taken to ground by the grounding conductor that should be installed on every pole in the system using butt-wrapping, terminating in a full diameter copper pole ground (butt) plate plus driven rod if possible. Properly constructed, with a basic insulation level (BIL) of not less than 30 kV and an overall installation resistance of less than 5 ohms, this protection scheme will generally perform as intended.

Grounding Tip 352: When using high-impulse, nonshielded construction, it is important to achieve a good low-resistance to ground that will be at ground potential of the earth located between the pole and devices.

HI-IMPULSE, NONSHIELDED CONSTRUCTION

In this type of construction, one ungrounded (phase) line conductor is located above two ungrounded (phase) line conductors which are placed below within a cone of protection.

Lightning strikes are directed to the upper ungrounded (phase) conductor and diverted to ground through a electrode path along the outer surface of the pole. This path is directed along the surface of the pole through adjacent air surrounding the electrodes.

For example, as voltage stress builds up due to a lightning strike, corona discharges on sharp edged metallic objects which are present on the poles and arms. Corona ionizes the pole surfaces and air, and due to this action, flashover of the high impulse voltage begins to divert down the lowest path. **(See Figure 12-17)**

Lightning Protection on Outside Overhead Lines

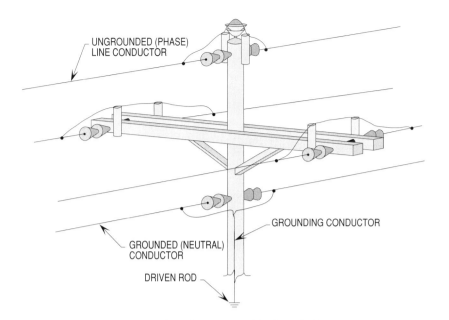

Figure 12-17. This illustration shows a detailed illustration of a hi-impulse nonshielded constructed ground protection scheme.

APPLICATION OF ARRESTERS

Lightning is an electrical discharge that occurs in the atmosphere between clouds or between clouds and earth. Very high-voltages are built up within a cloud. When this voltage becomes high enough, the insulating air between clouds or between clouds and earth is broken down and a charge of electricity passes from cloud to cloud or between cloud and earth.

Lightning strikes that hit overhead power lines and equipment can cause severe damage. Lightning arresters are used to clamp such surges of current and divert it to ground.

Lightning arresters are almost always applied between the ungrounded (phase) line conductors and ground because the lightning currents must be diverted to ground.

Distribution arresters are sensitive to the applied voltage and they will not interrupt the system power current if the system voltage exceeds the rating of the arrester.

It is essential that arresters be applied so the maximum system voltage that can appear across their terminals under conditions of use is not higher than their rated voltage. On some lines and equipment the normal voltage between ungrounded (phase) line conductors and ground can be exceeded during abnormal fault conditions.

It is most important that the selection of the arrester rating be made on the basis of past experience or reference to industry tables is recommended. If there is doubt as to the proper rating to be selected and applied, a manufacturer must be consulted.

See Figures 12-18(a) and (b) for the installation of lightning (surge) arresters and the grounding conductor connected to ground.

Grounding Tip 353: When designing, installing and maintaining lightning arresters to protect overhead lines and equipment, see Rules 190 through 193 in the NESC. Also, review the following IEEE standards:
- C62.1
- C62.11

Grounding Tip 354: For more information on designing and installing surge arresters, see **Chapter 14** in this book.

12-15

Figure 12-18(a). This illustration shows lightning arresters and how they are used to ground overhead lines.

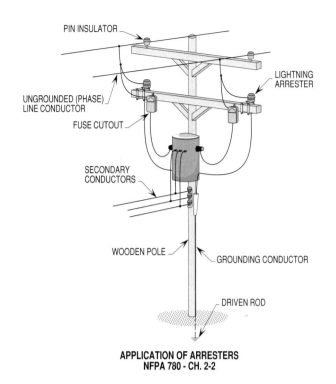

Figure 12-18(b). This illustration shows lightning arresters and how they are used to ground underground lines, service equipment and accessories.

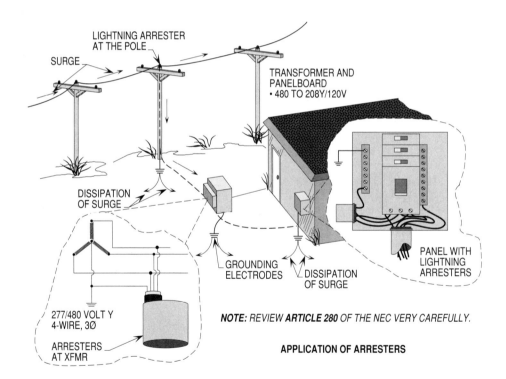

Name _____   Date _____

Chapter 12
Lightning Protection on Outside Overhead Lines

		Section	Answer

1. The shield conductor on lines supported by metal towers is not grounded at each tower.   _____   T   F

2. On distribution systems, the grounded (neutral) conductor may be installed above the phase line conductors to serve in the place of the shield conductor.   _____   T   F

3. A counterpoise grounding scheme is installed where the soil has a high-resistance.   _____   T   F

4. A grid grounding scheme consists of buried conductors and driven ground rods which are interconnected.   _____   T   F

5. The elementary lightning arrester consists of an air gap in parallel with a resistive element.   _____   T   F

6. Valve arresters consist primarily of a fiber expulsion chamber and a series of external and internal series gaps between lines and ground.   _____   T   F

7. Distribution class arresters can be used on electrical systems especially for the protection of oil insulated equipment.   _____   T   F

8. An overhead shield conductor can be used as a cone of protection for phase line conductors located below and within the zone of protection.   _____   T   F

9. Hi-impulse, nonshielded construction has a one phase line conductor located above two phase line conductors which are placed below within a cone of protection.   _____   T   F

10. Lightning arresters are used to clamp such surges (lightning strikes) of current and divert it to ground.   _____   T   F

11. Grounded neutral conductors used for primary and secondary lines and circuits must be grounded at each piece of electrical equipment terminated to such lines and with at least _____ grounding connections per mile along the line.   _____   _____

12. The size of the grounding conductor to the driven rod should be at least a _____ AWG copper conductor.   _____   _____

13. At least _____ turns of wire with a length of _____ ft. or greater provides a satisfactory grounding contact with the soil of the earth when using butt wrap grounds.   _____   _____

12-17

Section   Answer

_____ _____ **14.** If driven rods are used to ground a tower base, _____ rod is driven for each leg of the tower to ensure lower resistance to divert surges due to lightning strikes and switching procedures.

_____ _____ **15.** The area of protection for a grounding an object is a conical zone below and around such object that is based upon a _____ degree angle and _____ degree vertical respectively

13

LIGHTNING PROTECTION FOR BUILDINGS AND STRUCTURES

The Standard For The Installation of Lightning Protection Systems (NFPA 780 - 2000) is designed to provide guidelines in designing and installing a lightning protection system for intercepting a lightning strike and safely conducting the current to ground.

A basic lightning protection system consists of air terminals at all locations where lightning may directly strike a structure. Included in this system are properly designed electrical conductors connecting all air terminals and extending to ground with terminations at ground for distribution of the lightning discharge. Interconnection of grounded and ungrounded metal objects where a sideflash may occur shall be bonded to this lightning network.

AIR TERMINALS
NFPA 780, SEC.'s 2.1.1; 3.6

Air terminals shall be located on a structure at all locations considered where lightning strikes may occur. Since the ridge of a steep sloped roof or the outside edge of a flat or gently sloped roof is most often the target of lightning strikes, air terminals are placed at 20 ft. (6 m) to 50 ft. (15 m) intervals on these locations. Other obvious structural objects such as chimneys, gutters, flagpoles and ventilating equipment should be protected by air terminals. If exposed parts of a structure are constructed of metal with a thickness of 3/16 in. (4.8 mm), the installation of an air terminal is not required per NFPA 780, Sec. 3.6. The metal object, however, shall be interconnected to the system to provide a path to ground. **(See Figure 13-1)**

Figure 13-1. This illustration shows the procedure for designing and installing air terminals for a flat constructed roof area.

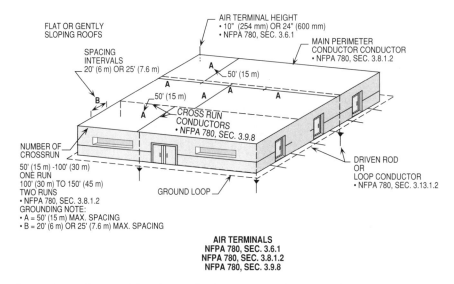

In determining the need for air terminals, some roof area and structural objects may fall within the influence of other air terminals. The area of influence is called the zone of protection. This zone of protection is considered to be about 150 ft. (45 m) final striking distance. Using a 150 ft. (45 m) radius arc allows the lightning protection designer to eliminate the need for air terminals on many lower roofs. **(See Figure 13-2(a))**

Figure 13-2(a). This illustration shows the zone of protection based on a 150 ft. (45 m) radius arc striking distance.

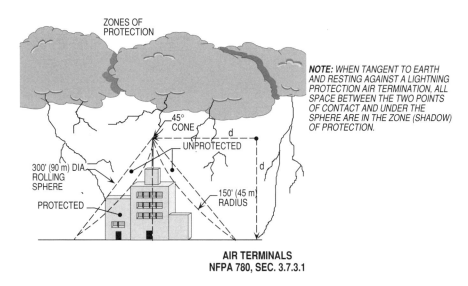

Structures not exceeding 25 ft. (7.6 m) above earth shall be considered to protect lower portions of a structure that is located in a one-to-two zone of protection. **(See Figure 13-2(b))**.

Figure 13-2(b). This illustration shows a 25 ft. (7.6 m) or less in height structure providing a one-to-two protection zone.

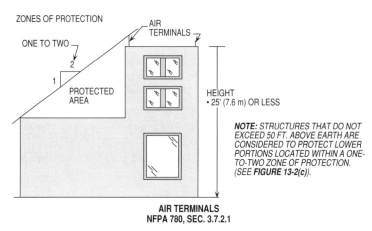

Structures not exceeding 50 ft. (15 m) above earth shall be considered to protect lower portions of a structure that is located within a one-to-one zone of protection. **(See Figure 13-2(c))**.

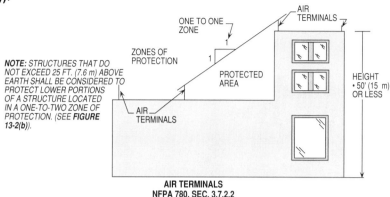

Figure 13-2(c). This illustration shows a 50 ft. (15 m) or less in height structure providing a one-to-one zone of protection.

Grounding Note: A six story building roof will offer protection for a four story level extending 24 ft. (7.3 m) from the higher roof.

For large flat or greatly sloping roof areas, air terminals shall be located so no unprotected area exceeds 50 ft. (15 m) in any dimension. **(See Figures 13-3(a) and (b))**

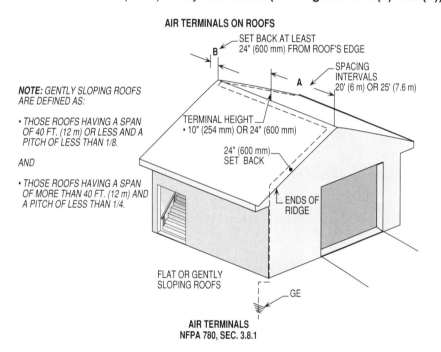

Figure 13-3(a). This illustration shows the requirements for spacing air terminals at 20 ft. (6 m) to 25 ft. (7.6 m) intervals on a gently sloping roof.

Grounding Tip 355: The building in **Figure 13-3(a)** has a two-way path to ground through the two grounding electrodes (GE). Note that one grounding electrode and down conductor are located at the left front of the building and the other is located at the back right side of the building.

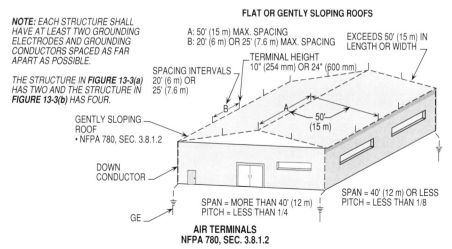

Figure 13-3(b). This illustration shows the requirements for spacing air terminals at a 50 ft. (15 m) maximum for a gently sloping roof that exceeds 50 ft. (15 m) in length or width.

HEIGHT OF TERMINALS
NFPA 780, SEC. 3.6.1; Figure 3.6.1

The height of air terminals shall be such as to bring the tip not less than 10 in. (254 mm) above the object to be protected for 20 ft. (6 m) maximum intervals and not less than 2 ft. above the object to be protected for 25 ft. (7.6 m) maximum intervals. **(See Figure 13-4)**

Figure 13-4. This illustration shows the height of air terminals based on the spacing intervals around the facility.

Grounding Tip 356: The height of air terminals shall be installed so that the tip is not less than 10 in. (254 mm) above the object to be protected for 20 ft. (6 m) maximum intervals and not less than 2 ft. (.6 m) above the object to be protected for 25 ft. (7.6 m) maximum intervals.

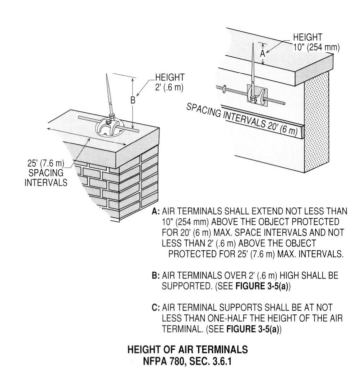

SUPPORT OF TERMINALS
NFPA 780, SEC. 3.6.2

Air terminals shall be secured against overturning either by attachment to the object to be protected or by means of braces which shall be permanently rigidly attached to the building. An air terminal exceeding 24 in. (600 mm) in height shall be supported at a point not less than one-half its height. **(See Figures 13-5(a) and (b))**

Figure 13-5(a). This illustration shows the requirements for supporting air terminals when their height is over 24 in. (600 mm).

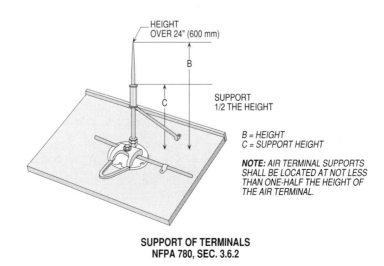

Lightning Protection for Buildings and Structures

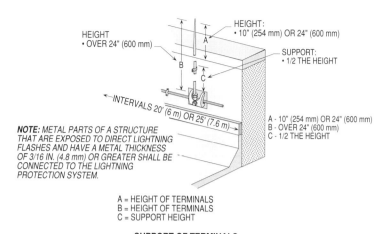

Figure 13-5(b). This illustration shows the supporting heights of air terminals based on their height and spacing intervals.

DOWNLEAD CONDUCTORS
NFPA 780, SEC. 3.9.10

Roof conductors shall interconnect all air terminals and provide a two way path from each air terminal to ground maintaining a horizontal or downward path. Note that there are exceptions to this requirement which are applied on roofs lower than the main roof and on dormers. Where a horizontal course cannot be maintained, a gradual rise not exceeding 3 in. (75 mm) per foot shall be permitted. Where the center roof conductors connect air terminals on large flat or gently sloping roofs, the conductor runs shall be cross connected to the roof perimeter every 150 ft. (46 m) of fraction thereof per NFPA 780, Sec. 3.9.8. **(See Figure 13-6)**

Grounding Tip 357: Structures exceeding 250 ft. (76 m) in perimeter shall have a down conductor for every 100 ft. (30 m) of perimeter or fraction thereof.

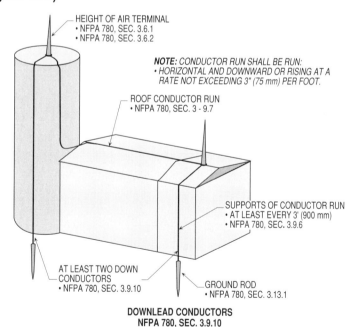

Figure 13-6. This illustration shows the requirements for designing and installing down-lead conductors to divert lightning strikes to earth through driven rods.

NUMBER OF DOWN CONDUCTORS
NFPA 780, SEC. 3.9.10

The down-lead conductors shall be as widely separated as possible. Preferably at the perimeter of the structure. All structures shall have a minimum of two paths to ground. Where protected roof perimeters exceed 250 ft. (76 m) additional down-leads shall be installed for every 100 ft. (30 m) of perimeter or fraction thereof. Structures with multiple levels of roof may require more down-leads in order to maintain two way paths from air terminals. **(See Figure 13-7)**

Figure 13-7. This illustration shows the procedure for determining the number of down conductors based on the width and length of the facility.

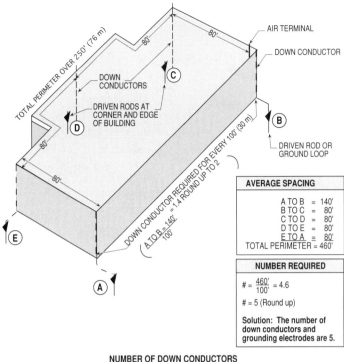

GROUNDING ELECTRODES
NFPA 780, SEC. 3.13 THRU 3.14

Each downlead conductor shall terminate in a ground terminal. These terminals may consist of a driven rod, multiple driven rods, metal plates, ground ring conductor or concrete-encase electrodes. A ground rod driven to a depth of 10 ft. (3 m) into undisturbed moist soil provides a proper ground in many cases. **(See Figure 13-8)**

The buried ground ring conductor if properly installed provides a good disturbed field for dissipating a lightning strike per NFPA 780, Sec. 3.13.3.

Figure 13-8. This illustration shows the requirements for installing a driven rod to be used as a grounding electrode for down conductors connected to air terminals on the roof.

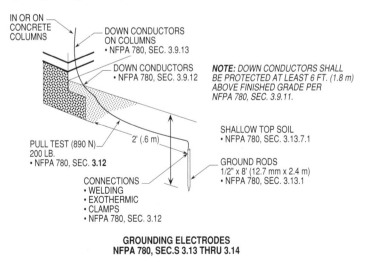

The use of concrete-encased conductors or rebars as a ground termination shall be used on new construction when the concrete foundation is in direct contact with earth per NFPA 780, Sec. 3.13.2. **(See Figure 13-9(a))**

See Figure 13-9(b) for a detailed illustration of how to install a buried counterpoise ground for connecting down conductors to earth ground.

Lightning Protection for Buildings and Structures

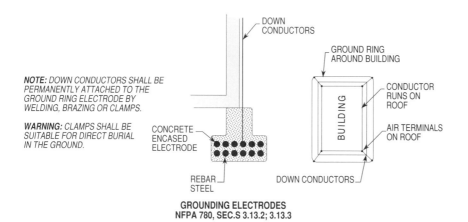

Figure 13-9(a). This illustration shows a concrete-encased electrode and ground ring used as grounding electrodes for connecting down conductors from air terminals on the roof.

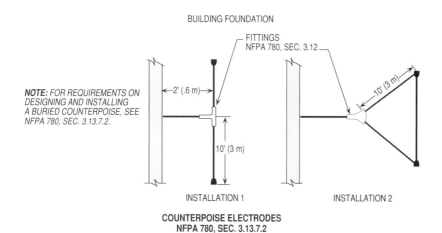

Figure 13-9(b). This illustration shows the requirements for designing and installing a counterpoise to be used as a grounding electrode for the lightning protection scheme.

CONDUCTOR BENDS
NFPA 780, SEC. 3.9.5

When routing a conductor, no bend of a conductor shall form an angle of less than 90 degrees, nor have a radius of bend of less than 8 in. (203 mm). **(See Figure 13-10)**

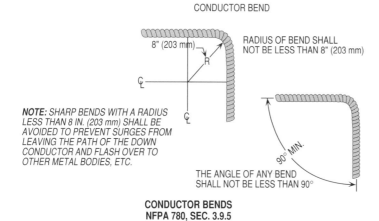

Figure 13-10. This illustration shows the requirements pertaining to the minimum bending radius and angle bend for down conductors.

CONDUCTOR SUPPORTS
NFPA 780 - SEC. 3.9.6

Conductors shall be permitted to be coursed through air without support for a distance of 3 ft. (.9 m) or less. Conductors that are coursed through air for longer distances shall be provided with a positive means of support that will prevent damage or displacement of the conductor.

Grounding Tip 358: All down conductors shall be connected to a low-resistance grounding electrode if lightning strikes are to be diverted properly.

ROOF CONDUCTORS
NFPA 780 - SEC. 3.9.7

Roof conductors that are coursed along ridges of gable, gambrel and hip roofs, around the perimeter of flat roofs, behind or on top of parapets and across flat or greatly sloping roof areas shall be interconnected to all air terminals. Conductors shall be coursed through or around obstructions in a horizontal plane with the main conductor. **(See Figure 13-11(a))**

See Figure 13-11(b) for a detailed illustration of how to course conductors on a chimney.

Figure 13-11(a). This illustration shows the types of gables that may need protection schemes designed and installed to protect from lightning strikes.

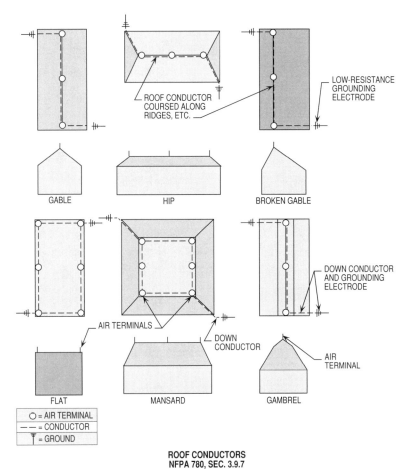

Figure 13-11(b). This illustration shows the requirements pertaining to the number of air terminals needed to protect a chimney from lightning strikes.

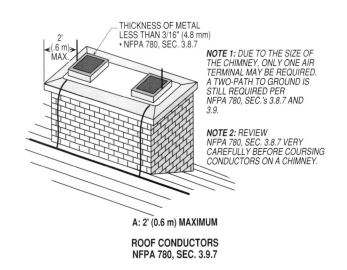

ONE WAY PATH
NFPA 780 - SEC. 3.9.1

Air terminals on a lower roof level that are interconnected by a conductor run from a higher roof level only require one horizontal or downward path to ground provided the lower level roof conductor run does not exceed 40 ft. (12 m). **(See Figure 13-12)**

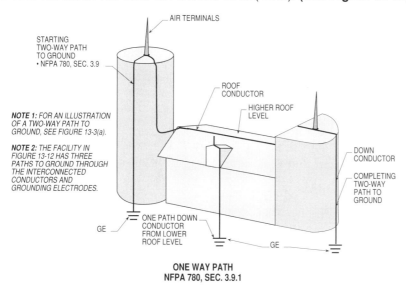

Figure 13-12. This illustration shows the requirements for a one-way path protection technique which is used in conjunction with the two-way path to ground protection scheme.

DEAD ENDS
NFPA 780 - SEC. 3.9.2

Air terminals shall be required to be "dead ended" with only one path to a main conductor on roofs below the main protected level. Note that this shall be permitted, provided the conductor run from the air terminal to a main conductor is not more than 16 ft. (4.9 m) in total length and maintains a horizontal or downward run. **(See Figure 13-13)**

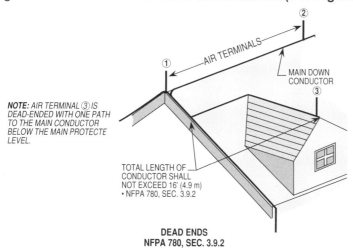

Figure 13-13. This illustration shows the requirements for designing and installing a dead-ended run conductor.

"U" OR "V" POCKETS
NFPA 780 - SEC. 3.9.4

Conductors shall maintain a horizontal or downward coursing free from "U" or "V" (down and up) pockets. These pockets are sometimes formed at low positioned chimneys, dormers or other projections on sloped roofs or at parapet walls, and shall be provided with a down conductor from the base of the pocket to ground or to an adjacent down-lead conductor. **(See Figure 13-14)**

Figure 13-14. This illustration shows the procedure in which the course conductors from the terminals on a chimney are installed to prevent "U" and "V" pockets from forming.

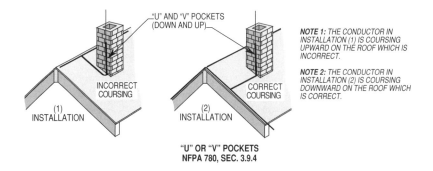

DOWN CONDUCTORS CONCEALED
NFPA 780, SEC.'s 3.15.1 THRU 3.15.4

Downlead conductors shall be permitted to be concealed within the building construction so only the air terminals are seen. On a reinforced concrete structure, the conductors shall be permitted to be placed directly in the roof, column or wall construction or may be provided with a metallic or nonmetallic raceway. In all cases where the conductor is encased, the rebar shall be interconnected at top and bottom of downlead conductors or at 100 ft. (30 m) intervals on roof. **(See Figure 13-15)**

Figure 13-15. This illustration shows the requirements for concealing down conductors within the building construction.

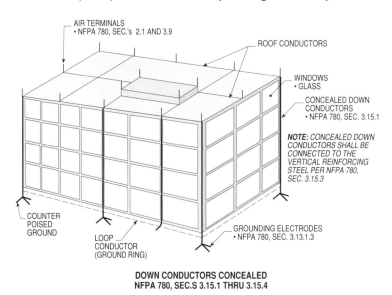

Grounding Tip 359: Piping such as water service, well casings and other underground conduits that are located within 25 ft. (7.6 m) of the structure shall be bonded and grounded per NFPA 780, Sec. 3-14.

CREATING A COMMON GROUNDING SYSTEM
NFPA 780, SEC. 3.14

Interconnection of all grounded systems within a structure shall be done to prevent fire hazards, structural or equipment damage and possible shock conditions. Systems considered grounded are electrical, communications, metal water piping, metal gas piping, antenna and roof items.

When providing a common ground with the metal water piping and metal gas piping, the connection shall be made on the owner's side of the meter. Where isolation joints are present in such service lines, the common protection shall be made on the owner's side of the joint and not the supplying utility.

The **NEC** and **NFPA 780** requires surge suppression devices be installed for protection of the structure only at entrances of electrical and telephone services and on radio and CATV lead-ins. Sometimes there is a need for additional surge suppression within a structure to safeguard specific types of equipment. **(See Figure 13-16)**

Lightning Protection for Buildings and Structures

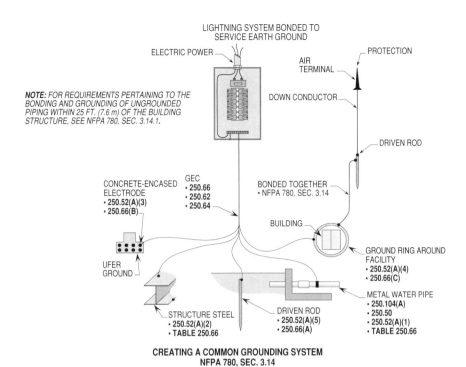

Figure 13-16. This illustration shows some of the items that shall be bonded and grounded together to form a common grounding system.

BONDING OF METAL
NFPA 780, SEC. 3.21

Structural components and equipment installed as part of a building may be affected by lightning currents, particularly, if they lead to ground and assist in providing a path to ground. The effects of these currents can be greatly reduced by bonding them together to form a common potential.

Bonding of metal bodies to the lightning protection system shall only be required where there is a possibility of a sideflash between them which can cause fire hazards, explosion of materials or personal injury. **(See Figure 13-17)**

Grounding Tip 360: For a technical discussion of lightning protection potential-equalization bonding procedures, see NFPA 780, Appendix J.

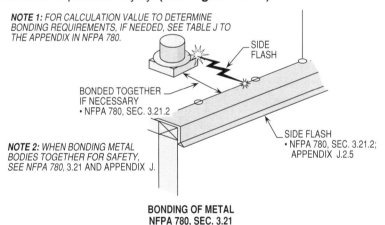

Figure 13-17. This illustration shows the requirements to consider when determining the distance between metal bodies that need to be bonded together to reduce the risk of flash over.

USING METAL FRAME OF BUILDING
NFPA 780 - SEC. 3.16

To reduce the necessity for bonding many bodies together, a system of potential equalization shall be applied to the structure. At or within 12 ft. (3.6 m) of ground level, all grounded bodies shall be interconnected per NFPA 780, Sec. 3.20.1. For structures more than 60 ft. (18 m) in height, the interconnection shall take the form of a ground loop conductor. On lower structures, the loop conductor is recommended, but not required by NFPA 780. At roof level, all grounded bodies shall be interconnected. **(See Figure 13-18)**

Figure 13-18. This illustration shows the structural steel shall be permitted to be used as the main conductor of a lightning protection system if electrically continuous or made to be electrically continuous.

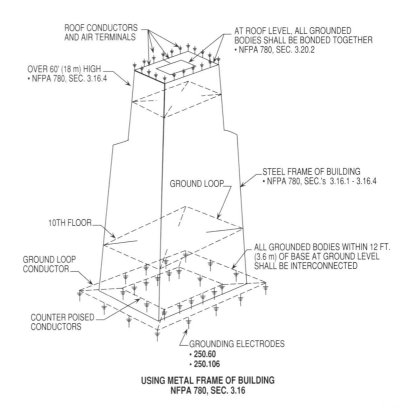

USING METAL FRAME OF BUILDING
NFPA 780, SEC. 3.16

Grounding Tip 361: When metal bodies, etc. are bonded together with a ground loop conductor, an equipotential plane is established that reduces voltage differences and provides safety from lightning surges.

If grounded systems such as metal water piping, metal gas piping and sprinkler piping have not been connected to the lightning system they shall be interconnected near the roof to form potential-equalization between metal grounded bodies.

LONG VERTICAL BODIES
NFPA 780, SEC. 3.16.1

Grounding Tip 362: The bonding of grounded or ungrounded long vertical metal bodies shall comply with the requirements in NFPA 780, Sec.'s 3.21.2 and 3.21.3.

Long vertical metal bodies, whether grounded or ungrounded are subject to high induced voltage when relatively near a lightning conductor. On steel framed structures, these long metal bodies exceeding 60 ft. (18 m) shall be bonded near each end when not bonded through construction, steel, etc. The same bonding shall be required for reinforced concrete structures.

If branches or previously bonded systems rise vertically 12 ft. (3.6 m) or greater, a formula shall be used to determine if more bonding is needed per NFPA 780, Sec. 3.21.2.

OVER 40 FT. (12m) OR 40 FT. OR LESS STRUCTURES
NFPA 780, SEC. 3.21.2(a); (b)

Grounding Tip 363: The following formula is applied for bonding requirements of structures not covered in NFPA 780, Sec. 3.21.2(a):

- $D = \dfrac{h}{6n}$ • km

For understanding and how to apply this formula, see NFPA 780, Sec.'s 3.21.2(a) and 3.21.2(b).

A different procedure is used and applied to the top of 60 ft. (18 m) or taller buildings including its lower portion. Separate procedures are applied to structures 40 ft. (12 m) or less in height. When grounded systems come within the calculated distance, for safety reasons per NFPA 780, they shall be bonded to the lightning protection system or to another grounded system that together forms a system of potential of equalization.

LIGHTNING PROTECTION FOR REINFORCED CONCRETE BUILDINGS
NFPA 780, SEC.'s 3.20.3(b); 3.21.1(b)

Lightning protection for reinforced concrete buildings can be easily installed during construction stages of the building. At this time, proper interconnection of the reinforcing steel and the lightning protection downlead cables can be made. Proper connections shall be made to the lightning conductor system at the highest and lowest points of the reinforcing steel. Note that the reinforcing steel shall not be used as the lightning conductor system unless such steel is made electrically continuous by welding and is of adequate size. Tie wires may not necessarily be capable of making the steel electrically continuous. **(See Figure 13-19)**

Grounding Tip 364: Column connections of ground terminals in Note 1 to **Figure 13-19** shall be made at columns having a surface of not less than 8 in. (200 mm).

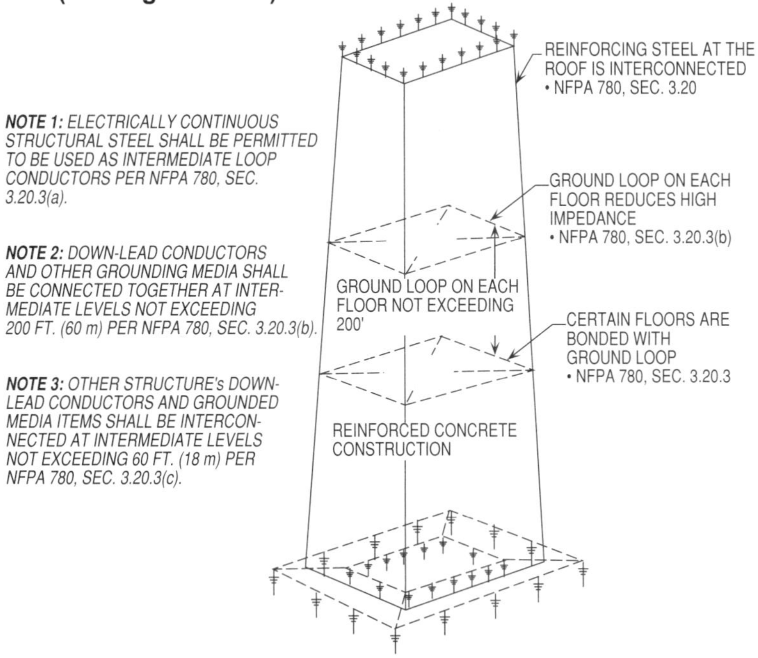

Figure 13-19. This illustration shows the requirements for using the reinforcing steel in concrete along with down conductors (if used) to form a reliable lightning protection scheme. (Buildings over 60 ft. (18 m) in height)

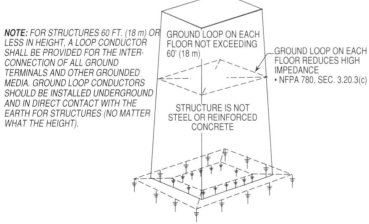

LIGHTNING PROTCTION FOR REINFORCED CONCRETE BUILDINGS
NFPA 780, SEC.'s 3.20.3(a) - 3.20.3(c)

Grounding Tip 365: A loop conductor is used for interconnecting down-lead conductors and other grounded media in Notes 1, 2 and 3 in **Figure 13-19**.

For structures other than steel or reinforced concrete, the grounded systems shall be interconnected with a loop conductor at levels not exceeding 60 ft. (18 m) **(See Figure 13-20)**

Figure 13-20. This illustration shows the bonding and grounding procedures for structures not exceeding 60 ft. (18 m) in height.

LIGHTNING PROTECTION FOR OTHER THAN STEEL OR REINFORCED CONCRETE
NFPA 780, SEC. 3.20.3(c)

LIGHTNING PROTECTION FOR STRUCTURAL STEEL BUILDINGS
NFPA 780, SEC.'s 3.21.1(a); 3.21.1(c)

In structural steel buildings, the steel columns of the building shall be permitted to be used as downleads. In such a design, every other column shall be grounded with the proper grounding electrode. For best results, in high-rise buildings, it is recommended to ground all columns and to connect all grounding electrodes with a buried counterpoise cable.

Grounding Tip 366: Metal parts of a structure shall be permitted to be used as a part of the lightning protection scheme in some cases.

For example, the structural metal framing having a cross-sectional area equal to the conductivity of main lightning conductors.

If it is electrically continuous, it shall be permitted to be used in lieu of separate down conductors. In such cases, air terminals shall be bonded to the framework at the top, and ground terminals shall also be provided at the bottom.

Figure 13-21. This illustration shows the procedure for bonding and grounding roof conductors, down conductors and structural steel to form a lightning protection system for high-rise structural steel buildings.

Roof air terminals should be connected to one another by a roof conductor that is bonded to the structural steel columns at 100 ft. (30 m) intervals or less.

On low-rise steel buildings, secondary bondings to objects of induction are not always required, due to these objects being inherently bonded to the building steel.

When a structure has a steel frame, the columns shall be permitted to be substituted for the down conductors. The roof circuit conductors shall be tied to the steel frame at intervals not exceeding 100 ft. (30 m). Ground terminals must be connected to every other steel column, but spaced not more than 60 ft. (18 m) apart. It shall be permitted to eliminate the roof conductors and connect each air terminal directly to the roof steel. **(See Figure 13-21)**

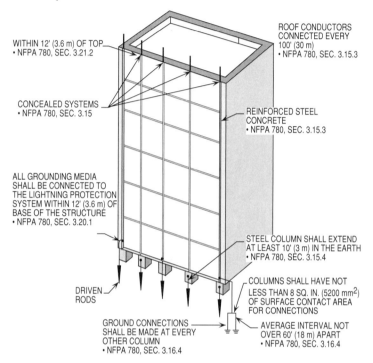

LIGHTNING PROTECTION FOR STRUCTURAL STEEL BUILDINGS
NFPA 780, SEC.'s 3.21.1(a) - 3.21.1(c)

For high-rise buildings, there are always concerns with high-resistance joints in metallic piping systems. This result can create high-impedance paths, it is recommended highly that a ground loop conductor be installed at certain levels throughout the height of the structure. Number and location of these installations shall be determined by the size and layout of the structure.

Grounding Tip 367: In cases where damage has occurred to a structure, the fault was usually due to additions or repairs to the building or to the deterioration or mechanical damage that was allowed to remain undetected and unrepaired, or both. Therefore, it is recommended that an annual visual inspection be made and the system be completely inspected every five years.

LIGHTNING PROTECTION FOR PRE-CAST CONCRETE BUILDINGS
NFPA 780, SEC.'s 3.16; 3.21

There are many buildings built today using pre-cast concrete construction. This type of construction contains varying amounts of reinforcing steel. During construction stages, the reinforcing steel shall be electrically connected to the lightning protection system in such a manner to provide dependable electrical continuity of the lightning protection scheme.

Tie plates or anchor plates should be welded to the reinforcing steel before these pre-cast units are ever poured. A connection to the anchor or tie plates during installation provides the same electrical potential as the lightning protection system for such building. **(See Figure 13-22)**

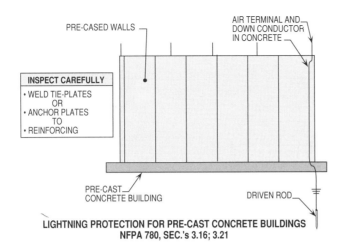

Figure 13-22. This illustration shows the procedure for installing a lightning protection scheme used for pre-cast concrete constructed buildings.

LIGHTNING PROTECTION FOR FOUNDATION AND SLAB BUILT ON SOLID ROCK
NFPA 780, SEC. 3.13.3

Where solid rock is supporting the foundation and slab, it is impossible to make a ground connection in the ordinary sense of the term because most kinds of rocks are insulated, or at least of high resistivity, and in order to obtain effective grounding, other and more elaborate means are necessary. The most effective means would be an extensive wire network laid on the surface of the rock surrounding the building to which the down conductors could be connected. The resistance to earth at some distant point of such an arrangement would be high but at the same time the potential distribution about the building would be substantially the same, as though it were resting on conducting soil, and the resulting protective effect also would be substantially the same. In other words, the steel in the foundation and slab as well as to the structural steel will be elevated at the same potential and tend to rise together during a surge. **(See Figure 13-23)**

Grounding Tip 368: Where soil is less than 1 ft. in (.3 m) depth, down conductors shall be connected to a loop conductor laid in a trench in the rock around the building. The loop conductor shall be at least equal to the main size lightning conductor. Plate electrodes can be attached to the loop conductor to provide a better contact with the earth.

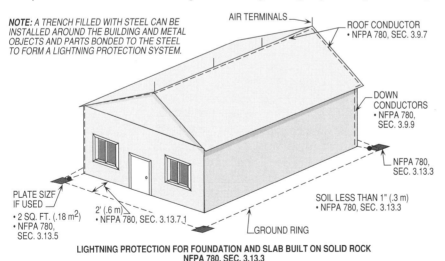

Figure 13-23. This illustration shows the procedure for installing a lightning protection system where ground rock is encountered.

LIGHTNING PROTECTION FOR INFORMATION TECHNOLOGY EQUIPMENT
NFPA 780, APPENDIX L

Lightning can adversely affect information technology equipment (ITE) systems in several ways. It can be responsible for interrupting power without excessive voltage or current surges ever reaching the ITE system. It can open distribution circuits or create voltage surges accompanied by current surges. These will enter conductive paths, following power conductors, communications lines, piping, structural steel and ducts until they pass through portions of the ITE hardware and its interconnections on its way to earth ground. The powerful electromagnetic pulse (EMP) created in the vicinity of a lightning stroke can induce significant noise impulse voltages and currents in the interconnections of an ITE system without having to flow through conductors to reach the system. **(See Figure 13-24)**

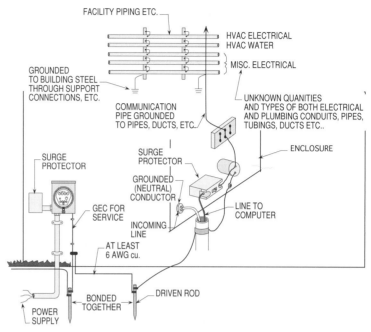

Figure 13-24. This illustration shows the procedure for designing and installing a lightning protection scheme for communications systems.

Grounding Tip 369: Lightning surges entering protected buildings on overhead or underground electrical power lines, or communications conductors or television or radio antenna wiring are not necessarily restricted to associated wiring systems and appliances. Therefore, such systems should be equipped with appropriate protective devices and bonded to assure a common potential to help reduce surges.

Grounding Tip 370: For filters to prevent power line interference from entering sensitive equipment, the frame of the filter must become part of the conductive enclosure shielding by bonding them together.

Grounding Note: For bonding and grounding rules pertaining to communications systems, see **250.60, 250.106** and **800.40(A)** through **(D)** in the NEC.

USE OF ARRESTERS AND FILTERS
NFPA 780, APPENDIX L.3.10

The noise signals from lightning can easily corrupt data signals. They can also be large enough to destroy insulation, conductors, contactors and other circuit elements and devices. Unless a direct stroke is encountered at or immediately outside the facility, the impulses which reach the ITE facility will most likely travel some distance along power or communications lines before they enter the facility building as illustrated in **Figure 13-25**. As they travel, they are fortunately attenuated and the voltage wave front becomes less steep. Arresters will have more time to respond to the more slowly rising voltage wave. Distance between the ITE system and the lightning stroke helps chances of survival.

The energy in lightning strokes is incredibly great. The energy needed to destroy or disrupt operation of a semiconductor is incredibly small. No single device can provide complete protection. The survival of ITE equipment and its chances or continuing operation without interruption or error during lightning storms in the vicinity depend upon a series of protective devices, starting with the protection of overhead power lines and ending with filters and grounding at the ITE hardware.

Lightning Protection for Buildings and Structures

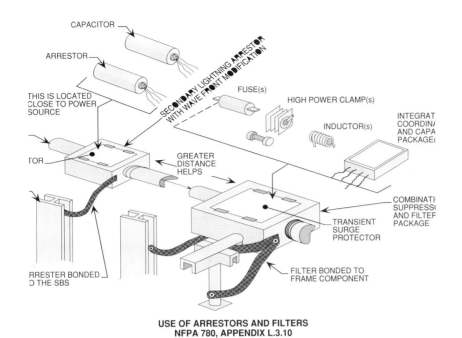

Figure 13-25. This illustration shows the filter and arrester bonded to the structural building steel and frame component of the equipment served or supporting post of a raised floor.

PRIMARY PROTECTION BELONGS OUTSIDE THE ITE ROOM
NEC 280.11

Protective devices are incapable of absorbing the energy of a lightning stroke or even a small part of that energy. Lightning arresters should more properly be called lightning diverters. Upon being exposed to an overvoltage, they become conducting and create a direct path to a grounding conductor leading to a driven earth ground rod or other earth electrodes. The conduction will cease when the current reaches zero and remains at or near zero long enough for the path to deionize and recover its insulating state. This may take several cycles and may involve a number of restrikes. **(See Figure 13-26)**

Grounding Tip 371: For the best protection possible, surge arresters shall be installed at the service equipment where only three will be needed. There shall be one provided for each phase-to-ground where the ground and grounded (neutral) conductor are common.

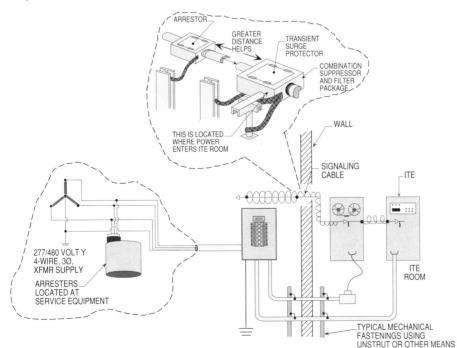

Figure 13-26. This illustration shows arresters and filters used for lightning protection located outside the ITE room.

13-17

TRAVELING WAVES
IEEE GREEN BOOK

Grounding Tip 372: To follow the current path of the lightning strike from the point of strike to the point of reflection and until such current path reaches the power center where the isolation transformer is installed, refer to **Figure 13-29**.

The traveling voltage waves on the overhead lines are created by the sudden appearance of large voltage drops through utility conductors as lightning currents pass through them on their journey to earth. Usually this is through the overhead grounded "static" wire which is usually strung above the ungrounded (phase) conductors. It is connected periodically to ground through wires on the pole and driven ground rods at their bases. In areas with frequent lightning storms, surge arresters between each ungrounded (phase) conductor and the ground wire plus a ground rod may be located at every pole.

When the traveling wave reaches any discontinuity in the line such as at the last pole where an overhead line may be terminated without a service tap, the voltage wave is reflected from the open end while voltage is momentarily doubled in the process. Service taps and overhead service drop conductors shall not be located at the end pole, but preferably 100 ft. (30 m) or more from it. **(See Figure 13-27)**

Figure 13-27. This illustration shows the current path for the lightning strike and the reflection wave occurring at point ③.

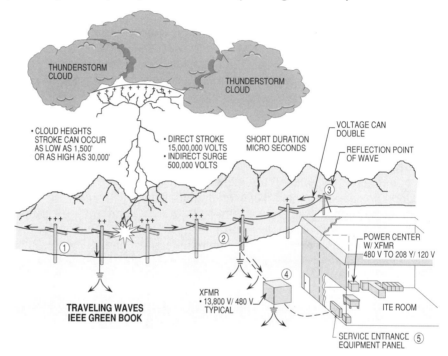

Grounding Tip 373: In Figure 13-27 arresters are installed at the utility pole ②, utility transformer ④ and also at the service equipment panel ⑤ to help divert surge current to ground.

USING ARRESTERS ON OVERHEAD LINES
NESC RULE 93(c)(4)

Although arresters on the overhead line will pass much current to ground, the voltage drop between the line and earth could easily be more than 100,000 volts peak at the first grounding point near a direct strike. This will attenuate as it travels along the line and it may attenuate more as it passes other surge arresters. At the point the surge enters an underground cable, the capacitance to ground will cause a further reduction in crest voltage. By the time it reaches the transformer primary for reduction to a typical voltage of 480Y/277 volts on the secondary, the magnitude of the lightning surge will have been reduced significantly. **(See Figure 13-28)**

GAP PROTECTORS INSIDE TRANSFORMERS

The input circuits to many transformers in this type of service may have self-contained gap protectors inside the unit. Whether or not the input fuses may blow, the purpose is to protect the transformer windings from insulation breakdown. The gaps shunt excess

Lightning Protection for Buildings and Structures

voltage to ground at the transformer if the gaps would spark over. By this time, the surge voltages and currents which have reached the transformer primary have been greatly reduced from the magnitude of the original lightning stroke. **(See Figure 13-28)**

The gaps may be supplemented by lightning arresters on the primary, but it is the decision of power companies to provide them if needed or owner of the overhead lines.

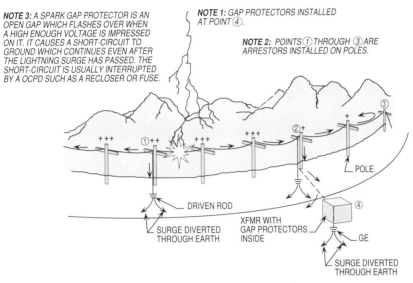

GAP PROTECTORS INSIDE TRANSFORMERS

Figure 13-28. This illustration shows arresters that are used on the utility poles at points ①, ② and ③ and gap protectors are used inside the utility transformer at point ④

RESIDUAL SURGE PROTECTION IN POWER AND COMMUNICATIONS LINES

As the lightning surge moves toward the load it is greatly attenuated by the separation of primary from secondary windings. Between the secondary winding and the service equipment for the building is an effective location to install supplementary low-voltage arresters where the potential is typically only 277 volts rms from each line to ground.

This is where the power actually enters the building, so arresters at this point will "condition" all the power entering the building and not just a portion which is directed to the ITE room. If only part of the circuits into the building have arresters, the surges could easily find other paths into the building and reach computer circuits by devious paths. **(See Figure 13-29)**

Grounding Tip 374: Because a lightning protection system is expected to remain in working condition for long periods of time with minimum attention, the mechanical construction should be strong, and the materials used should offer resistance to corrosion and mechanical injury.

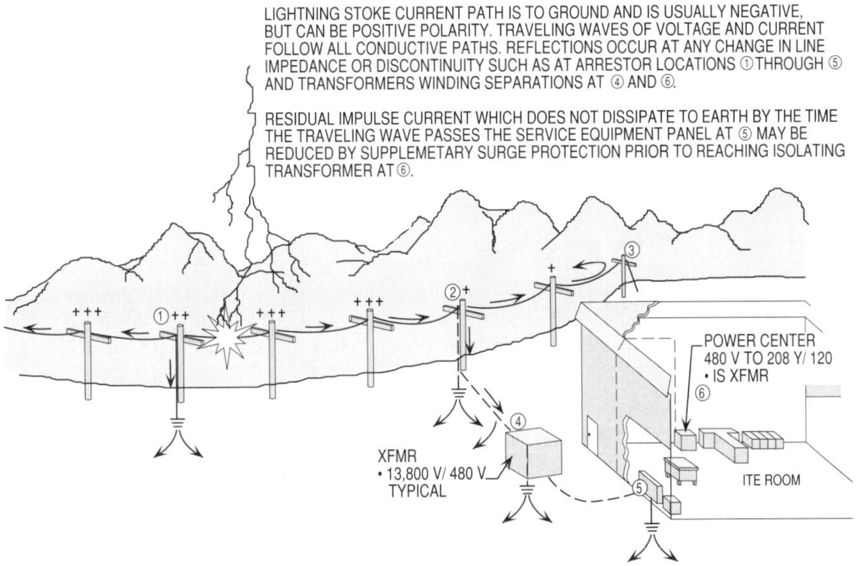

Figure 13-29. This illustration shows the location of the lightning strike at point ① and its path travels to point ⑥. Note that arresters are provided at points ①, ②, ③ and ⑤. Gap protectors are used at point ④ and surge suppressors are installed at point ⑥

RESIDUAL SURGE PROTECTION IN POWER AND COMMUNICATIONS LINES

13-19

Grounding Tip 375: Supplementary surge suppression devices can also be provided to protect power lines and associated equipment from both direct discharges and induced currents.

LIGHTNING PROTECTION SHOULD NOT BE INSTALLED IN ITE ROOM

The last place one wishes to have lightning surge currents is in the ITE room. To accomplish this, the current diversion must be accomplished as far from the ITE system as feasible. If the surge is allowed to reach to the ITE room and a lightning arrester is installed there, the only ground to which the surge current could be diverted at that point would be the ITE system ground. That could be disaster! However, supplementary surge protection to reduce residual impulse voltage from primary arresters is not a bad idea.

SURGE VOLTAGE AND CURRENT WAVES

In understanding surge voltage and current waves, one must remember that the energy within conductors can be neglected. All of the energy of the electromagnetic wave will be between conductors (electrostatic field) and around conductors (magnetic field). The conductors merely guide the wave. The coupling of those fields with other conductors which may parallel or cross them becomes a major path for energy to transfer to them.

LOCATING ARRESTERS AT SERVICE EQUIPMENT NEC 280.11

If a set of surge arresters is installed at the service equipment, only three will be needed, one for each ungrounded (phase) conductor to ground where ground and grounded (neutral) conductor are common. If located downstream where the grounded (neutral) conductor and ground are not a common conductor, an additional arrester would be needed for the grounded (neutral) conductor, if a neutral were in use. **(See Figure 13-30)**

Figure 13-30. This illustration shows arresters located at the service equipment. The path of the surge current is shown from points ① through ⑤

Grounding Tip 376: For a hook-up showing arresters located at service equipment, see **Figure 13-26** on page 13-17 in this chapter.

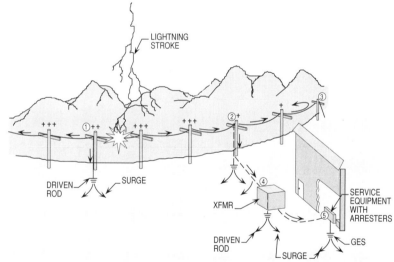

LOCATING ARRESTORS AT SERVICE EQUIPMENT

PROTECTION OF COMMUNICATIONS LINES

Communications lines need protection as much as power lines. Arresters should be located near the communications cable entrance to the building. For best protection, a coordinated combination of protective devices is needed. Carbon block gap or gas tube protectors should be the first elements to take the largest currents to ground. Going toward the data communications equipment, the communications conductors can be passed through such current limiting elements as resistors. On the protected equipment side of the current limiters, one can install devices which will limit the surge voltage to a much lower voltage with silicon protective elements which clamp on overvoltage and prevent further rise, yet become an open circuit at normal operation voltages. **(Refer to Figure 13-24)**

PROTECTION AT DEVICES

Residual surge protection can be placed directly at the input terminals of power and communications devices being protected. The longer the distance between the primary protection at the building entrance and the protected equipment, the more protection this distance will provide in attenuating the strength of the surge impulse from a lightning stroke.

PROTECTIVE DEVICE LOCATIONS

Protective device locations have been implied in the previous description. This primary protection devices should be located where the services enter the building. If the entrance points are not more than 25 (7.6 m) to 50 ft. (15 m) apart, the voltage occurring between their respective grounding points will not be difficult to limit. Voltage differences between grounding points at greater distances may require substantial interconnections of copper grounding busbar or strap in order to limit them. **(See Figure 13-31)**

Grounding Tip 377: Surge suppression devices should be installed on all wiring entering or leaving sensitive electronic equipment, usually power data or communications wiring.

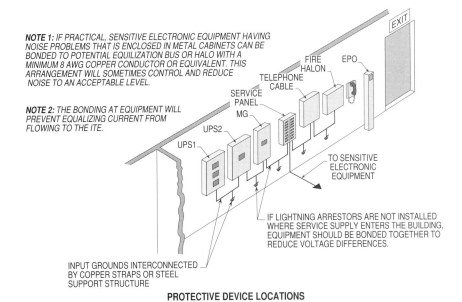

Figure 13-31. This illustration shows equipment bonded to form an equipotential plane to reduce voltage differences between pieces of equipment and help control noise problems.

UNWANTED PATH FOR SURGE ENTRY INTO THE ITE ROOM

Unwanted paths for surge entry into the computer room can be identified by a careful review of all conducting paths, whether by power conductors, communications lines, radio or television antenna cable, water pipe, ducts and structural members.

Grounding Tip 378: To reduce noise problems, all grounding conductors and conducting pipes which penetrate the ITE area shall be bonded together by short robust connections before they enter the computer room.

PATHS THROUGH EQUIPMENT

If air-conditioning equipment heat exchangers or condensers are located in a penthouse on the roof, make certain that the refrigerant or cooling pipes from penthouse to computer room are not a virtual lightning rod which when struck by lightning will guide the lightning into the computer room and its grounded cables as a path to the earth.

PROVIDING EQUIPOTENTIAL

In lightning storms areas, buildings of reinforced concrete should be constructed with steel reinforcing welded together for good electrical continuity before the concrete is poured. Any structural steel bolted to concrete reinforced columns should be electrically bonded to the reinforcing. When the building is completed, lightning protection should be installed in accordance with applicable codes. A perimeter conductor, to which all lightning protection downwires and driven ground rods are attached at or below ground level serves to help equalize the voltage gradient around the building at the time of a lightning stroke.

BONDING RODS
NFPA 780, SEC.'s 3.14.1; 3.14.2

A lightning protection ground rod system (alone) shall not be used for an electrical power service ground required by the NEC. However, this does not imply that they cannot be interconnected. NFPA 780 recommends that it be bonded to the grounding electrode system to reduce voltage differences.

Grounding Tip 379: For safety reasons, driven rods shall be bonded and grounded to the grounding electrode system and other system grounds.

BONDING PIPES AND ROOF EQUIPMENT
NFPA 780, SEC. 3.21.2; 3.21.3

Ducts, vents and pipes and all roof mounted electrical equipment shall be bonded to structural steel for electrical continuity wherever this is feasible to do so. The lightning protection system with sky-pointed lightning rods shall be coordinated with the electrical equipment grounding. The two systems should be complete in their own right, neither requiring use of the other system's conductors for integrity.

Within the building, conducting members such as pipes, ducts, conduits and steel structural members shall be bonded together electrically with short, direct, robust connections at intervals of about 12 ft. (3.6 m) or less. The structural steel bars for common support of multiple conduits, pipes and ducts usually accomplish this. The result is that large voltage differences will be safely equalized during a stroke of lightning current if it should enter the building. **(See Figure 13-32)**

Grounding Tip 380: Down conductors from roof-mounted air terminals (lightning rods) should not establish a path to ground through ITE circuits. Air-conditioning coolant pipes from the roof to computer room air handlers shall not bring in surges that can damage sensitive circuits.

Figure 13-32. This illustration shows metal pipes, conduits, cables and tubing shall be bonded together to reduce unwanted noise problems. This unwanted noise problem causes sensitive electronic circuits to malfunction and interrupt computer operation.

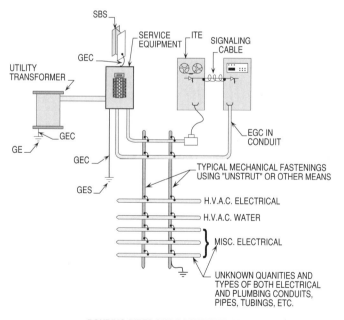

BONDING PIPES AND ROOF EQUIPMENT
NFPA 780, SEC.'s 3.14; 3.21.2; 3.21.3

METAL OBJECTS WITHIN 6 FT. (2 m) OF LIGHTNING CONDUCTORS
NFPA 780, APPENDIX J

Special attention should be given to conducting members or objects within the building which may be within, say, 6 ft. (2 m) of an exterior lightning down conductor to ground. To avoid the damage which could be caused by a side flash from the down conductor to

Lightning Protection for Buildings and Structures

the interior conducting members, an interconnection is sometimes made with or without a series spark gap inserted in that path. This predetermines the lightning current side path taken in the event that a surge is great enough to break down the gap. Without such a controlled path, alternate paths could create great damage. ITE interconnecting cables should be routed to avoid the possibility of their involvement in a direct or side flash lightning path. **(See Figure 13-33)**

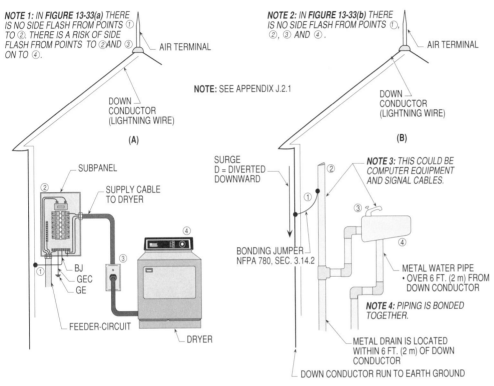

METAL OBJECTS WITHIN 6 FT. OF LIGHTNING CONDUCTORS
NFPA 780, APPENDIX J

Figure 13-33. This illustration shows where side flash could occur due to location of metal or conductive bodies, piping and wiring methods.

STATIC ELECTRICITY
NFPA 77

Electric charges, dislodged and trapped when insulating materials touch and are forcefully separated, can create significant voltages. The phenomenon is called "static electricity." Besides causing materials to cling to or repel each other, the charge supplies the necessary energy to produce a spark discharge when the distance between oppositely charged bodies becomes short enough. A 10,000 volt charge can create a spark approximately 0.5 in. long. **(See Figure 13-34)**

Grounding Tip 381: ITE systems can be selected which have been designed and tested to withstand electrostatic discharges up to 10 kV or greater without failure occurring.

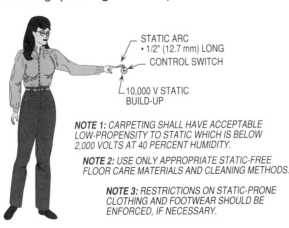

STATIC ELECTRICITY
NFPA 77

Figure 13-34. This illustration shows the requirements which will help reduce and control static electricity problems in ITE rooms and ITE areas.

Such discharges are in effect miniature lightning bolts. The total energy is far less than lightning strokes can deliver, but these discharges are much closer to the sensitive electronic equipment they are able to disturb. A discharge to any conducting surface creates very fast rise-time impulse noise current. All static discharges have become notorious for their ability to cause malfunctions and even destroy semiconductor circuits and devices.

ITE SUBJECTED TO STATIC CHARGES
NFPA 77, SEC. 6.4.1.3

Unfortunately, ITE sites often have static discharge problems without anyone being aware of it. Few subjects can feel a discharge of less than 3000 to 4000 volts from their own charged bodies, but this is more than enough to disrupt a computer circuit. It is not until discharge voltage exceeds 10,000 to 15,000 volts that subjects learn that discharges can be unpleasant or painful. Some techniques for controlling static problems are outlined below:

Grounding Tip 382: Humidity should not be less than 40 percent. High-humidity can cause problems such as corrosion, gold scavaging and ion migration on circuit-boards and connectors.

Grounding Tip 383: Static charges can be reduced, controlled or eliminated by:

- Controlling the relative humidity
- Using shielding on cables
- Using metallic frames or conductive paints
- Using an antistatic spray
- Using an ionizing air gun

HUMIDITY CONTROL
NFPA 77, SEC. 6.4.2

Humidity control to achieve 50 percent or higher relative humidity will greatly inhibit electrostatic problems. Too much humidity, however, will create corrosion problems and may make some paper products dimensionally unstable and too limp to handle. Most ITE equipment manufacturers recommended 40 to 60 percent relative humidity. Other say they can tolerate 20 to 80 percent. Correcting humidity (raising it in winter and trying to lower it in summer) requires much energy and is expensive. If other techniques are effective, the cost may be less. However, humidities below 30 percent and occasionally 40 percent may cause static electricity problems and dimensional change problems with printing paper.

CONDUCTIVITY OF FLOOR SURFACES
NFPA 77, SEC. 6.6.2

Conductivity of floor surfaces is a selectable characteristic which will prevent or limit static buildup on personnel who walk on them. Nylon carpeting can be among the worst. Vinyl-asbestos can be marginal. Pressure laminates are better but not perfect. Conducting synthetic rubber and other special purpose floor coverings for use in hospital operating rooms and munitions loading areas meet all conductivity requirements, but often lack attractive appearance and are costly. Floor polishes can make them unacceptable. It takes only a thin coating of carnauba wax to create a problem.

Typical floor surfaces which are acceptable at 50 percent relative humidity and which remain acceptable down to about 20 percent relative humidity include pressure laminates (unwaxed) and some carpeting with strands of conducting fiber in the yarn and semiconducting backing to spread the charge. The 20 percent is a lower humidity than most ITE equipment can tolerate without malfunctioning. Floor surfaces and their coverings can be tested by instrumentation illustrated in **Figure 13-35**.

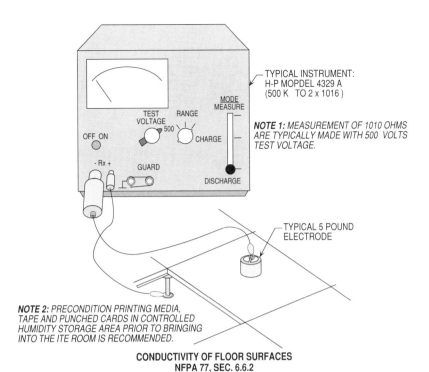

Figure 13-35. This illustration shows the procedure for performing a test to verify conductivity of computer room floor surface.

STATIC DRAIN PATHS
NFPA 77, SEC.'s 4.2; 4.3

Static drain paths from floor tiles or mats to the nearest grounded metal may be needed in heavy traffic areas. The charges from someone walking on the floor surface will spread through any conducting sheet, but a path to ground through conducting material is needed to dissipate the accumulated charge. This need not be a low resistance path. A resistance range from 0.5 to 20,000 megaohms at 50 percent relative humidity has been specified by a number of computer manufacturers, measured between a weighted electrode on the surface of the floor and the grounded subfloor. Another manufacturer specifies a maximum of 1000 megaohms. Many materials will qualify when humidity is 50 percent or better, but very few do so at 20 percent or less.

Metal contacts between removable floor panels and supporting metal structures are not necessarily needed. The fiber or synthetic cushions or gaskets which are sometimes used to make a floor quiet to walk on are acceptable if they are sufficiently conducting to meet the specified resistance range deemed adequate to drain static charge. If the facility is intended to operate at less than 50 percent relative humidity, the resistance should not exceed the 20,000 megaohms maximum at the lowest humidity to be encountered. These techniques will help reduce and control static charges.

A separate static drain path may not be necessary if there is sufficient conductivity of the floor surface and carpeting, if any, to remove static charge as it is generated. Resistances in the drain path of less than 0.5 megaohm are believed to increase the hazards of electric shock and the possibility of spark discharges which create unwanted electrical interference.

CARPETING WITH LOW-PROPENSITY TO STATIC ELECTRICITY
NFPA 77, SEC. 6.6.2

Carpeting with low-propensity to static electricity and suitable ITE installations is available in the form of mats, continuous material in customary widths and in squares which may be permanently attached to each of the removable floor tiles. **(See Figure 13-36)**

Grounding Tip 384: Appropriate floor coverings having pressure laminates with upper limit of 10^9 ohms on resistance to ground and surface resistivity is recommended. Carpeting should have acceptably low-propensity to static which is below 2000 volts at 40 percent relative humidity.

Note: To understand the fundamentals of static electricity, review the rules in Chapter 4 of NFPA 780.

Figure 13-36. This illustration shows carpet over tiles that works together to provide a low-propensity to static electricity in an ITE room or ITE area.

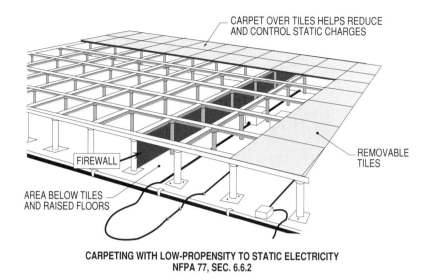

CARPETING WITH LOW-PROPENSITY TO STATIC ELECTRICITY
NFPA 77, SEC. 6.6.2

Where underfloor access is needed, the continuous material is not feasible. Some mats with heavy, flexible backing and bound edges will lie flat without being permanently attached to the floor. They can easily be lifted for underfloor access, cleaning or replacement. They do not appear to create a tripping hazard since the edges appear to remain on the floor.

Grounding Tip 385: Static electricity in ITE rooms and ITE areas can be reduced where its not a problem if appropriate static-free floor care materials and cleaning methods are used. Furniture surfaces and upholstery with low static propensity are highly recommended. Conducting glides, casters and wheels on legs and carts used should also be utilized.

FURNITURE AND UPHOLSTERY
NFPA 77, SEC. 6.6.1

Furniture and upholstery in ITE rooms and in offices where terminals are in use can be a contributory cause of static electricity. A person sitting in a chair and rising can generate a very large voltage. This can cause energetic spark discharge and cause corruption of data or even damage. The same materials used for computer room carpeting can be used for furniture upholstery and alleviate the problem. The casters on chairs and other movable office furniture should have metal or conducting rubber or plastic wheels. All stationary feet should be uninsulated. (Also review Chapter 4)

SHOES AND CLOTHING
NFPA 77, SEC. 6.6.2.4; 6.6.4

Shoes and clothing of personnel also generate static. Treating shoes and clothing with the antistatic preparations sold in markets appears to solve this problem.

Grounding Tip 386: Only qualified personnel with experience and knowledge of static electricity and its effects on ITE systems can help prevent such problems by following certain rules and procedures.

ION GENERATORS
NFPA 77, SEC. 6.5

Ion generators for neutralizing localized chronic static problems are a product commonly used in semiconductor assembly plant and in the printing industry to dissipate static charges. Without them, charges which accumulate on personnel and containers for semiconductor products become the source of spark discharges which damage or destroy the semiconductor product. In the handling or paper, static charges deflect the paper movement or cause it to stick.

DESENSITIZING ELECTRONIC CIRCUITS

Desensitizing electronic circuits highly susceptible to static discharge has become a specialized art. It is much easier to begin with the basic design of circuits, their layout and packaging than to fix a product which has already been manufactured and found to be overly sensitive. There are some steps which can be taken external to the product, however, such as installing an antistatic mat or installing balun filters in the data interface cables. It may also help to stop for a minute before touching an ITE unit, for during that interval any charge on a person may leak off and be reduced to an unacceptable level. **(See Figure 13-37)**

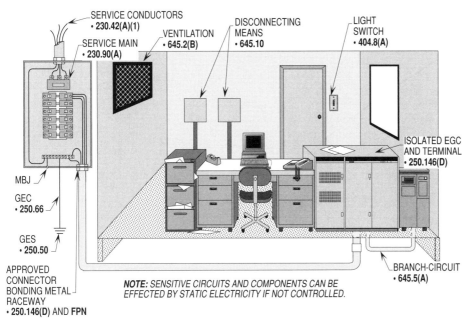

Figure 13-37. This illustration shows that static electricity can be a problem to electronic sensitive circuits if protection techniques are not employed.

Grounding Note: To help reduce the problems that effect ITE operation due to static electricity, review the recommendations covered in pages 13-24 through 13-27.

STATIC DISCHARGE SCENARIO

Static discharge on sensitive electronic equipment can create the electronic circuitry to malfunction or even cause permanent damage.

See Figures 13-38(a) and (b) for a detailed illustration of what happens to release a static charge.

Figure 13-38(a). This illustration shows a static charge built-up which is ready to be released when directly contacted.

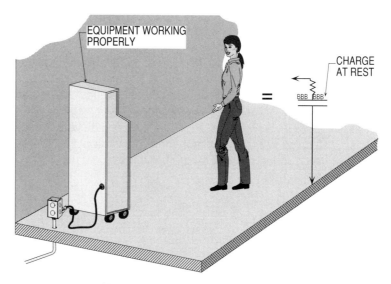

Figure 13-38(b). This illustration shows a static charge being released after direct contact has been made.

Name _____ Date _____

Chapter 13
Lightning Protection for Buildings and Structures

| | Section | Answer |

1. When metal parts of a structure are constructed of metal with a thickness of 3/16 in. metal, air terminals shall not be required to be installed. _____ T F

2. A lightning scheme that provides a protection arc of 150 ft. radius eliminates the need for air terminals on lower roof levels. _____ T F

3. Air terminal supports shall be not less than one-third the height of the air terminal. _____ T F

4. A buried ground ring if properly installed provides a good disturbed field for dissipating a lightning stroke. _____ T F

5. Conductors shall maintain a horizontal or downward coursing free from "U" or "V" pockets. _____ T F

6. The structural metal framing having a cross-sectional area equal to the conductivity of the main lightning conductor can be used instead of down conductors. _____ T F

7. A trench filled with steel shall not be installed around a building and used for lightning protection. _____ _____

8. For best protection possible, should filters be installed at the equipment to be protected. _____ _____

9. Arresters and filters shall not be used in combination for form a lightning protection scheme. _____ _____

10. Arresters installed at utility poles will help reduce the surge from a lightning strike. _____ _____

11. Lighting protection arresters should never be placed in computer rooms.

12. When waxing the floor of a computer room to protect from static charges, is static-free floor care wax required to be used? _____ _____

13. Buildings not exceeding _____ ft. above earth are considered to protect lower portions that are located in one-to-two zone of protection. _____ _____

   (a) 25 (b) 50 (c) 75 (d) 100

14. Buildings not exceeding _____ ft. above earth are considered to protect lower portions if they are located within a one-to-one zone of protection. _____ _____

   (a) 25 (b) 50 (c) 7 (d) 100

13-29

Section	Answer

15. Gently sloping roofs are defined as roofs having a span of _____ ft. or less with a pitch of less than 1/8.

    (a) 25   (b) 30   (c) 40   (d) 50

16. For large flat or greatly sloping roof areas, air terminals shall be located so that no unprotected area exceeds _____ ft. in any dimension.

    (a) 25   (b) 30   (c) 40   (d) 50

17. Each building shall have at least _____ grounding electrodes and grounding conductors spaced as far apart as possible.

    (a) one   (b) two   (c) all of the above   (d) none of the above

18. Air terminals shall extend not less than _____ in. above the object protected for 20 ft. maximum intervals.

    (a) 6   (b) 8   (c) 10   (d) 12

19. Air terminals over _____ in. in height shall be properly supported.

    (a) 12   (b) 24   (c) 36   (d) 48

20. The height of air terminals shall be such as to bring the tip not less than _____ ft. above the object to be protected for intervals of 25 ft. maximum.

    (a) 1   (b) 2   (c) 3   (d) 3 1/2

21. Buildings exceeding 250 ft. in perimeter shall have a down conductor for every _____ ft. of perimeter or fraction thereof.

    (a) 50   (b) 75   (c) 100   (d) all of the above

22. Down conductors shall be supported at least every _____ ft.

    (a) 3   (b) 4   (c) 5   (d) 6

23. Where a horizontal course cannot be maintained, a gradual rise not exceeding _____ in per foot is permitted.

    (a) 3   (b) 4   (c) 5   (d) 6

	Section	Answer

24. In many cases, a rod driven _____ ft. into undisturbed soil provides a proper ground for lightning surges.

    (a) 8   (b) 10 (c) all of the above   (d) none of the above

25. When routing a conductor, no bend of a conductor can form an angle of less than _____ degrees or have a radius of bend of less than 8 in.

    (a) 45   (b) 90 (c) all of the above   (d) none of the above

26. The total length of a dead-end conductor shall not exceed _____ ft. in length.

    (a) 16   (b) 18   (c) 20   (d) 24

27. Underground metal water piping located within _____ ft. of a building shall be bonded to the lightning protection system.

    (a) 25   (b) 50   (c) 75   (d) 100

28. It is recommended that bodies located within _____ ft. of down conductors shall be bonded to such conductor for safety.

    (a) 3   (b) 5   (c) 6   (d) all of the above

29. Where a building is over _____ ft. high, grounded bodies at roof level shall be bonded together.

    (a) 25   (b) 50   (c) 60   (d) all of the above

30. On high-rise buildings, down-lead conductors and other grounding media shall be connected together at intermediate levels not exceeding _____ ft.

    (a) 50   (b) 75   (c) 100   (d) 200

31. Ground terminals shall be connected to every other steel column and spaced not more than _____ ft. apart.

    (a) 30   (b) 40   (c) 50   (d) 60

32. The roof circuit conductors shall be tied to the steel frame at intervals not exceeding _____ ft.

    (a) 50   (b) 75   (c) 100   (d) 150

Section	Answer	

_____ _____ **33.** All grounding media shall be connected to the lightning protection system within _____ ft. of the base of the building.

        (a) 6   (b) 8   (c) 10   (d) 12

_____ _____ **34.** Steel columns used for grounding shall extend at least _____ ft. into the soil of the earth.

        (a) 8   (b) 10   (c) 12   (d) 15

_____ _____ **35.** Steel columns used for the termination of grounding conductors shall have at least _____ sq. in. of surface contact for making connections.

        (a) 8   (b) 10   (c) 12   (d) 14

_____ _____ **36.** Electrical equipment located at a single-point entry can be bonded together with (minimum) a _____ AWG copper conductor.

        (a) 10   (b) 8   (c) 6   (d) 4

_____ _____ **37.** A 10,000 volt electric static charge can produce a spark of about _____ in. long.

        (a) 0.5   (b) 1   (c) 2   (d) 3

_____ _____ **38.** For buildings _____ ft. or less in height, a loop conductor shall be provided for the interconnection of all ground terminals and other grounded media.

_____ _____ **39.** A direct lightning strike is capable of creating as much as _____ volts.

_____ _____ **40.** An indirect lightning stroke can create up to _____ volts.

14

SURGE ARRESTERS

A surge arrester is defined as a protective device for limiting surge voltage by discharging or bypassing surge current. It also prevents continued flow of follow current while remaining capable of repeating these functions as needed.

In elementary form it is simply a device having a small air gap with one side of the air gap connected to the circuit wire while the other side of the gap is connected to ground. The normal voltage of the circuit will not jump and cross the gap. However, the large voltage surges caused by lightning will arc across the gap and divert to ground limiting the voltage to the electrical system to a safe value.

The main function of surge arresters is to break down at voltages higher than the supply voltage, which allows the higher voltage and currents to flow to ground, and protects the elements and equipment on the electrical systems. When such surge passes to ground, the arrester resets, shutting off follow current from the supply system and allowing normal voltage and current to pass.

NUMBER REQUIRED
280.3

Where used on a supply circuit that supplies a single building or structure, a surge arrester is connected to each ungrounded (phase) conductor. Surge arresters are available in single units to connect to only one ungrounded (phase) conductor. In this case, one would be required for each ungrounded (phase) conductor. They also are available with three units in one enclosure, which will handle a three-phase supply.

If there are other supply conductors, such as supplied from a industrial site service pole, one set of surge arresters would be sufficient. Where there are a number of supply conductors, surge arresters shall be installed at the load end of such supply conductors. (**See Figure 14-1**)

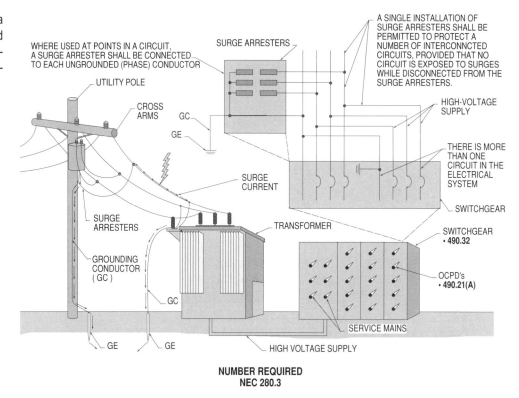

Figure 14-1. This illustration shows a single surge arrester shall be permitted to be used to protect a number of interconnected circuits, under certain conditions of use.

Grounding Tip 387: Low-voltage machines rated at 600 volts or less have relatively higher dielectric strength than machines rated over 600 volts. If these low-voltage machines are supplied by overhead lines that are exposed only to transformers which are equipped with adequate lightning protection on their primary, no additional lightning protection equipment is generally required.

Grounding Tip 388: On effectively grounded systems, arresters are usually rated at 80 percent of line-to-line voltage. On less than effectively grounded systems, arresters are usually rated at full line-to-line voltage.

SURGE ARRESTER SELECTION
280.4

Surge arresters shall be selected based upon the supply voltage and designed and installed to safely divert all types of surges that may be harmful to the electrical elements and equipment supplying power to the building, site, etc.

ON CIRCUITS OF LESS THAN 1000 VOLTS
280.4(A)

On electrical circuits of less than 1000 volts, it is required that the voltage rating of the surge arrester be equal to or greater than the maximum voltage of the phase-to-ground voltage present.

For example: A service of 480 volts-to-ground (rms), the maximum voltage can be calculated as follows:

Step 1: Finding formula
Max. V = V ÷ 0.707

Step 2: Applying formula
Max. V = 480 V ÷ 0.707
Max. V = 679 V

Solution: The maximum voltage calculated is 679 volts.

See **Figure 14-2** for the rules involved concerning the selection of surge arresters.

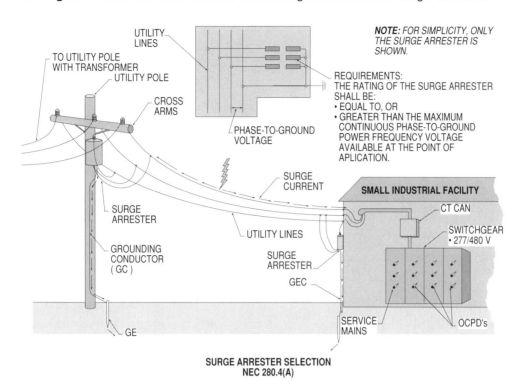

Figure 14-2. This illustration shows the rules that pertain to surge arrester selections, based upon circuits less than 1000 volts.

Grounding Tip 389: If the voltage-to-ground in **Figure 14-2** is 277 volts, the arresters shall be equal to or greater than this value.

ON CIRCUITS OF 1000 AND OVER
280.4(B)

This rule requires that the surge arrester shall have a rating of not less than 125 percent of the maximum circuit voltage rating, which is the phase-to-ground voltage.

Grounding Note: Do not get confused between rms voltage and maximum voltage (See Grounding Tip 390).

For Example: For a service of 2400 volts-to-ground (4160 V phase-to-phase) the maximum voltage-to-ground can be calculated as follows:

Step 1: Finding formula
Max. V = V x 125%

Step 2: Finding rating of arrester
Max. V = 2400 V x 125%
Max. V = 3000 volts

Solution: **The maximum voltage calculated is 3000 volts.**

See **Figure 14-3** for the requirements for computing the rating of surge arresters used on 1000 volts or less circuits.

The procedure for selecting surge arresters can be summed up as follows:

- For circuits of less than 1000 volts, the rating of the arrester shall be equal to the voltage-to-ground of the conductor.

- For circuits of 1000 volts or greater, 125 percent of the voltage-to-ground of the conductor.

Grounding Tip 390: The rms voltage is defined as being equal to the square root of the average value of the squares of all the instantaneous values of current or voltage during one-half cycle. Therefore, for a peak voltage of 679 volts, the rms AC voltage can be determined by applying the following multiplier times the peak voltage:

- 679 V peak
- x .707 multiplier
- 480.053 V rms

Solution: The root means square (rms) voltage is 480.76 volts.

Note: See calculation for finding maximum voltage in **"For Example"** on page 14-2.

Figure 14-3. The rating of silicon carbide-type surge arresters shall be calculated at not less than 125 percent of the maximum continuous phase-to-ground voltage (1000 volts or greater) available at the point of connection. (See **"For Example Problem"** on page 14-3 in this chapter for calculation procedure to select arrester ratings).

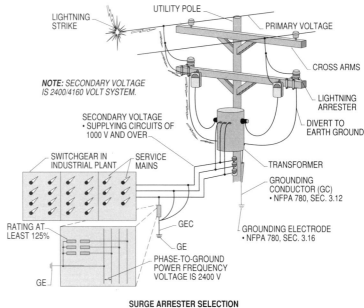

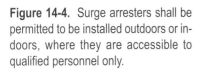

LOCATION
280.11

Surge arresters shall be permitted to be located indoors or outdoors and shall be made inaccessible to unqualified personnel. Whether or not to install surge arresters indoors or outdoors, the fact that some have exploded during operation must be taken into consideration. This is not a NEC requirement, but when possible they should be placed outdoors. However, **280.11** permits surge arresters listed for accessible locations to be located to unqualified personnel. **(See Figure 14-4)**

Figure 14-4. Surge arresters shall be permitted to be installed outdoors or indoors, where they are accessible to qualified personnel only.

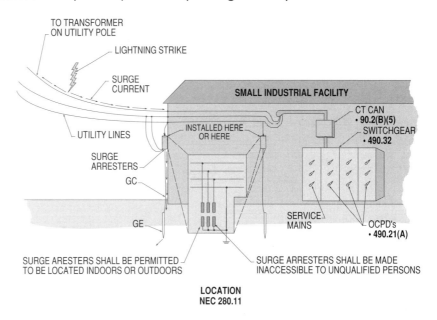

Standards to review
NEC 280.12
NESC 93.C(4)

ROUTING OF SURGE ARRESTERS CONNECTIONS
280.12

The connections from the supply system to the surge arresters shall be as short as possible. There shall be as few bends in the grounding conductors as possible. Lightning takes a direct path to the ground and moves so fast that it can't flow around sharp bends easily. The grounding conductors and their connection to ground are as follows:

- Conductors connecting arresters to ground or to the line shall be as short and direct as possible.
- Unnecessary bends shall not be permitted by the NEC for such conductors and connections.

See Figure 14-5 for the procedures of routing and connection of surge arresters.

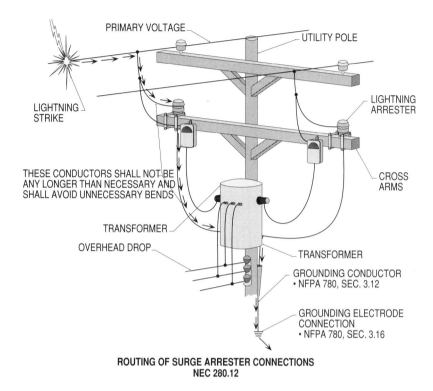

Figure 14-5. This illustration shows the conductors between the surge arrester line wire, bus and grounding connection, shall be routed without sharp bends.

INSTALLED AT SERVICES OF LESS THAN 1000 VOLTS 280.21

For surge arresters installed on the supply side of a service, the NEC allows four methods of connecting the ground side of the arrester to ground and they are as follows:

- They shall be permitted to be grounded to the "grounded service conductor" which may be a neutral.
- They shall be permitted to be grounded to the "Grounding electrode conductor" which connects to the grounding electrode system.
- They shall be permitted to be grounded to the grounding electrode for the service.
- They shall be permitted to be grounded to the equipment grounding terminal in the service equipment.

Grounding Tip 391: The ideal location of lightning arresters to obtain the best protection is to install them directly at the terminals of the apparatus being protected.

Grounding Note: The line and ground connecting conductors shall be required to be at least 14 AWG cu. or 12 AWG alu. **(See Figure 14-6)**

Figure 14-6. This illustration shows the line and grounding conductors shall be required to be at least 14 AWG copper or 12 AWG aluminum with as few bends as possible.

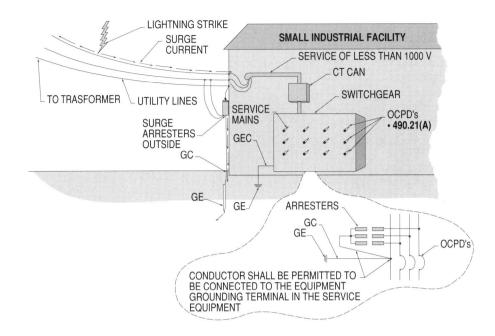

Grounding Tip 392: Arresters used on systems of 1000 volts or less, the arresters shall be equal to or greater than the maximum continuous phase-to-ground power frequency voltage available at the point of application.

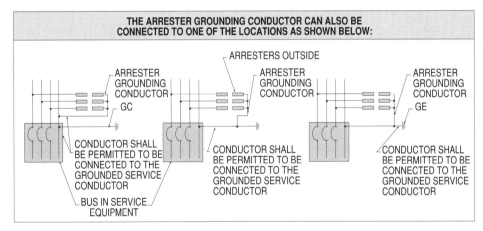

Grounding Tip 393: Lightning surges can't turn rapidly on a conductor, therefore its your responsibility to see that sharp bend in arrester grounding conductors are eliminated.

INSTALLATION ON THE LOAD SIDE OF SERVICES OF LESS THAN 1000 VOLTS
280.22

The line and grounding conductors shall not to be smaller than 14 AWG copper or 12 AWG aluminum. Care must be exercised when sizing these conductors and if possible, verify the proper size with the AHJ to ensure the correct size is selected. A surge arrester shall be permitted to be connected between any two conductors (ungrounded (phase) conductor(s), grounded (neutral) conductor and grounding conductor). The grounded (neutral) conductor and the grounding conductor shall be interconnected only by the normal operation of the surge arrester during a surge.

Again the shortest method of getting the surge to ground is always the best design. Avoid bends as much as possible to keep surges moving in a straight motion to ground.
(See Figure 14-7)

Surge Arresters

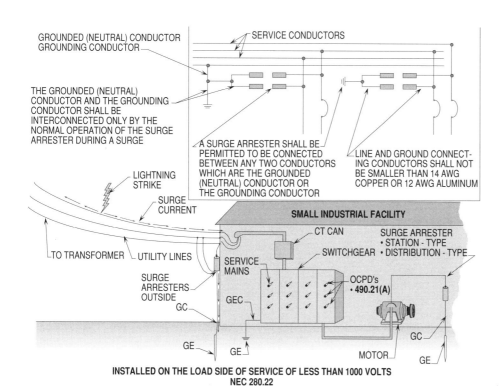

Figure 14-7. This illustration shows surge arresters shall be permitted to be connected on the load side of a service, to protect the load side equipment from surges of current.

Grounding Tip 394: Where the arrester voltage is required to be 3 to 15 kV, a choice between the distribution-type and the station-type arrester shall be made.

CIRCUITS OF 1000 VOLTS AND OVER - SURGE ARRESTER CONDUCTORS
280.23

For conductors connecting the surge arrester to both the ungrounded (phase) conductors and the ground, 6 AWG copper or aluminum is the smallest size that shall be permitted. **(See Figure 14-8)**

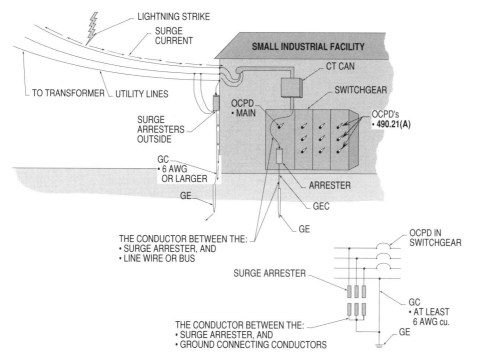

Figure 14-8. This illustration shows the grounding conductors used to interconnect surge arresters and ground them to earth shall be at least 6 AWG for copper or aluminum.

Grounding Tip 395: Where the switchgear is connected directly to the overhead line from roof bushings, lightning arresters should always be provided at the switchgear to obtain maximum protection. Sometimes, separation of arresters from the switchgear reduces the protection scheme.

CIRCUITS OF 1000 VOLTS AND OVER - INTERCONNECTIONS
280.24

The grounding conductor of a surge arrester protecting a transformer that supplies a secondary system shall be permitted to be interconnected as specified in Subsections **(A)** and **(B)** below. **(See Figure 14-9)**

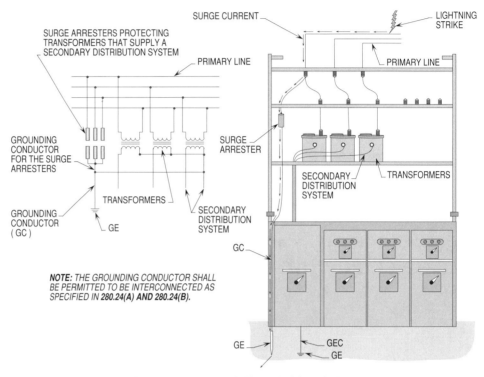

Figure 14-9. This illustration shows the connections of grounding conductors for surge arresters protecting transformers that supply a secondary distribution system.

METALLIC INTERCONNECTIONS
280.24(A)

Standards to review

NEC 280.24(A)
NESC 93.C(4)
NESC 192.

A metallic interconnection shall be permitted to be made to the secondary grounded (neutral) conductor, provided that this connection is addition to the direct grounding connection at the surge arrester. **(See Figure 14-10)**

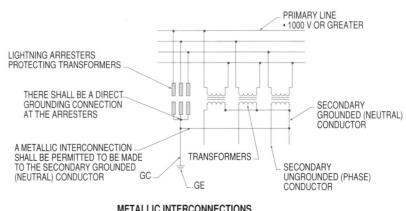

Figure 14-10. This illustration shows that under certain conditions of use, a metallic interconnection shall be permitted to be made to the secondary grounded (neutral) conductor for circuits operating at 1000 volts or greater.

CONNECTIONS
280.24(A)(1)

The conditions for permitting metallic interconnection, as covered in the paragraph above, are contingent upon meeting the requirements of both **280.24(A)(1)** and **(A)(2)**. It shall be permissible for an interconnection to be made between the surge arrester and the grounded (neutral) conductor of the secondary if the following rules are met:

- The secondary has the grounded (neutral) conductor connected elsewhere to a continuous metal underground metal piping system. (**See Figure 14-11**)

- If in urban areas there are a minimum of four water pipe grounding connections in a distance of one mile, the direct ground from the surge arrester may be eliminated and the secondary grounded (neutral) conductor used as the grounding for the surge arrester. (**See Figure 14-12**)

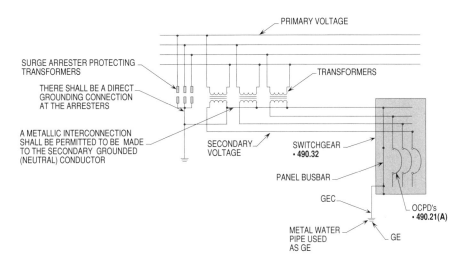

Figure 14-11. This illustration shows a continuous metal water pipe system shall be permitted to be used as a grounding electrode for the secondary system, if it complies with **250.50(A)(1)** and **250.104(A)**.

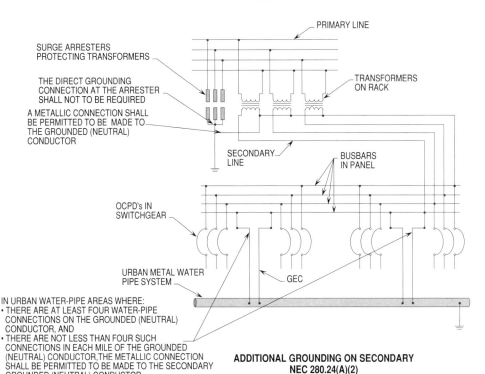

Figure 14-12. This illustration shows the requirements for installations where the neutral is multi-grounded. The direct ground from the surge arrester may be eliminated and the neutral used for the grounding of the arrester.

CONNECTIONS
280.24(A)(2)

In many cases the primary is four-wire wye, with the neutral grounded periodically. In this arrangement, the secondary grounded (neutral) conductor is usually interconnected with the primary grounded (neutral) conductor. Where the primary neutral is grounded in a minimum of four places in each mile, plus the secondary service ground, the surge arrester ground may be interconnected with the primary and secondary grounds in addition to the surge arrester grounding electrode. (**See Figure 14-13**)

Figure 14-13. This illustration shows the primary grounded (neutral) conductor of a metallic interconnected system shall have four grounds in each mile, which is in addition to the service grounds at the switchgear.

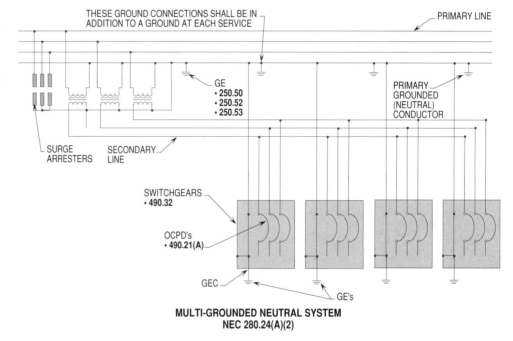

Grounding Tip 396: For requirements using multi-grounded neutral points, review Rules 96.A through 96.C in the NESC and **250.184(B)** in the NEC.

Grounding Tip 397: There are two types of protection for transformers to prevent insulation damage by lightning surges and they are:
- Spark gaps
- Valve arresters

A spark gap is an open gap which flashes over when a high enough voltage is impressed on it. It causes a short-circuit to ground which continues even after the lightning surge has passed. The short-circuit must be interrupted by an OCPD such as a recloser or fuse.

A valve type arrester consist of a series of small air gaps and a power current limiting resistive element. The small gaps are able to interrupt small currents but not high currents like the system fault current.

THROUGH SPARK GAP OR DEVICE
280.24(B)

A spark gap is an open gap that flashes over when a high enough voltage is impressed on it. It creates a short-circuit to ground which continues even after the lightning surge has passed. The short-circuit shall be interrupted by a overcurrent device such as a recloser or fuse.

If a surge arrester grounding conductor and secondary is not connected as in **280.24(A)** above but is otherwise grounded as in **250.50** and **250.52**, an interconnection shall be made through a spark gap or listed device as follows:

- For grounded or ungrounded primary systems, the spark gap or listed device shall have a 60-hertz breakdown voltage of at least two times the primary circuit voltage but not necessarily more than 10 kV, and there shall be at least one other ground on the grounded (neutral) conductor of the secondary not less than 20 ft. (6 m) distance from the surge arrester grounding electrode. (**See Figure 6-14**)

- For multiground neutral primary systems, the spark gap or listed device shall have a 60-hertz breakdown of not more than 3 kV, and there shall be at least one other ground on the grounded (neutral) conductor of the secondary not less than 20 ft. (6 m) distance from the surge arrester grounding electrode. (**See Figure 14-15**)

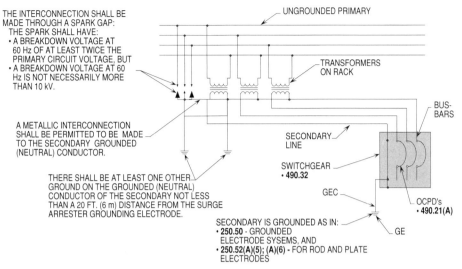

Figure 14-14. When the grounding conductors of surge arresters are grounded to a metal water piping system, as required by **250.50** and **250.52**, the rules in **280.24(A)(1)** shall not be applied.

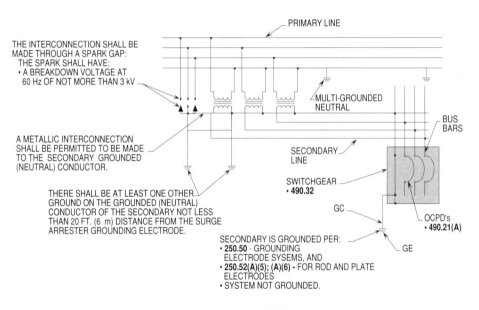

Figure 14-15. For multi-grounded neutral primary systems and for circuits of 1000 volts or greater, a metallic interconnection shall be permitted to be made through a spark gap. For the definition of a spark gap, see Grounding Tip 397 on page 14-10 in this book.

BY SPECIAL PERMISSION
280.24(C)

This permissive rule permits the authority having jurisdiction (AHJ) to grant special permission in writing to grant other than permitted in **280.24(A)** and **(B)** above.

Grounding Tip 398: When requesting a variance from a grounding requirement, the AHJ shall comply with **90.2(C)**, **90.4**, **110.2** and **110.3(B)** in the NEC.

Safety Tip - Grounding arresters

The ground side of an arrester for a transformer primary may be connected to the transformer secondary grounded (neutral) conductor as follows:
- **Method 1**
  Direct to the secondary grounded (neutral) conductor only.
- **Method 2**
  Direct to the secondary grounded (neutral) conductor and to an arrester ground.
- **Method 3**
  To the secondary grounded (neutral) conductor through a spark gap, and to an arrester ground.

Method 1 may be used only where the transformer secondaries serve four or more premises within a distance of one mile, and each service is grounded to a city water system.

Method 2 may be used when one or more premises are served by the transformer secondaries, and each service is grounded to a continuous water piping system.

Method 3 where the transformer serves a premises having a made electrode for a ground which may be a driven rod, pipe or plate, the primary arrester ground may be made to the secondary grounded (neutral) conductor, but only through a "spark gap" type arrester.

Grounding Note: A primary lightning arrester may be independently grounded without connection to a secondary grounded (neutral) conductor.

See Figure 14-16 for grounding when using special permission for a less stringent grounding scheme for surge arresters.

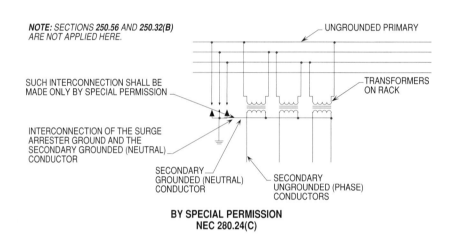

Figure 14-16. This illustration shows the procedure for installing and grounding surge arresters using special permission of the AHJ per **280.24(C)**. See **Grounding Tip 398** on page 14-11

Standards to review
NEC 280.25
NESC 93.C(4)
NESC 192.

GROUNDING
280.25

As usually as in the case of grounding, the grounding of surge arresters shall comply with the requirements in **Article 250** except as otherwise modified in this Article. Grounding conductors shall not be routed in metal enclosures unless they are bonded at both ends of such enclosures. **(See Figure 14-17)**

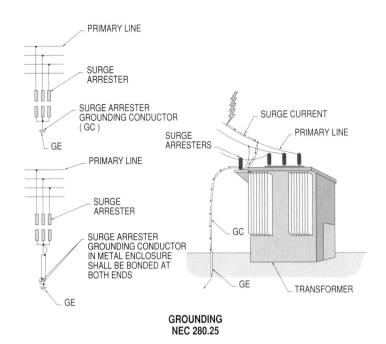

Figure 14-17. This illustration shows surge arresters shall be bonded and grounded by the requirements outlined in **Article 250**.

Name _____   Date _____

Chapter 14
Surge Arresters

	Section	Answer

1. Where used at a point on a circuit, a surge arrester shall be connected to each grounded conductor. _____ T F

2. Surge arresters shall be permitted to be located indoors and accessible to unqualified personnel if listed for the installation. _____ T F

3. The connections from the supply system to the surge arresters should be as short as possible. _____ T F

4. A spark gap is a series of small air gaps and a power current limiting resistive element. _____ T F

5. Grounding conductors for surge arresters shall be permitted to be routed in metal enclosures. _____ T F

6. On electrical circuits of less than _____ volts, it is required that the voltage rating of the surge arrester be equal to or greater than the maximum voltage of the phase-to-ground voltage present. _____ _____

7. Surge arresters on circuits of 1000 volts and over shall have a rating of not less than _____ percent of the maximum circuit voltage rating, which is the phase-to-ground voltage. _____ _____

8. The line and grounding conductors on the load side of services of less than 1000 volts shall not be smaller than _____ AWG copper. _____ _____

9. For conductors connecting the surge arrester to both the ungrounded conductors and the ground for circuits of 1000 volts and over, a _____ AWG copper is the smallest permitted size. _____ _____

10. For grounded and ungrounded primary systems, the spark gap or listed device shall have a 60-hertz breakdown voltage of at least two times the primary circuit voltage but not necessarily more than 10 kV, and there shall be at least one other ground on the grounded conductor of the secondary not less than _____ ft. distant from the surge arrester grounding electrode. _____ _____

15

TRANSIENT VOLTAGE SURGE SUPPRESSORS

A transient voltage surge suppressor is defined as a protective device for limiting transient voltages by diverting or limiting surge current. It also prevents continued flow of follow current while remaining capable of repeating these functions.

Transient voltage surge suppressors (TVSS) are devices that are installed on the load side of the service disconnect and protected by overcurrent protection devices. Note that **UL 1449, Std. Transient Voltage Surge Suppressors**, tests these devices at lower surge current levels than those for surge arresters covered in **Article 280** in the NEC and are evaluated under IEEE Standards. Requirements in this Article provide unique rules for TVSS devices and clarify the differences between these devices and surge arresters so that designers, installers and inspectors will not confuse the two in an electrical system.

Grounding Note: To aid in the protection of sensitive electronic equipment, the use of TVSS devices has greatly increased in residential, commercial and industrial installations. Previous editions of the NEC did not have requirements recognizing the installation of these devices which were being used everywhere. **Article 285** is equipped with the rules for installing these devices safely.

See Figure 15-1 for a detailed illustration pertaining TVSS installations.

Figure 15-1. This illustration shows the locations where TVSS devices are installed to protect sensitive electronic equipment from damaging power surges.

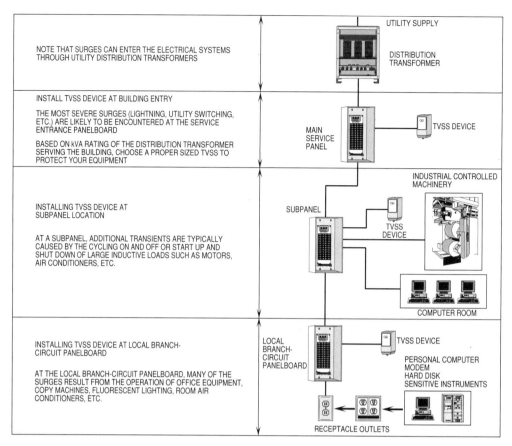

USE NOT PERMITTED
285.3

A TVSS shall not be used in the following:

- Circuits exceeding 600 volts

- Ungrounded electrical systems as permitted in **250.21**

- Where the rating of the TVSS is less than the maximum continuous phase-to-ground power frequency voltage available at the point of application.

FPN: For further information on TVSS's, see *NEMA LS 1-1992, Standard for Low Voltage Surge Suppression Devices*. The selection of a properly rated TVSS is based on criteria such as maximum continuous operating voltage, the magnitude and duration of overvoltages at the suppressor location as affected by phase-to-ground faults, system grounding techniques and switching surges

NUMBER REQUIRED
285.4

Where used on a point on a circuit, the TVSS shall be connected to each ungrounded (phase) conductor. Note that TVSS's shall be a listed device per **285.5** of the NEC.

SHORT CIRCUIT CURRENT RATING
285.6

The TVSS shall be marked with a short circuit current rating and shall not be installed at point on the system where the available fault current is in excess of that rating. This marking requirement shall not apply to receptacles.

See Figure 15-2 for specific rules pertaining to TVSS installation.

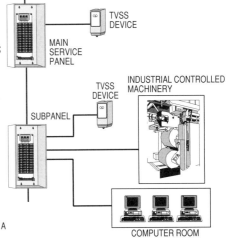

Figure 15-2. This illustration depicts certain installation rules pertaining to TVSS devices.

SCOPE.
THIS ARTICLE COVERS GENERAL REQUIREMENTS, INSTALLATION REQUIREMENTS AND CONNECTION REQUIREMENTS FOR TRANSIENT VOLTAGE SURGE SUPPRESSORS (TVSS) PERMANENTLY INSTALLED ON PREMISES WIRING SYSTEMS.
¥285.1

TRANSIENT VOLTAGE SURGE SUPPRESSOR (TVSS).
A PROTECTIVE DEVICE FOR LIMITING TRANSIENT VOLTAGES BY DIVERTING OR LIMITING SURGE CURRENT; IT ALSO PREVENT CONTINUED FLOW OF FOLLOW CURRENT WHILE REMAINING CAPABLE OF REPEATING THESE FUNCTIONS.
¥285.2

WHERE THE RATING OF THE TVSS IS LESS THAN THE MAXIMUM CONTINUOUS PHASE-TO-GROUND POWER FREQUENCY VOLTAGE AVAILABLE AT THE POINT OF APPLICATION.
¥285.3

NUMBER REQUIRED.
WHERE USED AT A POINT ON A CIRCUIT, THE TVSS SHALL BE CONNECTED TO EACH UNGROUNDED CONDUCTOR.
¥285.4

LISTING.
A TVSS SHALL BE A LISTED DEVICE.
¥285.5

SHORT CIRCUIT CURRENT RATING.
THE TVSS SHALL BE MARKED WITH A SHORT CIRCUIT CURRENT RATING AND SHALL NOT BE INSTALLED AT A POINT ON THE SYSTEM WHERE THE AVAILABLE FAULT CURRENT IS IN EXCESS OF THAT RATING. THIS MARKING REQUIREMENT SHALL NOT APPLY TO RECEPTACLES.
¥285.6

REQUIREMENTS FOR INSTALLING TVSS DEVICES
NEC 285.1 THRU 285.6

LOCATION
285.11

TVSS's shall be permitted to be located indoors or outdoors and shall be made inaccessible to unqualified persons, unless listed for installation in accessible locations.

ROUTING OF CONNECTIONS
285.12

The conductors used to connect the TVSS to the line or bus and to ground shall be no longer than necessary and shall avoid unnecessary bends.

See Figure 15-3 for specific rules pertaining to TVSS installations.

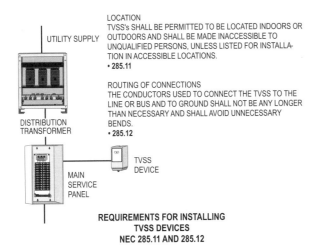

Figure 15-3. This illustration depicts certain installation rules pertaining to TVSS devices.

CONNECTIONS
285.21

Where a TVSS is installed, it shall be connected as follows:

SERVICE SUPPLIED BUILDING OR STRUCTURE
285.21(A)(1)

The transient voltage surge suppressor shall be connected on the load side of a service disconnect overcurrent device required in **230.91**.

FEEDER SUPPLIED BUILDING OR STRUCTURE
285.21(A)(2)

The transient voltage surge suppressor shall be connected on the load side of the first overcurrent device at the building or structure.

Grounding Note: Ex's to (1) and **(2)** - Where the TVSS is also listed as a surge arrestor, the connection shall be as permitted by **Article 280**.

SEPARATELY DERIVED SYSTEM
285.21(A)(3)

The TVSS shall be connected on the load side of the first overcurrent device in a separately derived system.

See Figure 15-4 for a detailed illustration of the requirements outlined in **285.21(A)(1) thru (A)(3)**.

CONDUCTOR SIZE
285.21(B)

Line and ground connecting conductors shall not be smaller than 14 AWG copper or 12 AWG aluminum.

CONNECTION BETWEEN CONDUCTORS
285.21(C)

A TVSS shall be permitted to be connected between any two conductors – ungrounded (phase) conductor(s), grounded (neutral) conductor and grounding conductor. The grounded (neutral) conductor and the grounding conductor shall be interconnected only by the normal operation of the TVSS during a surge.

See Figure 15-5 for a detailed illustration of the requirements outlined in **285.21(B)** and **285.21(C)**.

GROUNDING
285.25

Grounding conductors shall not be run in metal enclosures unless bonded to both ends of such enclosure.

See Figure 15-5 and 6 for the requirements pertaining to grounding and bonding TVSS devices.

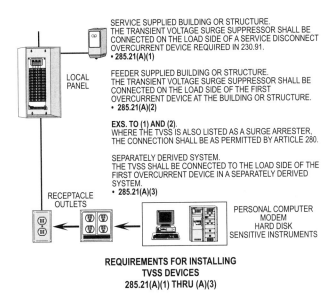

Figure 15-4. This illustration depicts certain installation rules pertaining to TVSS devices.

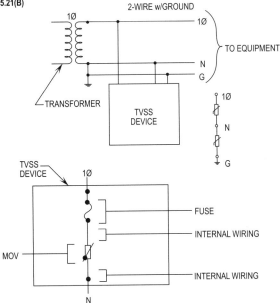

Figure 15-5. This illustration depicts certain installation rules that pertain to TVSS devices.

Figure 15-6. This illustration depicts the grounding and bonding rules that pertain to TVSS devices.

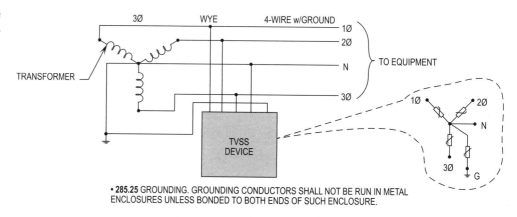

Name _____   Date _____

Chapter 15
Transient Voltage Surge Suppressors

	Section	Answer
1. A TVSS shall be permitted to be used for circuits exceeding 600 volts.	_____	T F
2. Line and ground connecting conductors shall not be smaller than 16 AWG copper.	_____	T F
3. Grounding conductors shall not be run in metal enclosures unless bonded to both ends of such enclosure.	_____	T F
4. Where used on a point on a circuit, the TVSS shall be connected to each ungrounded (phase) conductor.	_____	T F
5. TVSS's shall be permitted to be located indoors or outdoors and be accessible.	_____	T F

Appendix

This Appendix contains tables that can be used to size grounded (neutral) conductors, grounding electrode conductors and equipment bonding jumpers that are used for bonding and grounding the metal of enclosures on the supply side of the service disconnecting means.

Tables A, B and C are utilized for sizing the grounding electrode conductors, grounded (neutral) conductors, equipment bonding jumpers and main bonding jumpers on the supply side of the service equipment disconnecting means.

Tables D and E are utilized for sizing the equipment grounding conductors and equipment bonding jumpers on the load side of the service disconnecting means.

Table F is utilized for sizing conductors based on the circular mil rating.

Tables G, H and I are used to determine the maximum distance that a particular type conduit (with or without an equipment grounding conductor) can be run and used as an equipment grounding means to open an OCPD to deenergize power on a feeder or branch-circuit during a ground-fault condition.

Tables J, K and L are tips utilized when designing and installing a grounding, lightning and static electricity protection scheme.

Table M is a checklist chart that is utilized for grounding and bonding of electrical systems to help provide guidance to the user so as to maintain a safe and reliable grounding system.

Table N is used as a reference of other codes and standards that are used with the NEC to design and ensure power quality, safety and operational integrity of sensitive electronic equipment.

Appendix A

Use the Table below to size the minimum size grounded conductor (neutral or phase), the minimum size equipment bonding jumper and the minimum size main bonding jumper.

The grounding electrode conductor is also sized from this table when it's connected to the metal water pipe or building steel. **Note:** This table is used when there is no OCPD ahead of the conductors supplying the equipment. Such conductors are service conductors and transformer secondary conductors. For example, to ground a 400 KCMIL copper service to structural building steel, A 1/0 AWG copper grounding electrode conductor is required per **Table 250.66**. (See Appendix B and C)

TABLE 250.66. GROUNDING ELECTRODE CONDUCTOR FOR ALTERNATING-CURRENT SYSTEMS			
SIZE OF LARGEST SERVICE-ENTRANCE CONDUCTOR OR EQUIVALENT AREA FOR PARALLEL CONDUCTORS[1]		SIZE OF GROUNDING ELECTRODE CONDUCTOR	
COPPER	ALUMINUM OR COPPER-CLAD ALUMINUM	COPPER	ALUMINUM OR COPPER-CLAD ALUMINUM[2]
2 OR SMALLER	1/2 OR SMALLER	8	6
1 OR 1/0	2/0 OR 3/0	6	4
2/0 OR 3/0	4/0 OR 250 KCMIL	4	2
OVER 3/0 THROUGH 350 KCMIL	OVER 250 KCMIL THROUGH 500 KCMIL	2	1/0
OVER 350 KCMIL THROUGH 600 KCMIL	OVER 500 KCMIL THROUGH 900 KCMIL	1/0	3/0
OVER 600 KCMIL THROUGH 1100 KCMIL	OVER 900 KCMIL THROUGH 1750 KCMIL	2/0	4/0
OVER 1100 KCMIL	OVER 1750 KCMIL	3/0	250 KCMIL

Appendix B

Determining the grounding elements at the service equipment based on Table 250.66.

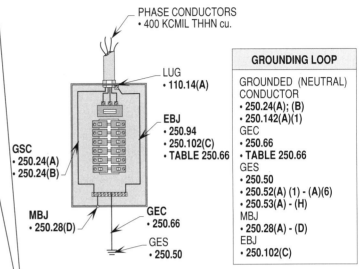

ELEMENTS OF THE SERVICE EQUIPMENT

Using copper, size the grounding electrode conductor from **Table 250.66** based on the size of the largest service conductor.

Step 1: Sizing GEC
Table 250.66
400 Kcmil requires 1/0 AWG cu.

Solution: 1/0 AWG cu. is the minimum size grounding electrode conductor.

Using copper, size the grounded (neutral) conductor (GSC) per **250.24(B)(1)** and **Table 250.66** based on the size of the grounding electrode conductor per Step 1 above.

Step 2: Sizing GSC
250.24(B)(1); Table 250.66
1/0 AWG cu. GEC requires 1/0 AWG cu.

Solution: 1/0 AWG cu. is the minimum size grounded (neutral) conductor.

Using copper, size the equipment bonding jumper per **250.102(C)** and **Table 250.66** based on the size of the grounding electrode conductor per Step 1 above.

Step 3: Sizing EBJ
260.102(C); Table 250.66
1/0 AWG cu. requires 1/0 AWG cu.

Solution: 1/0 AWG cu. is the minimum size equipment bonding jumper.

Using copper, size the main bonding jumper per **250.28(D)** and **Table 250.66** based on the size of the grounding electrode conductor per Step 1 above.

Step 4: Sizing MBJ
250.28(D); Table 250.66
1/0 AWG requires 1/0 AWG cu.

Solution: 1/0 AWG cu. is the minimum size main bonding jumper.

Sizing grounding electrode conductor per Step 1 and **Table 260.66** requires 1/0 AWG cu.

Grounding Note 1: The box shows the minimum size grounding elements for the service above.

TABLE 250.66. GROUNDING ELECTRODE CONDUCTOR FOR ALTERNATING-CURRENT SYSTEMS

SIZE OF LARGEST SERVICE-ENTRANCE CONDUCTOR OR EQUIVALENT AREA FOR PARALLEL CONDUCTORS[1]		SIZE OF GROUNDING ELECTRODE CONDUCTOR	
COPPER	ALUMINUM OR COPPER-CLAD ALUMINUM	COPPER	ALUMINUM OR COPPER-CLAD ALUMINUM[2]
2 OR SMALLER	1/2 OR SMALLER	8	6
1 OR 1/0	2/0 OR 3/0	6	4
2/0 OR 3/0	4/0 OR 250 KCMIL	4	2
OVER 3/0 THROUGH 350 KCMIL	OVER 250 KCMIL THROUGH 500 KCMIL	2	1/0
OVER 350 KCMIL THROUGH 600 KCMIL	OVER 500 KCMIL THROUGH 900 KCMIL	1/0	3/0
OVER 600 KCMIL THROUGH 1100 KCMIL	OVER 900 KCMIL THROUGH 1750 KCMIL	2/0	4/0
OVER 1100 KCMIL	OVER 1750 KCMIL	3/0	250 KCMIL

Grounding Note 2: A 1/0 AWG cu. grounding electrode conductor per **Table 250.66** is required because the 400 KCMIL cu. of the service falls between 350 KCMIL and 600 KCMIL.

Appendix C

Determining the size grounding electrode conductor based on the type of electrodes that are available per 250.50(A) through (D) in the NEC.

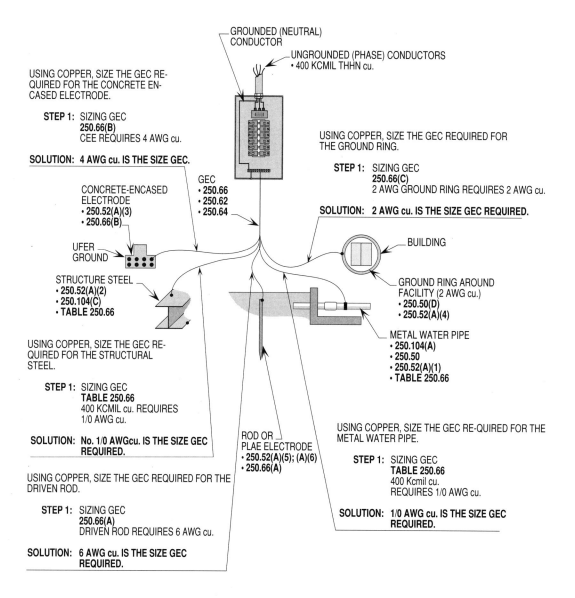

Appendix D

Use the Table below to size the minimum size equipment grounding conductor and the minimum size equipment bonding jumper to connect the noncurrent-carrying metal parts of electrical equipment enclosures to earth ground. **Note:** This table is used when there is an OCPD ahead of the conductors supplying the equipment.

For example, the size EGC required to ground the noncurrent-carrying parts of a electrical enclosures with a 225 OCPD ahead of the circuit conductors is a 4 AWG copper conductor per **Table 250.122**.

TABLE 250.122. MINIMUM SIZE EQUIPMENT GROUNDING CONDUCTORS FOR GROUNDING RACEWAY AND EQUIPMENT

RATING OR SETTING OF AUTOMATIC OVERCURRENT DEVICE IN CIRCUIT AHEAD OF EQUIPMENT, CONDUIT, ETC., NOT EXCEEDING (AMPERES)	SIZE (AWG OR kcmil)	
	COPPER	ALUMINUM OR COPPER-CLAD ALUMINUM*
15	14	12
20	12	10
30	10	8
40	10	8
60	10	8
100	8	6
200	6	4
300	4	2
400	3	1
500	2	1/0
600	1	2/0
800	1/0	3/0
1000	2/0	4/0
1200	3/0	250
1600	4/0	350
2000	250	400
2500	350	600
3000	400	600
4000	500	800
5000	700	1200
6000	800	1200

225 A FALLS BETWEEN 200 A AND 300 A

Note: Because the 225 amp OCPD exceeds the 200 amp rating in **Table 250.122** and falls between the 200 amp and 300 amp ratings, the size EGC must be sized at 4 AWG copper and not 6 AWG copper.

Appendix E

Determining the size equipment grounding conductor based on the size OCPD protecting the circuit conductors per Table 250.122.

Using copper, size the equipment grounding conductor based on the 300 amp OCPD protecting the circuit conductors.

Step 1: Sizing EGC
Table 150.122
300 A OCPD requires 4 AWG cu.

Solution: 4 AWG cu. is the size equipment grounding conductor required.

Using copper, size equipment bonding jumper per **250.102(D)** and **Table 250.122** based on the size of the equipment grounding conductor per Step 1 above.

Step 2: Sizing EBJ
250.102(D); Table 250.122
300 A OCPD requires 4 AWG cu.

Solution: 4 AWG cu. is the size equipment bonding jumper required.

Solution: The size of the equipment grounding conductor and equipment bonding jumper are required to be at least the same size per 250.122 and 250.102(D).

> Sizing equipment grounding conductor per step 1 and **Table 250.122** is 4 AWG cu. and is required.

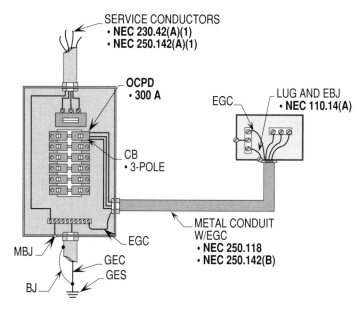

TABLE 250.122		
TABLE 250.122. MINIMUM SIZE EQUIPMENT GROUNDING CONDUCTORS FOR GROUNDING RACEWAY AND EQUIPMENT		
RATING OR SETTING OF AUTOMATIC OVERCURRENT DEVICE IN CIRCUIT AHEAD OF EQUIPMENT, CONDUIT, ETC., NOT EXCEEDING (AMPERES)	SIZE (AWG OR kcmil)	
	COPPER	ALUMINUM OR COPPER-CLAD ALUMINUM*
15	14	12
20	12	10
30	10	8
40	10	8
60	10	8
100	8	6
200	6	4
300	4	2
400	3	1
500	2	1/0
600	1	2/0
800	1/0	3/0
1000	2/0	4/0
1200	3/0	250
1600	4/0	350
2000	250	400
2500	350	600
3000	400	600
4000	500	800
5000	700	1200
6000	800	1200

Grounding Note: A 4 AWG cu. equipment grounding conductor is required based on the 300 amp OCPD protecting the circuit conductors.

Appendix F

When selecting various values of grounding conductors to be used in an electrical system, use the Table below. When selecting various values related to a particular size, circular mils (CM) is used for conductor size 18 AWG through 4/0 AWG and for conductor sizes 250 to 2000, KCMIL's is used. For example, the DC resistance of a coated 4/0 AWG copper conductor is .0626 ohms.

Size AWG/ KCMIL	Area Cir. Mills	Conductors Standing Quantity	Standing Diam. In.	Overall Diam. In.	Overall Area Sq. In.^{2}	DC Resistance at 75°C (167°F) Copper Uncoated ohm/kFT	Copper Coated ohm/kFT	Aluminum ohm/kFT
18	1620	1	—	0.040	0.001	7.77	8.08	12.8
18	1620	7	0.015	0.046	0.002	7.95	8.45	13.1
16	2580	1	—	0.051	0.002	4.89	5.08	8.05
16	2580	7	0.019	0.058	0.003	4.99	5.29	8.21
14	4110	1	—	0.064	0.003	3.07	3.19	5.06
14	4110	7	0.024	0.073	0.004	3.14	3.26	5.17
12	6530	1	—	0.081	0.005	1.93	2.01	3.18
12	6530	7	0.030	0.092	0.006	1.98	2.05	3.25
10	10380	1	—	0.102	0.008	1.21	1.26	2.00
10	10380	7	0.038	0.116	0.011	1.24	1.29	2.04
8	16510	1	—	0.128	0.013	0.764	0.786	1.26
8	16510	7	0.049	0.146	0.017	0.778	0.809	1.28
6	26240	7	0.061	0.184	0.027	0.491	0.510	0.808
4	41740	7	0.077	0.232	0.042	0.308	0.321	0.508
3	52620	7	0.087	0.260	0.053	0.245	0.254	0.403
2	66360	7	0.097	0.292	0.067	0.194	0.201	0.319
1	83690	19	0.066	0.332	0.087	0.154	0.160	0.253
1/0	105600	19	0.074	0.373	0.109	0.122	0.127	0.201
2/0	133100	19	0.084	0.419	0.138	0.0967	0.101	0.159
3/0	167800	19	0.094	0.470	0.173	0.0766	0.0797	0.126
4/0	211600	19	0.106	0.528	0.219	0.0608	0.0626	0.100
250	—	37	0.082	0.575	0.260	0.0515	0.0535	0.0847
300	—	37	0.090	0.630	0.312	0.0429	0.0446	0.0707
350	—	37	0.097	0.681	0.364	0.0367	0.0382	0.0605
400	—	37	0.104	0.728	0.416	0.0321	0.0331	0.0529
500	—	37	0.116	0.813	0.519	0.0258	0.0265	0.0424
600	—	61	0.099	0.893	0.626	0.0214	0.0223	0.0353
700	—	61	0.107	0.964	0.730	0.0184	0.0189	0.0303
750	—	61	0.111	0.998	0.782	0.0171	0.0176	0.0282
800	—	61	0.114	1.03	0.834	0.0161	0.0166	0.0265
900	—	61	0.122	1.09	0.940	0.0143	0.0147	0.0235
1000	—	61	0.128	1.15	1.04	0.0129	0.0132	0.0212
1250	—	91	0.117	1.29	1.30	0.0103	0.0106	0.0169
1500	—	91	0.128	1.41	1.57	0.00858	0.00883	0.0141
1750	—	127	0.117	1.52	1.83	0.00735	0.00756	0.0121
2000	—	127	0.126	1.63	2.09	0.00643	0.00662	0.0106

Appendix G

Maximum length of an equipment grounding conductor that may safely be used as an equipment grounding circuit conductor. Based on a ground-fault current of 400 percent of the overcurrent device rating. Circuit 120 volts to ground; 40 volts drop at the point of fault. Ambient Temperature 25°C.

COLUMN 1	COLUMN 2	COLUMN 3	COLUMN 4	COLUMN 5	COLUMN 6	COLUMN 7	COLUMN 8
COPPER EQUIPMENT GROUNDING CONDUCTOR SIZE ***	COPPER CIRCUIT CONDUCTORS	MAXIMUM LENGTH OF RUN (IN FT.) USING COPPER EQUIPMENT GROUND CONDUCTOR	ALUMINUM EQUIPMENT GROUNDING CONDUCTOR SIZE ***	ALUMINUM CIRCUIT CONDUCTORS	MAXIMUM LENGTH OF RUN (IN FT.) USING ALUMINUM EQUIPMENT GROUND CONDUCTOR	FOR COPPER AND ALUMINUM OVERCURRENT DEVICE RATING AMPS. 75°C **	FAULT CLEARING CURRENT 400% O. C. DEVICE RATING AMPS.
14 AWG	14 AWG	253	12 AWG	12 AWG	244	15	60
12 AWG	12 AWG	300	10 AWG	12 AWG	226	20	80
10 AWG	10 AWG	319	8 AWG	8 AWG	310	30	120
10 AWG	8 AWG	294	8 AWG	8 AWG	232	40	160
10 AWG	6 AWG	228	8 AWG	4 AWG	221	60	240
8 AWG	3 AWG	229	6 AWG	1 AWG	222	100	400
6 AWG	3/0 AWG	201	4 AWG	250 KCMIL	195	200	800
4 AWG	350 kcm	210	2 AWG	500 KCMIL	204	300	1200
3 AWG	600 kcm	195	1 AWG	900 KCMIL	192	400	1600
2 AWG	2-4/0 AWG	160	1/0 AWG	2-400 KCMIL	163	500	2000
1 AWG	2-300 KCMIL	160	2/0 AWG	2-500 KCMIL	161	600	2400
1/0 AWG	3-300 KCMIL	134	3/0 AWG	3-400 KCMIL	131	800	3200
2/0 AWG	4-250 KCMIL	114	4/0 AWG	4-400 KCMIL	115	1000	4000
3/0 AWG	4-300 KCMIL	106	250 kcm	4-500 KCMIL	107	1200	4800
4/0 AWG	4-600 KCMIL	93	350 kcm	4-900 KCMIL	97	1600	6400
250 KCMIL	5-600 KCMIL	78	400 kcm	5-800 KCMIL	79	2000	8000
350 KCMIL	6-600 KCMIL	*	600 kcm	6-900 KCMIL	*	2500	10,000
400 kcm	8-500 KCMIL	*	600 kcm	8-750 KCMIL	*	3000	15,000
500 KCMIL	8-1000 KCMIL	*	800 kcm	8-1500 KCMIL	*	4000	16,000
700 KCMIL	10-1000 KCMIL	*	1200 kcm	10-1500 KCMIL	*	5000	20,000
800 KCMIL	12-1000 KCMIL	*	1200 kcm	12-1500 KCMIL	*	6000	24,000

* Calculations Necessary
** 60°C for 20- and 30-ampere devices
*** Based on NEC Chapter 9, Table 8

Tables G, H and I are
Based on Georgia Tech Model

Appendix H

Using an equipment grounding conductor with or without metal conduit on 120 volt to ground circuits

COLUMN 1	COLUMN 2	COLUMN 3	COLUMN 4	COLUMN 5	COLUMN 6	COLUMN 7	COLUMN 8	COLUMN 9	COLUMN 10
OVERCURRENT DEVICE RATING AMPERES (75°C)	400% (4IP) OVERCURRENT DEVICE RATING AMPERES	CIRCUIT CONDUCTOR SIZE AWG-kcmil COPPER OR ALUMINUM	STEEL EMT, IMC GRC TRADE SIZE	(1) EQUIPMENT GROUNDING CONDUCTOR SIZE COPPER OR ALUMINUM	LENGTH OF EMT RUN COMPUTED MAXIMUM	LENGTH OF IMC RUN COMPUTED MAXIMUM	LENGTH OF GRC RUN COMPUTED MAXIMUM	COPPER GROUNDING CONDUCTOR W/O STEEL CONDUIT MAXIMUM RUN	ALUMINUM GROUNDING CONDUCTOR W/O STEEL CONDUIT MAXIMUM RUN
					(IN FEET)	(IN FEET)	(IN FEET)		
20	80	12	1/2	—	395	398	384	—	—
20	80	12	—	12	—	—	—	300	—
20	80	10 alu.	—	10 alu.	—	—	—	—	293
30	120	10	1/2	—	358	383	364	—	—
30	120	10	3/4	—	404	399	386	—	—
30	120	10	—	10	—	—	—	319	—
30	120	8 alu.	—	8 alu.	—	—	—	—	310
40	160	8	3/4	—	407	414	395	—	—
40	160	8	1	—	447	431	418	—	—
40	160	8	—	10	—	—	—	294	—
40	160	8 alu.	—	8 alu.	—	—	—	—	232
60	240	6	3/4	—	350	383	363	—	—
60	240	6	1	—	404	400	382	—	—
60	240	6	—	10	—	—	—	228	—
60	240	4 alu.	—	8 alu.	—	—	—	—	221
100	400	3	1 1/4	—	402	397(4)	373	—	—
100	400	3	—	8	—	—	—	229	—
100	400	1 alu.	—	8 alu.	—	—	—	—	222
200	800	3/0	2	—	390	389	363	—	—
200	800	3/0	—	6	—	—	—	201	—
200	800	250 alu.	—	4 alu.	—	—	—	—	195

Examples of Maximum Length Equipment Grounding Conductor (Steel EMT, IMC, GRC, and Copper or Aluminum Wire) computed as a safe return fault path to overcurrent device based on 1997 Georgia Tech Software Version (GEMI Windows, 2000) with an ARC voltage of 40 and 4 IP at 25°C ambient based on a circuit voltage of 120 volts to ground THHN/THWN insulation.

Appendix I

Using an equipment grounding conductor with or without metal conduit on 277 volt to ground circuits

COLUMN 1	COLUMN 2	COLUMN 3	COLUMN 4	COLUMN 5	COLUMN 6	COLUMN 7	COLUMN 8	COLUMN 9	COLUMN 10
OVERCURRENT DEVICE RATING AMPERES (75°C)	400% (4IP) OVERCURRENT DEVICE RATING AMPERES	CIRCUIT CONDUCTOR SIZE AWG-kcmil COPPER OR ALUMINUM	STEEL EMT, IMC GRC TRADE SIZE	(1) EQUIPMENT GROUNDING CONDUCTOR SIZE COPPER OR ALUMINUM	LENGTH OF EMT RUN COMPUTED MAXIMUM (IN FEET)	LENGTH OF IMC RUN COMPUTED MAXIMUM (IN FEET)	LENGTH OF GRC RUN COMPUTED MAXIMUM (IN FEET)	COPPER GROUNDING CONDUCTOR W/O STEEL CONDUIT MAXIMUM RUN	ALUMINUM GROUNDING CONDUCTOR W/O STEEL CONDUIT MAXIMUM RUN
20	80	12	1/2	—	1170	1179	1140	—	—
20	80	12	—	12	—	—	—	800	—
20	80	10 alu.	—	10 alu.	—	—	—	—	870
30	120	10	1/2	—	1199	1135(4)	1143	—	—
30	120	10	3/4	—	—	1182	—	—	—
30	120	10	—	10	—	—	—	946	—
30	120	8 alu.	—	8 alu.	—	—	—	—	920
40	160	8	3/4	—	1208	1228	1170	—	—
40	160	8	1	—	1326	1276	1239	—	—
40	160	8	—	10	—	—	—	871	—
40	160	8 alu.	—	8 alu.	—	—	—	—	690
60	240	6	3/4	—	1039	1134	1075	—	—
60	240	6	1	—	1197	1186	1131	—	—
60	240	6	—	10	—	—	—	676	—
60	240	4 alu.	—	8 alu.	—	—	—	—	657
100	400	3	1 1/4	—	1192	1176(4)	1107	—	—
100	400	3	—	8	—	—	—	680	—
100	400	1 alu.	—	8 alu.	—	—	—	—	659
200	800	3/0	2	—	1157	1155	1077	—	—
200	800	3/0	—	6	—	—	—	598	—
200	800	250 alu.	—	4 alu.	—	—	—	—	578

Examples of Maximum Length Equipment Grounding Conductor (Steel EMT, IMC, GRC, and Copper or Aluminum Wire) computed as a safe return fault path to overcurrent device based on 1997 Georgia Tech Software Version (GEMI Windows, 2000) with an ARC voltage of 40 and 4 IP at 25°C ambient based on a circuit voltage of 277 volts to ground THHN/THWN insulation.

Appendix J

Grounding Tips

When noise problems interrupt computer operations, use the following tips below as a troubleshooting guide to help locate the problem and make approximate corrections.

1. Are grounding requirements of the ITE manufacturer consistent with NEC requirements? Differences and underlying rationale should be discussed and agreed to. There may be good reasons for differences.

2. Is an isolating transformer required or recommended? This affects the point where computer logic ground conductors and power source neutral grounding points will come together at a common point.

3. Where will the ITE system's central grounding point be located? If a modular power center (power peripheral) is used, it may be located there.

4. Will the computer room raised floor structure be specified with interconnecting bolted horizontal struts, suitable for use as a zero signal reference grid? This could save alot of money compared with construction using copper conductors or straps, and could enhance performance.

5. Are the ground conductors for non-ITE equipment separated from ITE grounding conductors except at some upstream common connections, typically at the building service equipment or other common separately derived power source (such as a transformer)?

6. Will the communications and power grounding systems in the building be bonded together at an appropriate upstream common grounding point? This is needed for safety and to minimize noise voltage differences without providing conducting paths through the ground conductors of the ITE system. Use modems or baluns to accomodate interconnection of systems having voltage differences between their respective grounds.

7. Are all grounding conductors and conducting pipes which penetrate the ITE area bonded together by short, robust connections *before* they enter the computer room? Equalizing currents need not flow in the ITE room if this practice is followed.

8. Are all ITE units and their accessories listed or approved by UL or other acceptable safety testing laboratory which is recognized by the state, county or municipality in which the ITE units are to be installed?

9. Does the ITE premise's wiring meet the local/National Elecrical Code requirements?

10. If relocatable power taps are used to cord-and-plug connected computers and their associated equipment, there should'nt be any electric heaters or clocks plugged into the outlets of the power tap assembly.

11. Are filters installed to limit noise problems, and if so, are they grounded properly?

Appendix K

Lightning Protection Tips

When designing, installing or inspecting a lightning protection system, use the following tips 1 through 8 to verify that computer systems are protected from lightning surges. Naturally, the location in which the computer system is installed makes a big difference in how many protective devices are needed.

1. Does the building structure have lightning protection? (refer to Underwriters' Laboratories, Inc. Subject 96A and National Fire Protection Association, Lightning Code NFPA 780)

2. Buildings in which structural steel is bonded together by welding(as oposed to reinforcing steel in concrete which is electrically discontinuous or merely touching) offers better lightning protection of circuits within the building.

3. Down-wires from roof-mounted air terminals (lightning rods) should not establish a path to ground through ITE circuits. Air conditioning coolant pipes from the roof to computer room air handlers must not become a direct path for lightning surges to reach the ITE circuits.

4. Lightning protection down-wires to ground should be separated by at least 6 ft. from ITE power or communication circuits to avoid a side flash.

5. All incoming power and communications equipment conductors should be protected by surge protection devices having shunt overvoltage paths to ground and series impedances, if necessary, to limit surge currents. Short ground connections (a few inches or less, if possible) between each conductor and the surge protectors (line-to-line, line-to-neutral, and neutral-to-ground conductor) make surge arrestors much more effective on fast rise-time impulses than if leads are a foot or more in length. The place for primary lightning protectin is at the building entrance. Supplementary protection may be placed at the input and output of load devices such as rectifier/charges for UPS installations, motor-generators, voltage regulators, isolating transformers and filters.

6. Metal oxide varistors (MOVs) are effective in shunting impulse voltages, but they have been known to deteriorate with repeated exposure to impulses, and sometimes fail in a short-circuit mode which will throw flaming material about. MOVs must be properly protected in enclosures and use of interrupting devices for sustained currents. There are also other devices with similar characteristics, each with advantages and limitations.

7. Down-conductors that are grounded to driven rods used to divert lightning surges must be bonded to the grounding electrode system of the service equipment in order to form an equipotential plane and reduce voltage differences.

8. Is the required ground loop on each floor of a high-rise building bonded per NFPA 780?

Appendix L

Static Electricity Tips

When protecting computer systems and personal from static electricity charges, use the following tips 1 through 7 to help alleviate such problems in the work area.

1. Humidity not less than 40 percent. (High humidity can cause other problems such as corrosion, gold scavaging and ion migration on circuit boards and connectors)

2. Appropriate floor coverings (pressure laminates with upper limit of 10^9 ohms on resistamce to ground and surface resistivity).Carpeting to have acceptably low propensity to static (below 2000 V at 40 percent relative humidity.

3. Appropriate static-free floor care materials and cleaning methods.

4. Furniture surfaces and upholstery with low static propensity. Use conducting glides, casters and wheels on legs and carts used in ITE room.

5. Restrictions on static-prone clothing and footwear.

6. Precondition printing media, tape and punched cards in controlled humidity storage area prior to bringing into the ITE room.

7. Select ITE equipment which has been designed and tested to withstand electrostatic discharge (ESD) to 10 kV or more without failure.

Appendix M

The checklist below can be used to design, install or inspect the grounding and bonding of an electrical system for safety.

GROUNDING CHECKLIST

- Using the grounded (neutral) conductor

    ____ NEC 250.142(A) - supply-side
    ____ NEC 250.142(B) - load-side
    ____ NEC 250.140 - frames of ranges, ovens, etc.

- Types of grounding electrode systems

    ____ NEC 250.52(A)(1) - metal water pipe
    ____ NEC 250.52(A)(2) - structural steel
    ____ NEC 250.52(A)(3) - concrete encased
    ____ NEC 250.52(A)(4) - ground ring
    ____ NEC 250.52(A)(5) - driven rod
    ____ NEC 250.52(A)(6) - plate

- Check sizing of GEC's and EGC's

    ____ NEC Table 250.66 - sizing to water pipe
    ____ NEC 250.66(B) - sizing to rebar
    ____ NEC 250.66(A) - sizing to driven rod
    ____ NEC Table 250.122 - min. size EGC's

- Check point of attachment for GEC's

    ____ NEC 250.10 - attachment
    ____ NEC 250.68 - accessible
    ____ NEC 250.70 - connection

- Type of grounded fittings for connection

    ____ NEC 250.70 - exothermic welding
    ____ NEC 250.70 - listed lugs
    ____ NEC 250.70 - pressure connectors
    ____ NEC 250.70 - listed clamps

- Check for bonding of service equipment

    ____ NEC 250.92(A)(1) - service raceways
    ____ NEC 250.92(A)(2) - meter base enclosure
    ____ NEC 250.92(A)(3) - service equipment
    ____ NEC 250.92B) - all other enclosures

- Check method of bonding service equipment

    ____ NEC 250.92(B)(1) - grounded service
    ____ NEC 250.92(B)(2) - threaded connections
    ____ NEC 250.92(B)(3) - threaded couplings
    ____ NEC 250.92(B) - bonding jumpers
    ____ NEC 250.92(B)4) - other devices

- Check grounded terminal bar

    ____ NEC 250.2(D) - grounding path
    ____ NEC 250.28 - main bonding jumper
    ____ NEC 250.28(A) - material
    ____ NEC 250.28(B) - color and visible
    ____ NEC 250.28(D) - sizing

- Check bonding utilization equipment

    ____ NEC 250.102(D) - sizing
    ____ NEC 250.102(E) - installation
    ____ NEC 250.97 - over 250 volts
    ____ NEC Table 250.122 - min. size EGC's

- Check equipment grounding conductors

    ____ NEC 250.134(B) - circuit conductors
    ____ NEC Table 250.122 - min. size EGC's

- Check grounding of terminal bar

    ____ NEC 408.20 - grounding of panelboard
    ____ NEC 250.130 - EGC connections
    ____ NEC 250.50 - grounding electrode system
    ____ NEC 250.96(B) - connecting receptacles
    ____ NEC 250.142(B) - load size equipment

- Check for grounding of two or more buildings

    ____ NEC 250.32(B) - grounded conductor
    ____ NEC 250.32(C) - ungrounded systems
    ____ NEC 250.32(D) - disconnecting means
    ____ NEC 250.32(F) - grounding conductor

- Check bonding and grounding of swimming pools

    ____ NEC 680.26(B) - bonded parts
    ____ NEC 680.26(C) - common bonding grid
    ____ NEC 680.6 - grounding
    ____ NEC 680.23(F)(2) - methods of grounding

- Check bonding and grounding of hot tub, spas, etc.

    ____ NEC 680.43(D) - bonding
    ____ NEC 680.43(E) - methods of bonding
    ____ NEC 680.43(F) - grounding
    ____ NEC 680.71 - protection
    ____ NEC 680.72 - other electric equipment
    ____ NEC 680.53 - bonding
    ____ NEC 680.54 & 55 - grounding

Appendix N

Other Codes and Standards

There are many different codes and standards which are used with the NEC to design and ensure power quality, safety and operational integrity of sensitive electronic equipment and they are as follows:

Standard	Subject Covered
NFPA 50	Computer Rules Interface
NFPA 75	Life/Fire Safety
NFPA 780	Lightning and Surge Protection
UL 467	Grounding and Bonding
UL 1449	Surge Protection
UL 1363	Transient Surge Protection
UL 1950	Computer Rules Interface
NIST-SP768	Disturbances
NEMA-UPS	Equipment Interface
IEEE C62	Surge Protection
IEEE C84.1	Voltage Levels
IEEE STD-141	Industrial Powering
IEEE STD-142	Grounding, Powering
IEEE STD-446	Interface Equipment
IEEE STD-493	Reliability
IEEE STD-519	Harmonics
IEEE STD-929	Harmonics, Utility Interface
IEEE STD-1001	Harmonics, Utility Interface
IEEE STD-1035	Harmonics, Utility Interface
IEEE P-487	Life/Fire Safety
IEEE P-1100	Grounding, Powering, Disturbances, Interface Equipment, Monitoring, Load Susceptibility
IEEE P-1159	Disturbances, Monitoring

Appendix O

Tables used to size down conductors from air terminals to grounding earth electrode(s)

An ordinary structure is any structure which is used for ordinary purposes whether commercial, industrial, farm, institutional or residential. Such structures not exceeding 75 ft. (23 m) in height must be protected with Class I materials as outlined in Table O-1. Structures greater than 75 ft. (23 m) in height must be protected with Class II materials as outlined in Table O-2.

For structure parts exceeding 75 ft. in height and other parts that do not exceed 75 ft. in height, use Class II air terminals and conductors for those areas exceeding 75 ft. in height. Note that Class II conductors from the higher parts must be extended to ground and interconnected with the balance of the system.

TABLE O-1 MINIMUM CLASS I MATERIAL REQUIREMENTS

Type of Conductor		Copper		Aluminum	
		Standard	Metric	Standard	Metric
Air Terminal, Solid	Diameter	3/8 in.	9.5 mm	1/2 in.	12.7 mm
Air Terminal, Tubular	Diameter	5/8 in.	15.9 mm	5/8 in.	15.9 mm
	Wall Thickness	0.033 in.	0.8 mm	0.064 in.	1.6 mm
Main Conductor, Cable	Size ea. Strand	17 AWG		14 AWG	
	Wgt. per Length	187 lb/1000 ft.	278 g/m	95 lb/1000 ft.	141 g/m
	Cross Sect. Area	57,400 CM	29 mm^2	98,600 CM	50 mm^2
Main Conductor, Solid Strip	Thickness	0.051 in.	1.30 mm	0.064 in.	1.63 mm
	Width	1 in.	25.4 mm	1 in.	25.4 mm
Bonding Conductor, Cable (solid or stranded)	Size ea. Strand	17 AWG		14 AWG	
	Cross Sect. Area	26,240 CM		41,100 CM	
Bonding Conductor, Solid Strip	Thickness	0.051 in.	1.30 mm	0.064 in.	1.63 mm
	Width	1/2 in.	12.7 mm	1/2 in.	12.7 mm

TABLE O-2 MINIMUM CLASS II MATERIAL REQUIREMENTS

Type of Conductor		Copper		Aluminum	
		Standard	Metric	Standard	Metric
Air Terminal, Solid	Diameter	1/2 in.	12.7 mm	5/8 in.	15.9 mm
Main Conductor, Cable	Size ea. Strand	15 AWG		13 AWG	
	Wgt. per Length	375 lb/1000 ft.	558 g/m	190 lb/1000 ft.	283 g/m
	Cross Sect. Area	115,000 CM	58 mm^2	192,000 CM	97 mm^2
Bonding Conductor, Cable (solid or stranded)	Size ea. Strand	17 AWG		14 AWG	
	Cross Sect. Area	26,240 CM		41,100 CM	
Bonding Conductor, Solid Strip	Thickness	0.051 in.	1.30 mm	0.064 in.	1.63 mm
	Width	1/2 in.	12.7 mm	1/2 in.	12.7 mm

Appendix P

The average number of thunderstorm days per year (Isokeraunic Levels). The observation range for each point is about 62 square miles.

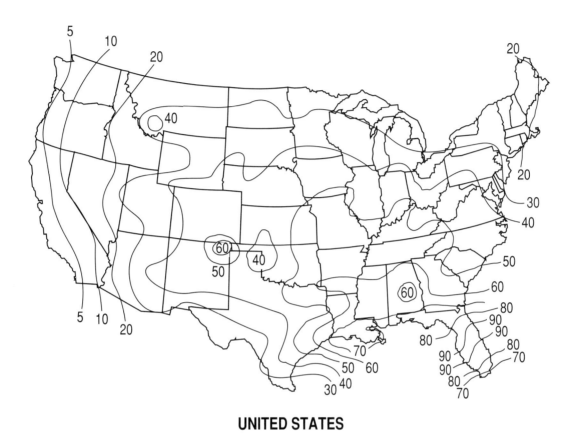

UNITED STATES

A-Q

The above shows the areas in the United States where the resistance of the ground is high. Note that areas of high ground resistance requires a proper designed and installed grouinding electrode system(s).

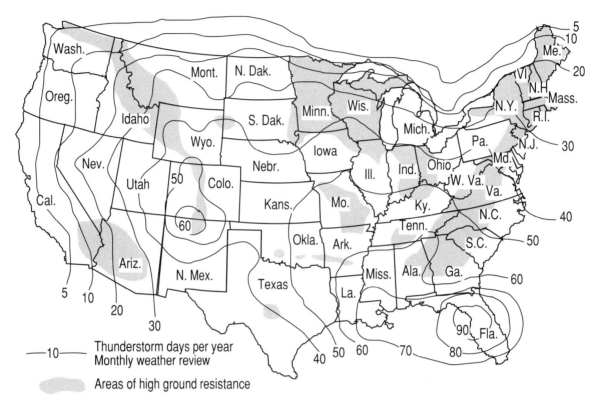

UNITED STATES

Abbreviations

A
A - Amps
AC - Alternating-current
AC - Armored cable
A/C - Air-conditioning
AHJ - Authority having jurisdiction
alu. - Aluminum
AWG - American wire gauge

B
BC - Branch-circuit
BJ - Bonding jumper

C
CB - Circuit breaker
CEE - Concrete-encased electrode
CM - circular mil
cu. - Copper

D
D - Distance
DC - Direct-current
DPCB - Double-pole circuit breaker

E
E - Voltage
EBJ - Equipment bonding jumper
EGC - Equipment grounding conductor
EMT - Electrical metallic tubing

F
FC - Feeder-circuit
FMC - Flexible metal conduit
FMT - Flexible metallic tubing
FNMC - Flexible nonmetallic conduit
ft. - foot

G
G - Ground
GC - Grounding conductor
GE - Grounding electrode
GEC - Grounding electrode conductor
GES - Grounding electrode system
GFCI - Ground-fault circuit-interrupter
GR - Ground
GSC - Grounded service conductor

H
H - Hots

I
IEGC - Isolated equipment grounding conductor
IMC - Intermediate metal conduit
ITE - Information technology equipment

K
kV - Kilo-volts
kVA - Kilo-volt amps
kW - Kilo-watts

L
LFMC - Listed flexible metal conduit
LTFMC - Liquidtight flexible metal conduit

M
Max. - Maximum
MBJ - Main bonding jumper
MC - Metal-clad
MI - Mineral-insulated
Min. - Minimum

MOLPD - Motor overload protective device
MPC - Modular power centers
MWP - Metal water pipe

N
N - Netural
NEC - National Electrical Code
NFPA - National Fire Protection Association
NMC - Nonmetallic conduit
No. - Number

O
OCPD - Overcurrent protection device
OSHA - Occupational Safety and Health Administration

R
R - Resistance
RMC - Rigid metal conduit
rms - root means square
RNMC - Rigid nometallic conduit

S
SBS - Structural building steel
SDS - Separately derived system
SE - Service-entrance
Sec. - Section
SGE - Supplementary grounding electrode
sq. ft. - square foot
sq. in. - square inches

T
TV - Television
TVSS - Transient voltage surge suppressors

U
UF - Underground feeder
UL - Underwriters Laboratories
UPS - Uninterrupted power supply

V
V - Volts
VA - Volt amps
VD - Voltage drop

W
W - Watts
W - wire

X
XFMR - Transformer

Z
Z - Impedance
ZSRG - Zero signal reference grid

Glossary

A

AC Power Interface. The electrical points where the TVSS is electrically connected to the AC power system.

Approved. Acceptable to the authority having jurisdiction.

B

Balun Transformer. A longitudinal transformer in which inductive effects are canceled and common-mode noise is sufficiently restricted.

Bonding (Bonded). The permanent joining of metallic parts to form an electrically conductive path that will ensure electrical continuity and the capacity to conduct safely any current likely to be imposed.

Bonding Jumper. A reliable conductor to ensure the required electrical conductivity between metal parts required to be electrically connected.

Bonding Jumper, Equipment. The connection between two or more portions of the equipment grounding conductor.

Bonding Jumper, Main. The connection between the grounded circuit conductor and the equipment grounding conductor at the service.

C

Combination Wave. (Also called "combination surge") A surge delivered by a generator which has the inherent capability of applying a 1.2/50 μs voltage wave across an open circuit, and delivering an 8/20 μs current wave into a short-circuit. The exact wave that is delivered by the generator and instantaneous impedance to which the combination surge is applied.

Component – Transient Voltage Surge Suppressor. A TVSS intended solely for factory installation in another component, device or product. Examples are discrete devices such as gas discharge tubes, metal oxide varistors (MOV's), and avalanche junction diodes, or combinations of such devices, that are not provided with enclosures, have enclosures that are incomplete, or are otherwise unsuitable for direct field installation or direct connection to a branch-circuit.

Connector, Pressure (Solderless). A device that establishes a connection between two or more conductors or between one or more conductors and a terminal by means of mechanical pressure and without the use of solder.

Cord Connected (CC). Any TVSS provided with a power-supply cord terminating in an attachment plug for connection of the device to a receptacle in the AC power circuit.

Crest (Peak) Value (of a Wave, Surge or Impulse). The maximum value that a wave, surge or impulse attains.

Coupling. Circuit element(s) or network that may be considered common to the input mesh and the output mesh and through which energy may be transferred from on to the other.

D

Direct Plug-In (DPI). Any TVSS incorporating integral blades for direct insertion into a standard wall receptacle.

Dropout. A loss of equipment operation due to noise, sag or interruption.

E

Electromagnetic Interference Filter. An EMI filter is intended to attenuate unwanted radio-frequency signals (such as noise or interference) generated from electromagnetic sources. EMI filters consist of capacitors and inductors used alone or in combination with each other and may be provided with resistors.

Equipment. A general term including material, fittings, devices, appliances, fixtures, apparatus and the like used as a part of, or in connection with, an electrical installation.

F

Fault Current. The current from the connected power system that flows in a short-circuit.

Fitting. An accessory such as a locknut, bushing or other part of a wiring system that is intended primarily to perform a mechanical rather than an electrical function.

Follow (Power) Current. The current from the connected power source that flows through a TVSS during and following the passage of discharge current. Examples of devices that permit follow current are gas discharge tubes and thyristors.

Frequency Deviations. Power frequency increases or decreases lasting from several cycles to several hours.

G

Ground. A conducting connection, whether intentional or accidental, between an electrical circuit or equipment and the earth, or to some conducting body that serves in place of the earth.

Grounded. Connected to earth or to some conducting body that serves in place of the earth.

Grounded, Effectively. Intentionally connected to earth through a ground connection or connections of sufficiently low-impedance and having sufficient current-carrying capacity to prevent the buildup of voltages that may result in undue hazards to connected equipment or to persons.

Grounded Conductor. A system or circuit conductor that is intentionally grounded.

Grounding Conductor. A conductor used to connect equipment or the grounded circuit of a wiring system to a grounding electrode or electrodes.

Grounding Conductor, Equipment. The conductor used to connect the noncurrent-carrying metal parts or equipment, raceways and other enclosures to the system grounded conductor, the grounding electrode conductor, or both, at the service equipment or at the source of a separately derived system.

Grounding Electrode Conductor. The conductor used to connect the grounding electrode to the equipment grounding conductor, to the grounded conductor, or to both, of the circuit at the service equipment or at the source of a separately derived system.

Ground-Fault Circuit Interrupter. A device intended for the protection of personnel that functions to deenergize a circuit or portion thereof within an established period of time when a current to ground exceeds some predetermined value that is less than that required to operate the overcurrent protective device of the supply circuit.

Ground-Fault Protection of Equipment. A system intended to provide protection of equipment from damaging line-to-ground fault currents by operating to cause a disconnecting means to open all ungrounded conductors of the faulted circuit. This protection is provided at current levels less than those required to protect conductors from damage through the operation of a supply circuit overcurrent device.

Ground Loop. A condition created when two or more points in an electrical system that are nominally at ground potential are connected by a conducting path such that either or both points are not at the same potential.

Glossary

H

Harmonics. Power frequency harmonics are periodic distortions of the time-domain voltage or current waveforms seen as integer multiples of the fundamental frequency.

I

Identified (as applied to equipment). Recognizable as suitable for the specific purpose, function, use, environment, application, etc. where described in a particular Code requirement.

Interruption. The complete loss of voltage for a period time.

Isolated Secondary Circuit. A circuit derived from an isolating source (such as a transformer, optical isolator, limiting impedance or electromechanical relay) and having no direct connection back to the primary circuit (other than through the grounding means). A secondary circuit that has a direct connection back to the primary circuit is considered part of the primary circuit.

L

Labeled. Equipment or materials to which has been attached a label, symbol or other identifying mark of an organization that is acceptable to the authority having jurisdiction and concerned with product evaluation, that maintains periodic inspection of production of labeled equipment or materials, and by whose labeling the manufacturer indicates compliance with appropriate standards or performance in a specified manner.

Listed. Equipment, materials or services included in a list published by an organization that is acceptable to the authority having jurisdiction and concerned with evaluation of products or services, that maintains periodic inspection of production of listed equipment or materials or periodic evaluation of services, and whose listing states that either the equipment, material or services meets identified standards or has been tested and found suitable for a specified purpose.

N

Noise. Unwanted wideband disturbances superimposed upon a useful signal that tend to obscure its information content.

Noise, Common Mode. The noise that appears equally and in-phase between each signal conductor and ground.

Noise, Differential Mode. See Noise, Transverse Mode.

Noise, Normal Mode. See Noise, Transverse Mode.

Noise, Transverse Mode. Noise signals measurable between or among active circuit conductors feeding the subject load, but not between the equipment grounding conductor or associated signal reference structure and the active circuit conductors.

O

Off-Line Operation. Pertaining to UPS systems whereby an inverter is off during normal oepration conditions.

On-Line Operation. Pertaining to UPS systems whereby an inverter is operating and supplying the load during normal operation conditions.

Overcurrent. Any current in excess of the rated current of equipment or the ampacity of a conductor. It may result from overload, short-circuit and ground fault.

Overload. Operation of equipment in excess of normal, full-load rating, or of a conductor in excess of rated ampacity that, when it persists for a sufficient length of time, would cause damage or dangerous overheating. A fault, such as a short-circuit or ground fault, is not an overload.

P

Power Conditioner. A device designed to reduce or eliminate electrical disturbance or electrical noise on AC power lines supplying a critical or sensitive load.

Power Disturbance. Any deviation from the nominal value or selected thresholds (based on load tolerance) of the input AC power characteristics.

Power Quality. A concept of powering and grounding sensitive electronic equipment in a manner that is suitable to the operation of that equipment.

R

Raceway. An enclosed channel of metal or nonmetallic materials designed expressly for holding wires, cables or busbars, with additional functions as permitted in the NEC. Raceways include, but are not limited to, rigid metal conduit, rigid nonmetallic conduit, intermediate metal conduit, liquidtight flexible conduit, flexible metallic tubing, flexible metal conduit, electrical metallic tubing, underfloor raceways, cellular concrete floor raceways, cellular metal floor raceways, surface raceways, wireways and busways.

Relocatable Power Tap (RPT). A relocatable power tap consists of an attachment plug cap and a length of flexible cord terminated in an enclosure in which are mounted one or more receptacles. A relocatable power tap may be provided with suitable supplementary overcurrent protection, switches and indicator lights singly or in any combination. A relocatable power tap containing three or more receptacles may also employ a transient voltage surge suppressor (TVSS) and/or an electromagnetic interference (EMI) filter.

S

Sag. A momentary (0.5 to 120 cycles) reduction in voltage at the power frequency beyond a particular piece of equipment's voltage tolerance.

Separately Derived System. A premises wiring system whose power is derived for a battery, a solar photovoltaic system, or from a generator, transformer or converter windings, and that has no direct electrical connection, including a solidly connected grounded circuit conductor, to supply conductors originating in another system.

Shield. A conductive sheath (usually metallic) applied over the insulation of a conductor or conductors to provide means to reduce electrostatic coupling between the conductors so shielded and other conductors that may be susceptible to or that may be generating unwanted electrostatic or electromagnetic fields (noise).

Shielding. The use of a conducting and/or ferromagnetic (permeable) barrier between a potentially disturbing noise source and sensitive circuitry. It is used to protect cables (data and power) and electronic circuits.

Supplementary Protection Device. A device intended for use as overcurrent, over-temperature or over and under-voltage protection within a RFP where branch-circuit overcurrent protection is already provided.

Surge. See Transient.

Surge Suppressor. A device or network of devices that reduces the voltage of an incoming transient.

T

Transformer, Isolation. A transformer of the multiple winding type, with the primary and secondary windings physically arrange separated, that inductively couples its secondary winding to the ground feeder systems that energizes its primary winding, thereby preventing primary circuit potential from being impressed on the circuits.

Touch Potential. The voltage difference between any two conductive surfaces that can be touched by an individual.

U

Undervoltage. Any long-term change below the prescribed input voltage range for a given piece of equipment.

UPS, On-Line. UPS system which, in normal operation, supplies the critical load from the output of its inverter.

V

Voltage Distortion. Any deviation from the nominal sine wave form of the AC line voltage.

Voltage Regulation. The degree of control or stability of the voltage waveform at the load. The ability of the source to provide effective constant voltage (EMF) at the load.

Voltage to Ground. For grounded circuits, the voltages between the given conductor and that point or conductor of the circuit that is grounded; for ungrounded circuits, the greatest voltage between the given conductor and any other conductor of the circuit.

Topic Index

A

AC Circuits and Systems to be Grounded	2-1
AC Circuits of 50 to 1000 Volts	2-3
AC Circuits of Less Than 50 Volts	2-2
AC Systems of 1 kV and Over	2-4
AC Systems of 50 Volts to 1000 Volts Not Required to be Grounded	2-5
Accessibility	3-21
Accessibility not Required	3-22
Additional Bonding Jumper	2-25
Adjustment for Voltage Drop	6-37
Agricultural Buildings or Structures	2-34
Air Terminals	13-1
Aircraft Hangers	6-6, 6-26
Alternate Grid	11-25
Alternative to Rewiring Building	11-17
Aluminum or Copper-Clad Aluminum Conductors	3-15
Application of Arresters	12-9, 12-15
Approval of Equipment	11-7

B

Baluns and Filters	11-39
Bonded Network	11-1
Bonding	5-1
Bonding for Over 250 Volts	5-16
Bonding in Hazardous (Classified) Locations	5-18
Bonding Jumper (Main)	2-21
Bonding Jumper - Class 1, Class 2 or Class 3 Circuits	2-25
Bonding Loosely Joint Metal Raceways	5-18
Bonding of Electrically Conductive Materials and Other Equipment	1-3
Bonding of Metal	13-11
Bonding of Piping Systems and Exposed Structural Steel	5-25
Bonding of Services	5-1
Bonding Other Enclosures — General	5-14
Bonding Pipes and Roof Equipment	13-22
Bonding Rods	13-22
Bonding for Other Systems	5-5
Braided Cables	11-43
Bulk Storage Plants	6-8, 6-27
Butt Wrap Ground	12-5
By Means of a Separate Flexible Wire or Strap	7-7
By Means of an Equipment Grounding Conductor	7-7
By Special Permission	14-11

C

Carpeting With low Propensity to Static Electricity	13-25
Cases of Instruments, Meters and Relays - Operating Voltage 1 kV and Over	9-5

Central Grounding Point 11-8
Circuits Leaving ITE Room 11-29
Circuits not to be Grounded 2-6
Circuits of 1000 Volts and Over
   Interconnections 14-8
   Surge Arrester Conductors 14-7
Class I, Divisions 1 and 2 .... 6-3, 6-21
Class I, Zones 0, 1 and 2 6-5
Class II, Divisions 1 and 2 .... 6-4, 6-24
Class III, Divisions 1 and 2 .... 6-4, 6-25
Temporary Currents not Classified
   as Objectionable Currents 1-8
Clean Surfaces 1-11
Coaxial Cables 11-41
Commercial Garages, Repair and Storage 6-6, 6-26
Common Grounding Electrode 3-12
Communications Systems -
   Bonding of Electrodes 5-8
   Electrode 5-7
   Electrode Connection 5-8
   Grounding Conductor 5-6
Community Antenna Television and Radio Distribution
   Bonding of Electrodes 5-14
   Electrode 5-12
   Electrode Connection 5-13
   Grounding Conductor 5-11
Compliance with Safety Codes 11-6
Concept of Single Point Entry 11-10
Concrete-Encased Electrode 3-5, 3-20
Conductivity of Floor Surfaces 13-24
Conductor Bend 13-7
Conductor to be Grounded — AC System 2-18
Conductor Supports 13-7
Conductors in Parallel 6-40
Conductors Larger Than 6 AWG 6-33
Cone of Protection 12-11
Connected to a Concrete-Encased Electrode 8-7
Connected to a Ground Ring 8-8
Connecting Receptacle Grounding Terminal to Box 7-11
Connections 14-9
Connection of Grounding and Bonding Equipment ... 1-10
Connection of ITE with Straps 11-26
Connections 14-10
Contact Devices or Yokes 7-11
Continuity and Attachment of Equipment
   Grounding Conductors to Boxes 7-14
Continuous 3-16
Controlling Harmonics 11-32
Cord-and-Plug Connected Equipment 7-7
Counterpoise Ground 12-6
Creating a Common Grounding System 13-10

D

DC Fire Alarm Circuits 8-3
Dead Ends 13-9
Derived Neutral Systems 10-2
Desensitizing Electronic Circuits 13-27
Design of Grid 11-25
Direct-Current Bonding Jumper 8-9
Direct-Current Circuits and Systems to be Grounded ... 8-1
Dirty Grounds 11-19
Disadvantage of Standby Power Systems 11-37

Disconnecting Means 2-41
Distribution Apparatus 6-12
Distribution Class Arrester 12-9
Don't be Confused by Term Isolating Transformer 11-17
Down Conductors Concealed 13-10
Downlead Conductors 13-5
Driven Rods 12-3

E

Earth Ground Connection 11-13
Eccentric or Concentric Knockouts 5-16
Effective Grounding Path 3-22
Electric Signs 6-16
Electrical Circuits 5-28
Electrical Contact with Metal 6-2
Elevators and Cranes 6-15
Enclosures for Grounding Electrode Conductors 3-18
Enclosures for Motor Controllers 6-15
Equipment Bonding Jumper - Installation 5-24
Equipment Bonding Jumper on Supply Side
   of Service 5-20
Equipment Bonding Jumper —
   Material and Attachment 5-20
Equipment Connected by Cord-and-Plug 6-21
Equipment Considered Effectively Grounded 7-5
Equipment Fastened-in-Place or Connected by
   Permanent Wiring Methods (Fixed) 6-1
Equipment Fastened-in-Place or Connected by
   Permanent Wiring Methods (Fixed) - Grounding 7-4
Equipment Grounding Conductor 2-37
   Connections 7-1
   Continuity 6-41
   Installation 6-35
   Types 7-4
Equipment Grounding Conductors 10-5
Equipment Secured to Grounded Metal Supports 7-6
Exposed Noncurrent-Carrying Metal Parts 10-7
Expulsion-Type Arrester 12-8

F

Fastened-in-Place or Connected by
Permanent Wiring Methods (Fixed) Specific 6-13
Feeders and Branch-Circuits 11-34
Filters 11-38
Fixed Equipment not Required to be Grounded 6-12
Flat-Top Construction 12-13
Floor Boxes 7-12
Foil Shield Cables 11-43
For Grounded Systems 7-2
For Ungrounded Systems 7-2
Frames of Ranges and Clothes Dryers 7-8
Frequency Checks 11-22
Frequency of Noise Signals 11-9
Furniture and Upholstery 13-26

G

Gap Protectors Inside Transformers 13-18
Gapless Type Arresters 12-9
Garages, Theaters and Motion Picture Studios 6-15
Gasoline Dispensing and Service Stations 6-7
General Requirements for Grounding and Bonding ... 1-1

Ground Currents 11-2
Ground Detectors 8-2
Ground Grid 12-6
Ground Ring 3-5, 3-21
Ground Voltage Equalization 11-4
Ground-Fault Current 10-7
Ground-Fault Detection and Relaying 10-8
Ground-Fault Return Path 11-4
Grounded Conductor Brought
   to Service Equipment 2-12
Grounded Service Conductor 5-3
Grounding and Bonding for Cable Trays 7-17
Grounding Electrode -
Interior Metal Water Pipe 2-31
Grounding Electrode - One Branch-Circuit 2-36
Grounding Electrode -
Used in Service Equipment 2-31
Grounding Electrode Conductor 2-25, 2-34, 3-15
Grounding Electrode
   Conductor - Class 1, Class 2 or Class 3 2-28
Grounding Electrode Conductor
Connection to Grounding Electrodes 3-21
Grounding Electrode Conductor Installation 3-15
Grounding Electrode Conductor Location 2-46
Grounding Electrode Conductor Material 3-15
Grounding Electrode Conductor Taps 3-17
Grounding Electrode System 3-1
Grounding Electrodes 13-6
Grounding Facts 11-1
Grounding Higher Frequencies 11-9
Grounding Impedance Location 2-45
Grounding Instrument Transformer Cases 9-3
Grounding Myths Related to ITE 11-18
Grounding of Electrical Systems 1-1, 1-2
Grounding of Equipment 10-9
Grounding of Systems Supplying
   Portable or Mobile 10-6
Grounding Separately Derived AC Systems 2-21
Grounding Service - Supplied AC Systems 2-8
Grounding Systems 12-3
Grounding Tips 11-14

H

Health Care Facilities 6-9, 6-28
Height of Terminals 13-4
Hi-Impulse, Nonshielded Construction 12-14
High-Impedance 2-17
High-Impedance Grounded Neutral Systems 2-44
How to Control Noise with Filters 11-40
Humidity Control 13-24

I

Identification of Equipment Grounding Conductors 6-33
Identification of Wiring Device Terminals 6-41
Identified and Insulated 10-4
Impedance Grounded Neutral Systems 10-4
In Other Than Residential Occupancies 6-29
In Residential Occupancies 6-28
Independent Grounding Prohibited 11-15
Industrial and Commercial Buildings 3-4
Install Isolation Transformer Close to ITE 11-12

Installation on the Load Side of
   Services of Less than 1000 Volts 14-6
Installed at Services of Less Than 1000 Volts ... 14-5
Installing Additional Driven Rods 11-21
Instrument Grounding Conductor 9-6
Instrument Transformer Cases 9-2
Instrument Transformer Circuits 9-1
Interconnected Grounds Between Systems 11-22
Intermediate Class Arresters 12-10
Intrinsically Safe Systems 6-5
Ion Generators 13-26
Isolated Grounding Circuits 5-15
Isolated Receptacles 7-13
Isolating Transformers 11-31
Isolating Transformers not Always Required 11-34
Isolation 10-8
Isolation Grounds 11-15
Isolation of Objectionable Direct-Current
   Ground Currents 1-9
ITE Subjected to Static Charges 13-24

L

Length of Strap 11-27
Lightning Arresters 12-7
Lightning Protection Should not be
   Installed In ITE Room 13-20
Lightning Protection for Foundation and
   Slab Built on Solid Rock 13-15
Lightning Protection for Information
   Technology Equipment 13-15
Lightning Protection for Pre-Cast
   Concrete Buildings 13-14
Lightning Protection for Reinforced
   Concrete Buildings 13-12
Lightning Protection for Structural Steel Buildings 13-13
Lightning Protection Systems 5-28
Limitations to Permissible Alterations 1-6
Line Voltage Regulators 11-40
Listed Equipment 6-12, 6-30
Load-Side Equipment 7-10
Locating Arresters at Service Equipment 13-20
Location 10-4, 14-4
Long Feeder-Circuit Runs 11-30
Long Vertical Bodies 13-12
Low-Impedance Paths 11-19
Low-Resistance Connection not a Myth 11-18
Luminaires (Lighting Fixtures) 6.19

M

Magnetic Coupling Controlled Voltage Regulators 11-41
Main Bonding Jumper - Construction 2-20
Main Bonding Jumper - Attachment 2-20
Main Bonding Jumper - Size 2-21
Matched Filters 11-40
Metal Boxes 7-14
Metal Car Frames 7-6
Metal Frame of the Building or Structure 3-6
Metal Frames of Electrically
   Heated Appliances 6-12, 6-30
Metal Gas Piping 5-27

Entry	Page
Metal Objects Within 6 ft. of Lightning Conductors	13-22
Metal Underground Gas Piping Systems	5-27
Metal Underground Water Pipe	3-3
Metal Underground Water Pipe - Continuity	3-3
Metal Underground Water Pipe - Supplemental Electrode Required	3-3
Metal Water Pipe	3-3
Metal Water Piping - Buildings of Multiple Occupancy	5-25
Metal Water Piping - Multiple Buildings or Structures Supplied From a Common Service	5-26
Metal Water Piping - Separately Derived Systems	5-27
Metal Well Casings	6-21
Metallic Interconnection	14-8
Methods of Bonding at the Service	5-2
Methods of Grounding Mobile Homes	7-15
Methods of Grounding and Bonding Conductor Connections to Electrodes	3-23
Mobile Home Grounding	7-15
Modular Power Centers are Sometimes Used	11-34
Motion Picture Projection Equipment	6-18
Motor Circuits	6-40
Motor Frames	6-14
Motor-Operated Water Pumps	6-20
Motors - Where Guarded	6-30
Multiconductor Cable	6-34
Multiple Circuit Connections	7-10
Multiple Circuits	6-38
Multiple Grounding	10-3
Multiple Separated Grounds	11-38

N

Entry	Page
Neutral Conductor	2-45, 10-3
Neutral Conductor Routing	2-45
Neutral Grounding Conductor	10-3
No Modifications of Equipment	11-35
Nonelectric Equipment	6-31
Nongrounding Receptacle Replacement or Branch-Circuit Extensions	7-2
Nonmetallic Boxes	7-14
Not on Switchboards	9-3
Not Required to be Grounded	4-2, 4-4, 6-29, 10-10
Not Smaller Than the Largest Conductor	8-6
Not Smaller Than the Neutral Conductor	8-5
Number of Down Conductors	13-5
Number Required	14-1

O

Entry	Page
Objectionable Current Alterations To Stop	12-11
Over Grounding Conductors	1-6
Off-Premises Source	8-4
On Circuits of 1000 and Over	14-3
On Circuits of Less Than 1000 Volts	14-2
On Dead-Front Switchboards	9-4
On Live-Front Switchboards	9-4
On-Premises Source	8-4
One Locknut	5-17
One Way Path	13-9
OSHA, NFPA, UL and Local Codes	11-7
Other Conductor Enclosures and Raceway	4-4
Other Devices	5-4
Other Local Metal Underground Systems and Structures	3-7
Other Metal Piping	5-27
Over 150 Volts-to-Ground	6-11, 6-28
Over 40 ft. or 40 ft. or Less Structures	13-12
Overhead Lines Below Cone	12-11

P

Entry	Page
Path for Fault Current	1-6
Paths Through Equipment	13-21
Pipe Organs	6-14
Plate Electrodes	3-10
Point of Connection For Direct-Current Systems	8-4
Portable and Vehicle-Mounted Generators	2-39
Portable Generators	2-39
Portable or Mobile Equipment	10-6
Power-Limited Remote-Control Signaling and Fire Alarm Circuits	6-18
Primary Protection Belongs Outside the ITE Room	13-17
Proper Grounding Techniques	11-20
Protection at Devices	13-21
Protection of Communications Lines	13-20
Protection of Ground Clamps and Fittings	1-10
Protective Device Locations	13-21
Providing Equipotential	13-21

R

Entry	Page
Radio and Television Equipment - Bonding of Electrodes (Receiving Stations)	5-9
Radio and Television Equipment - Electrode (Receiving Stations)	5-10
Radio and Television Equipment - Grounding Conductor (Receiving Stations)	5-5
Rectifier-Derived DC Systems	8-2
Requirements of NEC and Other Codes and Standards	11-6
Residual Surge Protection in Power and Communications Lines	13-19
Resistance of Made Electrodes	3-10
Results of Independent Grounds	11-20
Ribbon Cable	11-43
Ride Through Periods	11-37
Rod and Pipe Electrodes	3-9
Rod-to-Earth Resistance	11-21
Rod, Pipe or Plate Electrodes	3-5, 3-8, 3-9, 3-20, 8-6
Roof Conductors	13-8
Routing	2-12
Routing of Surge Arresters Connections	14-4

S

Entry	Page
Separately Derived System	2-4
Service Raceways and Enclosures	4-1
Severe Harmonic Distortion	11-40
Shield and Grounded Neutral Conductor	12-2
Shield Static Conductor	12-2
Shielded Construction	12-13
Shielding Requirements	11-5

Shields Connected at Both Ends 11-42
Shields Connected at One End 11-42
Shoes and Clothing 13-26
Short Sections of Raceways 7-4
Signaling Cables Create Ground Loops 11-23
Single Point Entry Bonding 11-29
Size of Direct-Current Grounding
   Electrode Conductor 8-5
Size — Equipment Bonding Jumper
   on Load Side of Service 5-22
Size of Equipment Grounding Conductors 6-36
Size of AC Grounding Electrode Conductor 3-19
Skid-Mounted Equipment 6-19
Solidly Grounded Neutral Systems 10-2
Splices in Busbars 3-17
Spray Application, Dipping and Coating Processes 6-9
Stand-Alone Units 11-14
Static Drain Paths 13-25
Static Electricity 13-23
Station Class Arresters 12-10
Stopping Objectionable Current 1-6
Structural Building Steel 3-19
Structural Steel 5-28
Supplementary Grounding Electrodes 3-10
Supply-Side Equipment 7-9
Support of Terminals 13-4
Surface-Mounted Boxes 7-11
Surge Arrester Selection 14-2
Surge Voltage and Current Waves 13-20
Surges 12-1
Switchboard Frames and Structures 6-13
Swimming Pool Bonding 7-19
System Neutral Connection 2-45, 10-5

T

Techniques of Grounding 11-2
Techniques of Grounding Safely 11-6
Techniques of Grounding Signaling Cables 11-41
Terminating Shields to Clean Surfaces 11-44
Threaded Connections 5-3
Threadless Couplings and Connectors 5-3
Threadless Fittings 5-17
Three-Wire, Direct-Current Systems 8-3
Through Spark Gap or Device 14-10
Tools and Portable Handlamps 6-30
Touch Voltage 11-3
Tower Base Ground 12-6
Trailing Cable and Couplers 10-9
Transformer Characteristics of Use 11-32
Transformer Impedance 11-33
Transformer With Shields 11-33
Transformer Without a Shield 11-33
Transformers - Where to Locate Them 11-30
Transformer Centrally Located 11-30
Traveling Waves 13-18
Two Locknuts 5-17
Two-Wire, Direct-Current Systems 8-1
Types of Arresters 12-8
Types of Construction 12-13
Types of Equipment Grounding Conductors 6-31

U

"U" OR "V" Pockets 13-9
Underground Service Cable 4-3
Underground Service Cable or Conduit 4-3
Underground Service Conduit Containing Cable 4-3
Ungrounded Direct-Current
   Separately Derived Systems 8-12
Ungrounded Separately Derived AC Systems 2-33
Uninterruptable Power Systems (UPS) 11-36
Unwanted Path for Surge
   Entry into the ITE Room 13-21
Use of Arresters and Filters 13-16
Use of Grounded Circuit Conductor for
   Grounding Equipment 7-9
Use of Air Terminals 3-13
Use of Lightning Rods 5-28
Using Arresters on Overhead Lines 13-18
Using Filters 11-11
Using Isolation Transformers 11-11
Using Metal Frame of Building 13-11
Using Multiple Shields to Control Noise 11-43

V

Valve Type Arresters 12-9
Vehicle-Mounted Generators 2-40
Vertically or Horizontally 6-2

W

Water Pipe Minimum Size 3-21
Wet or Damp Locations 6-2
Wiring Methods With Equipment
   Grounding Conductor 6-11
Wiring Systems and Their Effect on ITE 11-28
With Circuit Conductors 7-5

Z

Zero Signal Reference Grid 11-24
Zero Signal Reference Grid not a Substitute
   for Equipment Grounding Conductor 11-27